P9-DMO-936

Chemical Fate and Transport in the Environment

SECOND EDITION

Chemical Fate and Transport in the Environment

SECOND EDITION

Harold F. Hemond

Ralph M. Parsons Laboratory
Department of Civil and Environmental Engineering
Massachusetts Institute of Technology
Cambridge, Massachusetts

Elizabeth J. Fechner-Levy

Abt Associates, Inc.
Bethesda, Maryland

ACADEMIC PRESS
An Imprint of Elsevier

San Diego London Boston New York Sydney Tokyo Toronto

Cover image courtesy of Harold F. Hemond.

This book is printed on acid-free paper. ♾

Academic Press
An Imprint of Elsevier
525 B Street, Suite 1900, San Diego, California 92101-4495, USA
http://www.academicpress.com

Academic Press
84 Theobald's Road, London WC1X 8RR, UK
http://www.academicpress.com

Library of Congress Catalog Card Number: 99-65407

ISBN-13: 978-0-12-340275-2
ISBN-10: 0-12-340275-1

PRINTED IN THE UNITED STATES OF AMERICA
08 09 10 11 12 9 8

CONTENTS

PREFACE

This textbook is expanded and extensively revised from the first edition of *Chemical Fate and Transport in the Environment*. It is intended for a one-semester course covering the basic principles of chemical behavior in the environment. The approach is designed to include students who are not necessarily pursuing a degree in environmental science, but whose work may require a basic literacy in environmental transport and fate processes.

Written as a survey text suitable for graduate students of diverse backgrounds, this book may also be appropriate for use in some undergraduate curricula in environmental engineering. Concepts are developed from the beginning, assuming only prior familiarity with basic freshman chemistry, physics, and math. Certain simplifications are made, and the material is intentionally presented in an intuitive fashion rather than in a rigorously mathematical framework. Nevertheless, the goal is to teach students not only to understand concepts but also to work practical, quantitative problems dealing with chemical fate and transport.

Depending on the nature of the class, the instructor may wish to spend more time with the basics, such as the mass balance concept, chemical equilibria, and simple transport scenarios; more advanced material, such as transient well dynamics, superposition, temperature dependencies, activity coefficients, redox energetics, and Monod kinetics, can be skipped. Similarly, by omitting Chapter 4, an instructor can use the text for a water-only course. In the case of a more advanced class, the instructor is encouraged to expand on the material; suggested additions include more rigorous derivation of the transport equations, discussions of chemical reaction mechanisms, introduction of quantitative models for atmospheric chemical transformations, use of computer software for more complex groundwater transport simulations, and inclusion of case studies and additional exercises. References are provided with each chapter to assist the more advanced student in seeking additional material.

This book was originally based on notes for a class titled Chemicals in the Environment: Fate and Transport, which the first author has taught for 10 years at the Massachusetts Institute of Technology. Several classes have now used the first edition of the textbook; each time, we have benefited from thoughtful feedback from students and teaching assistants, and we have included many of their suggestions in this second edition. We hope to hear of the experiences of others, students and instructors alike, who use this text in the coming years. We hope you find the book helpful, even enjoyable, and come away sharing both our enthusiasm for the fascinating environment we inhabit and our desire to treat the environment from a basis of appreciation and understanding.

Harry Hemond
Liz Fechner-Levy

CHAPTER 1

Basic Concepts

1.1 INTRODUCTION

"By sensible definition any by-product of a chemical operation for which there is no profitable use is a waste. The most convenient, least expensive way of disposing of said waste—up the chimney or down the river—is the best" (Haynes, 1954). This quote, describing once common industrial waste disposal practices, reflects the perception at that time that dispersal of chemical waste into air or water off the factory site meant that the chemical waste was gone for good. For much of the 20th century, many industries freely broadcast chemical waste into the environment as a means of disposal. Other human activities, including use of pesticides and disposal of household waste in landfills, also contributed enormous loads of anthropogenic chemicals to the environment.

Today, however, not only have the gross pollution effects of emissions from stacks, pipes, and dumps become evident, but more subtle and less predictable effects of chemical usage and disposal have also manifested themselves. Some lakes, acidified by atmospheric deposition from power plants, smelters, and automobiles, have lost fish populations, while other lakes have burgeoned with unwanted algal growth stimulated by detergent disposal, septic leachate, and urban and agricultural runoff. Municipal wells have been shut down due to chemical contamination emanating from landfills. Populations of several species of birds of prey have been decimated by pesticides that have become concentrated in their tissues and have adversely affected their reproduction. Humans have been poisoned by polychlorinated biphenyls (PCBs) and by mercury acquired from the environment via the food chain. Even nontoxic, seemingly harmless chlorofluorocarbons (CFCs) from spray cans and refrigerators have threatened the well-being of humans as well as the functioning of ecosystems by creating a hole in Earth's protective ozone shield.

Nevertheless, it is neither possible nor desirable for modern societies to stop all usage or environmental release of chemicals. Even in prehistoric times, tribes of troglodytes roasting hunks of meat over their fires were unknowingly releasing complex mixtures of chemicals into the environment. It is imperative, however, that modern societies understand their environment in sufficient detail so that accurate assessments can be made about the environmental behavior and effects of chemicals that they use. This includes an understanding of both chemical *transport,* referring to processes that move chemicals through the environment, and chemical *fate,* referring to the eventual disposition—either destruction or long-term storage—of chemicals. It then can be hoped that societies will make intelligent, informed decisions that will protect both human health and the environment, while allowing human beings to enjoy the benefits of modern technology.

This book presents the principles that govern the fate and transport of many classes of chemicals in three major environmental media: surface waters, soil and groundwater (the subsurface), and the atmosphere. These several media are treated in one book for three related reasons. First, this is primarily an introductory textbook, and a broad scope is most appropriate to the student who has not specialized in a particular environmental medium. Second, chemicals released into the environment do not respect the boundaries between air, water, and soil any more than they respect political boundaries. Because exchanges among these media are common, modeling a chemical in any single medium is unlikely to be adequate for obtaining a full description of the chemical's fate. Third, a great deal of insight can be gained by comparing and contrasting chemical behavior in surface waters, soil and groundwater, and the atmosphere. For example, although the fact is not immediately apparent in much of the literature, the mathematics describing physical transport in each medium are almost identical; the transport equation that models the mixing of industrial effluent into a river is also useful for describing the movement of contaminants in groundwater or the mixing of air pollutants in the atmosphere. Contrasts are also instructive; for example, the dominant fate process for a chemical in the atmosphere may be photode-

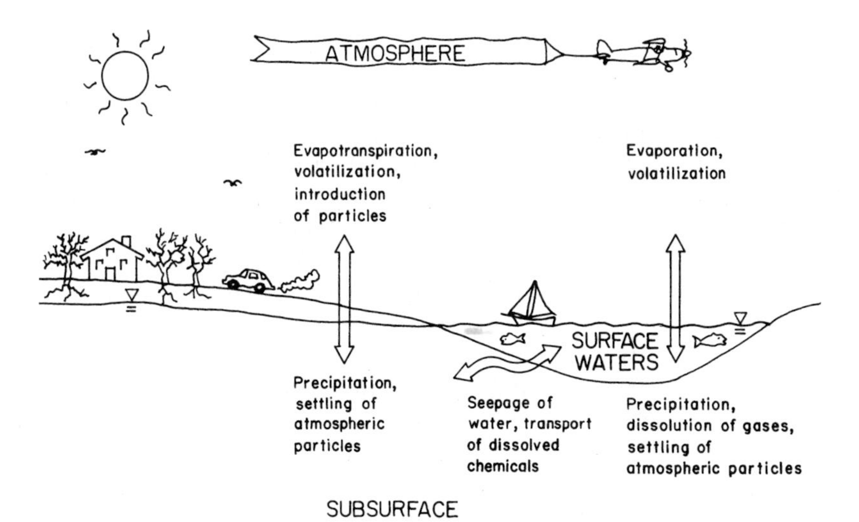

FIGURE 1-1 Three major environmental media: surface waters, the subsurface (soil and groundwater), and the atmosphere. Although each medium has its own distinct characteristics, there are also many similarities among them. Few chemicals are restricted in their movement to only one medium; thus chemical exchanges among the media must be considered. Several very generalized exchange processes between media are shown.

gradation, whereas in the subsurface environment, biodegradation of the chemical may dominate.

Knowledge of the principles underlying the fate and transport of chemicals in the environment allows problems ranging from local to global scales to be defined and analyzed. This first chapter presents fundamental concepts that apply universally to any environmental medium. The subsequent three chapters focus on surface waters, the subsurface environment, and the atmosphere, respectively; see Fig. 1-1 for a diagram of some of the interrelationships among these media. In each chapter, each medium is discussed in terms of its basic physical, chemical, and biological attributes; then the fate and the transport of introduced chemicals are considered.

1.2 CHEMICAL CONCENTRATION

Perhaps the single most important parameter in environmental fate and transport studies is chemical *concentration* (C). The concentration of a chemical is a measure of the amount of that chemical in a specific volume or mass of air, water, soil, or other material. Not only is concentration a key quantity in fate and transport equations; a chemical's concentration in an environmental medium also in part determines the magnitude of its biological effect.

Most laboratory analysis methods measure concentration. The choice of units for concentration depends in part on the medium and in part on the process that is being measured or described. In water, a common expression of concentration is mass of chemical per unit volume of water. Many naturally occurring chemicals in water are present at levels of a few milligrams per liter (mg/liter). The fundamental dimensions associated with such a measurement are [M/L^3]. The letters M, L, and T in square brackets refer to the fundamental dimensions of mass, length, and time, which are discussed further in the Appendix. For clarity in this book, specific units, such as (cm/hr) or (g/m^3), either are free-standing or are indicated in parentheses, not in square brackets.

Another common unit of concentration in water is *molarity*. Recall that a *mole* of a chemical substance is composed of 6.02×10^{23} atoms or molecules of that substance. Molarity refers to the number of moles per liter of solution and is denoted *M*, with neither parentheses nor square brackets around it in this book.

A related unit, *normality* (*N*), refers to the number of *equivalents* of a chemical per liter of water. An equivalent is the amount of a chemical that either possesses, or is capable of transferring in a given reaction, 1 mol of electronic charge. If a chemical has two electronic charge units per molecule, 1 mol of the chemical constitutes two equivalents [e.g., a mole of sulfate (SO_4^{2-}) is equal to two equivalents, and a one molar (1 *M*) solution of sodium sulfate (Na_2SO_4) is two normal (2 *N*)].

In soil, a chemical's concentration may be measured in units such as milligrams per cubic centimeter (mg/cm^3). Expressing concentration as mass per unit volume for soil, however, carries the possibility of ambiguity; soils undergo volume changes if they are compressed or expanded. Accordingly, it is often more useful to express soil concentration as mass per unit mass, such as (mg/kg) because the mass of soil does not vary with changes in the degree of compaction of the soil.

For air, which is highly compressible, expression of chemical concentration as mass per volume is even more ambiguous. The volume of a given mass of air changes significantly with changes in pressure and temperature, and thus the chemical concentration, when expressed as mass per volume, also changes. Consider the following situation: if initially there is 1 μg of benzene vapor (C_6H_6) per cubic meter of air and the barometric pressure decreases by 5%, the concentration of benzene as expressed in units of mass per volume also decreases by 5%, because the air expands while the mass of benzene remains constant. Expressing the benzene concentration as mass of benzene per mass of air removes all ambiguity because changes in temperature and pressure do not affect the mass of air present.

Ambiguity can also arise when a laboratory reports the concentration of a chemical that can exist in more than one form *(species)* without reference to a particular species. For example, if the concentration of nitrogen in a water sample is reported simply as 5 mg/liter, it is unclear whether 1 liter contains 5 mg of nitrogen atoms (N) or 5 mg of one of the nitrogen species present, such as nitrate (NO_3^-), nitrite (NO_2^-), ammonia (NH_3), or ammonium (NH_4^+). In such a situation, clarification from the laboratory performing the measurements must be obtained to understand the actual chemical mass present. Furthermore, if the species associated with a concentration is not known, a conversion from mass to moles is not possible. [Recall that to convert mass to moles, the molecular weight of the species measured must be known; the number of moles is then equal to the mass (in grams) divided by the molecular weight.]

Numerous other options exist for specifying a concentration; common ones are parts per thousand (ppt or ‰), parts per million (ppm), or parts per billion (ppb). For the soil and air cases just mentioned, ppm on a mass basis is numerically equal to milligrams (mg) of chemical per kilogram (kg) of soil or air. Parts per million is also sometimes used on a volume basis. This may be inferred from context or made clear by the term ppm(v); 1 ppm(v) of helium in air would correspond to 1 ml of helium in 1000 liters (1 m^3) of air. For water, the density of which is approximately 1 g/cm^3, parts per million corresponds to milligrams of chemical per liter of water (mg/liter) in dilute solutions.

No matter which units are used, however, concentration is the measure of interest for predictions of a chemical's effects on an organism or the environment. Concentration is also critical in one of the most important concepts of environmental fate and transport: the bookkeeping of chemical mass in the environment.

1.3 MASS BALANCE AND UNITS

1.3.1 MASS BALANCE AND THE CONTROL VOLUME

Three possible outcomes exist for a chemical present at a specific location in the environment at a particular time: the chemical can remain in that location, can be carried elsewhere by a *transport* process, or can be eliminated through *transformation* into another chemical. This very simple observation is known as *mass balance* or *mass conservation.* Mass balance is a concept around which an analysis of the fate and transport of any environmental chemical can be organized; mass balance also serves as a check on the completeness of knowledge of a chemical's behavior. If, at a later time in an analysis, the original mass of a chemical cannot be fully accounted for, then there is an incomplete understanding of how transformation and transport processes are affecting that chemical. Accurate fate and transport modeling results from an understanding of every process contributing to the mass balance of a chemical.

Implicit in the application of the mass balance concept is the need to choose a *control volume*. A control volume is any closed volume, across whose boundaries we propose to account for all transport of a chemical, and within whose boundaries we propose to account for all the chemical initially present (stored), as well as all processes (*sources* or *sinks*) that produce or consume the chemical. The *mass balance expression* for any chemical in any control volume during any time interval can be written as

Change in storage of mass = mass transported in − mass transported out

+ mass produced by sources − mass eliminated by sinks. [1-1a]

The mass balance expression in a control volume can also be written in terms of rates, that is, mass per time [M/T]:

Rate of change in storage of mass = mass transport rate in

− mass transport rate out + mass production rate by sources

− mass elimination rate by sinks. [1-1b]

Control volumes are chosen to be convenient and useful. While the choice of a good control volume is somewhat of an art and depends on both the chemicals and the environmental locations that are of interest, the control volume boundaries are almost always chosen to simplify the problem of determining chemical transport into and out of the control volume.

As an example of an environmental pollution problem requiring the choice of a control volume, consider a lake that is receiving industrial effluent from a discharge pipe. To establish a useful volume within which we could describe the fate and transport of chemicals in the effluent, we might choose the entire lake, as shown in Fig. 1-2. The upper boundary of the control volume is then the lake surface; transport across this boundary is described by the general principles that govern chemical transport between water and air. For some *nonvolatile* chemicals (i.e., chemicals that do not rapidly move from a dissolved phase in the water to a gaseous phase in the air), this transport rate is negligible. For many other chemicals, enough knowledge exists to make reasonable estimates of the air/water exchange rate (based on factors such as the volatility of the chemical, the rate of chemical diffusion in water, and the amount of turbulence in the lake). The lower boundary of the control volume

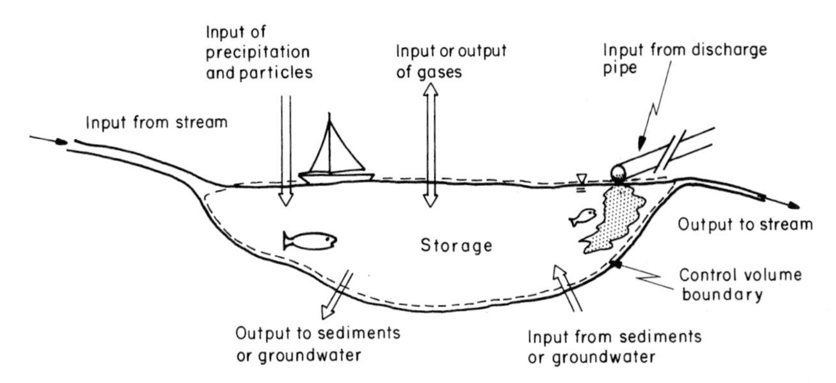

FIGURE 1-2 An example of a control volume that is useful for analyzing chemical behavior in a lake because it facilitates the measurement or estimation of chemical inputs and outputs. Chemicals may enter the control volume via the inflowing stream and the industrial discharge pipe and may leave the control volume across the air–water interface and sediment–water interface as well as via the outflowing stream.

might be chosen to be just above the uppermost layers of lake bottom sediments. Water seeping into or out of the lake sediments is then one transport mechanism by which a dissolved chemical could cross the lower boundary of the control volume. The lake control volume might also receive chemical input from an inflowing stream; the rate at which a chemical enters the lake from the stream could be estimated by multiplying the streamflow by the concentration of the chemical in the stream.

To complete a mass balance on a chemical within the control volume, internal consumption, production, and storage of the chemical also must be quantified. In some cases source and sink strengths can be estimated based on knowledge of chemical and biological composition and physical attributes of the lake. If the lake is well mixed, storage of the chemical in the lake control volume at a given time can be estimated as the product of the chemical concentration in the lake water and the total volume of water in the lake.

Note that the left-hand sides of Eqs. [1-1a] and [1-1b] are zero if storage does not change with time. This is one example of *steady state*, a description that applies to any problem in which quantities do not change with time. (Mathematically, all derivatives with respect to time are zero in steady-state systems.) Steady-state assumptions often simplify the analysis of a problem but should not be invoked when a *transient* (time-varying) situation exists.

If all terms but one are known in the mass balance expressions of Eqs. [1-1a] and [1-1b], the control volume can be used to estimate an otherwise unmeasurable transport, source, or sink term, as shown in Example 1-1.

EXAMPLE 1-1

Consider the lake shown in Fig. 1-2. For this example, assume that the discharge pipe releases to the lake small amounts of various alcohols from an industrial fermentation process, and it is desired to estimate the rate at which alcohols are degraded in the lake. One of them, butanol (C_4H_9OH), is released to the lake at the rate of 20 kg/day. Butanol is measured in the lake water on several occasions at a concentration of 10^{-4} kg/m^3; no butanol is detected in the inflowing stream. Average streamflow at the outlet of the lake is measured to be 3×10^4 m^3/day. What is the magnitude of internal sinks of butanol?

By using the mass balance equation of Eq. [1-1b]:

Mass elimination rate by sinks = mass transport rate in

− mass transport rate out + mass production rate by sources

− rate of change in storage of mass.

By assuming no butanol exchange with the atmosphere, this equation can be rewritten more explicitly for the mass balance of butanol:

Internal sink rate = discharge pipe input rate + stream input rate

− stream output rate + internal source rate − rate of change in storage.

By considering the equation term by term:

- The discharge pipe input rate of butanol is 20 kg/day.
- The stream input rate is zero.
- By assuming that the lake is well mixed (i.e., the butanol concentration is the same everywhere in the lake), the stream output rate is $(3 \times 10^4 \text{ m}^3/\text{day}) \cdot (10^{-4} \text{ kg/m}^3) = 3$ kg/day.
- Assume there are no internal sources of butanol, so the internal source rate is zero.
- Given that the butanol concentration in the lake water seems to be at steady state at 10^{-4} kg/m^3, the rate of change in storage is zero.

Therefore, the mass balance is

$$\text{Internal sink rate} = 20 \text{ kg/day} + 0 - 3 \text{ kg/day} + 0 - 0 = 17 \text{ kg/day}.$$

The internal sink appears to consume 17 kg of butanol per day, although it is not known by what *processes* this consumption occurs (e.g., through biodegradation or through consumption by fish). Three other limitations of this mass balance analysis are (1) the calculated internal sink rate may actually be an overestimate because atmospheric exchange is being neglected. There is probably not much butanol in precipitation, but the assumption of no *volatilization* (transfer to the air) should be tested and perhaps an atmospheric output rate term added to the mass balance equation (this is discussed in Section 2.3). (2) The well-mixed lake assumption may not be appropriate; the measurements of butanol concentration may not be representative of the butanol concentration in water leaving the lake in the stream outflow. (3) If the lake processes are not at steady state, the rate of change in storage term may be nonzero, thereby affecting the calculated internal sink rate.

A lake can also illustrate a theoretically valid, but *not* useful, control volume. Consider a control volume that comprised only the northern half of the lake; the southern boundary of the control volume would then be a surface cutting vertically across the entire lake from the water surface to the lake sediments. Measurement of chemical transport across this boundary would be immensely difficult; it would require detailed water flow measurements at an impossibly large number of sites, given that the speed and direction (i.e., the velocity) of water currents in a lake typically vary from place to place and time to time. Such a control volume would not simplify estimates of chemical inputs and outputs.

A little reflection on a variety of other environmental pollution situations suggests any number of relevant control volumes having convenient, useful, and well-defined boundaries. Three typical examples are shown in Fig. 1-3. If

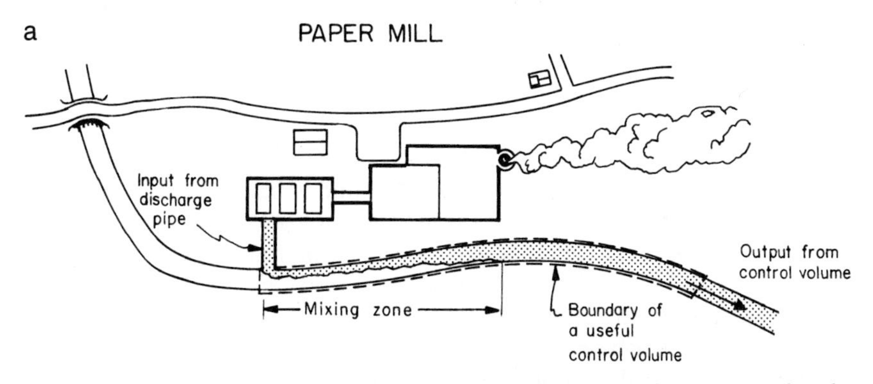

FIGURE 1-3 a Examples of useful control volumes for three principal environmental media. Control volume (a) would be practical if we were studying the various processes that remove a contaminant from a river; the difference between the input and output fluxes would represent internal sinks in the river or volatilization loss to the air *(Figure continues)*.

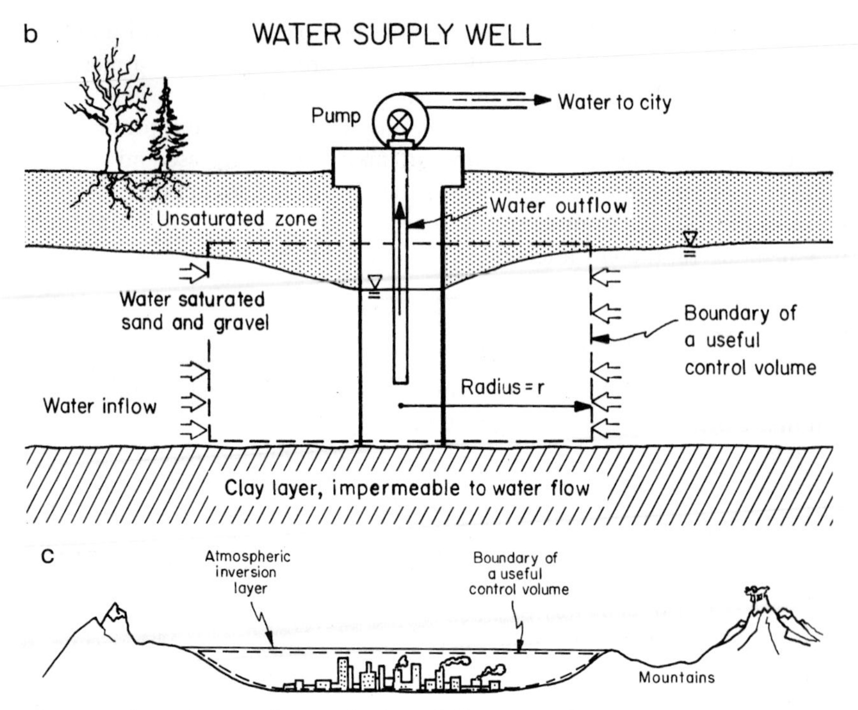

FIGURE 1-3 b–c *(Continued)* Control volume (b) has water inflow perpendicular to the boundary; by equating an expression for groundwater flow into the control volume to expressions for change in water storage and for the removal of water by the pump, we can derive equations that describe the hydraulic behavior of the well and the movement of chemicals associated with the groundwater. Control volume (c) would be useful when atmospheric conditions, such as an inversion layer, prevent the upward transport of an air pollutant over a city. Knowledge of the air volume into which a pollutant is mixed allows an estimation of the rate of change of the air pollutant concentration (i.e., the rate per unit volume at which the pollutant is being stored in the air), if the rate of pollutant release into the air volume is known.

the fate of river pollution in the vicinity of an industrial outfall is of concern, a specific reach of river beginning just upstream of the outfall and extending downstream to some location where the pollutant has become fully mixed across the river could constitute a good control volume. For analysis of the movement of water to a groundwater well, a cylindrical volume containing a portion of the water-bearing formation from which water is drawn into the well serves as a useful control volume (as illustrated further in Chapter 3). In situations pertaining to urban air quality, an imaginary "bubble" above a city might be a useful control volume to consider, especially if air flow patterns and natural barriers such as hills hinder transport of airborne chemicals across the wall of the imaginary bubble. Depending on the particulars of a situation, more than one practical control volume may be defined.

1.3.2 Consistency of Units

Anyone working in the environmental sciences must become familiar with the basic physical dimensions and units, many of which are described in the Appendix. A rigorous check for consistency of units is an excellent device for catching errors in expressions used in the modeling of chemical fate and transport. To confirm that an answer has the correct dimensions, one should express units along with each quantity that enters a mathematical expression. Not only does this often give insight into the mathematical expression, but it also highlights missing or superfluous terms that lead to spurious units and erroneous answers. For example, reconsider the lake control volume described in Section 1.3.1. If the rate at which a certain chemical was advected into the lake by the stream (mass per unit time) needed to be determined, the concentration of the chemical in the stream and the average velocity and the cross-sectional area of the stream would be multiplied together. Without containing actual numbers, such a calculation might look like

$$\text{Rate of chemical inflow} = \text{velocity (m/sec)} \cdot \text{area (m}^2\text{)} \cdot \text{concentration (g/m}^3\text{)}.$$

The final units of the answer would be (g/sec) with dimensions of [M/T], entirely appropriate to express the rate of chemical inflow to the lake. Alternatively, if (ft/sec) had been used for river velocity, the units of the answer would have been (g · ft/m · sec), a very good sign that a consistent set of units had not been used in the original expression. If the units for velocity had been omitted, the answer would have had the units of (g/m), which are clearly incorrect, in part because there is no time unit.

1.4 PHYSICAL TRANSPORT OF CHEMICALS

Most physical transport of chemicals in the environment occurs in the fluids air and water. There are primarily two kinds of physical processes by which chemicals are transported in these fluids: bulk movement of fluids from one location to another, and random (or seemingly random) mixing processes within the fluids. Both types of transport processes are implicitly included in the input and output terms of Eqs. [1-1a] and [1-1b]. (Biological transport, such as the swimming of a contaminated fish, is less amenable to analysis by the methods of physics—a fish's agenda depends on feeding and avoiding predators!) The first type of process, *advection*, is due to bulk, large-scale

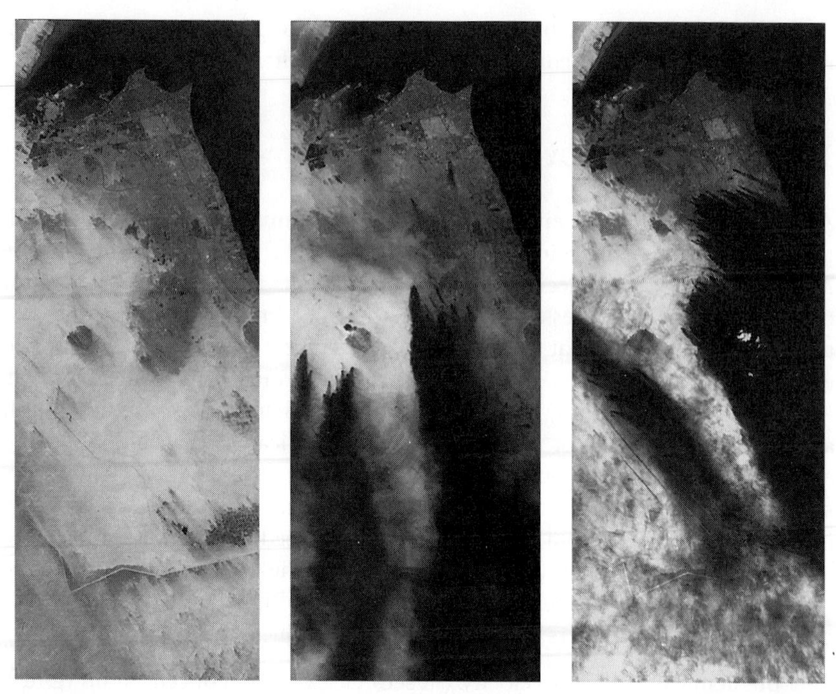

FIGURE 1-4 An example of pollutant advection and diffusion in the atmosphere. Smoke from multiple burning oil wells in Kuwait is carried downwind by advection. At the same time, the plumes of smoke widen because of diffusive transport, one of the major Fickian transport processes. Imagery courtesy of Space Imaging, Thornton, Colorado, USA.

movement of air or water, as seen in blowing wind and flowing streams. Figure 1-4 shows the advective transport of smoke from burning oil wells in Kuwait. (*Convection,* a similar term, often implies vertical advection of air or water resulting from density differences.) A chemical present in air or water is passively carried by this bulk advective movement, resulting in chemical transport.

In the second type of transport process, a chemical moves from one location in the air or water where its concentration is relatively high to another location where its concentration is lower, due to random motion of the chemical molecules *(molecular diffusion),* random motion of the air or water that carries the chemical *(turbulent diffusion),* or a combination of the two. Transport by such random motions, also called *diffusive* transport, is often modeled as being *Fickian.* Sometimes the motions of the fluid are not entirely random; they have a discernible pattern, but it is too complex to characterize. In this situation, the mass transport process is called *dispersion,* and it is also commonly treated as a Fickian process, even though in some situations it may only approximate true Fickian transport. In a given amount of time, the distances over which mass is carried by Fickian transport (molecular diffusion, turbulent diffusion, and dispersion) are usually not as great as those covered by advection.

1.4.1 QUANTIFICATION OF ADVECTIVE TRANSPORT

The bulk motion of fluid is common throughout the environment; this advective motion is described mathematically by the *direction* and the *magnitude* of its velocity. If a chemical is introduced into flowing air or water, the chemical is transported at the same velocity as the fluid. While "spreading" due to Fickian transport may occur at the same time, as described in the next section, the *center of mass* of the chemical moves by advection at the average fluid velocity.

The rate at which a chemical is transported per unit area is often expressed in terms of *flux density.* Flux density is the mass of chemical transported across an imaginary surface of unit area per unit of time (Fig. 1-5) and is often given the symbol J. Note that the imaginary surface may be one of the boundaries of a control volume. Flux density due to advection is equal to the product of a chemical's concentration in the fluid and the velocity of the air or water,

$$J = CV, \tag{1-2}$$

where J is the flux density [M/L^2T], C is the chemical concentration [M/L^3], and V is the fluid velocity [L/T].

The velocities of air and water frequently vary with time, as is evident to anyone who has stood in a gusty wind or swum in a turbulent river. Consequently, any estimate of flux density due to advection by a turbulent fluid flow must involve a time period over which flow variations and corresponding fluctuations of chemical concentration are averaged. Often the fluctuations in time are faster than the instruments for determining velocity and chemical concentration can follow, and the instruments inherently provide averaged values. In other situations, instruments can easily detect and measure the

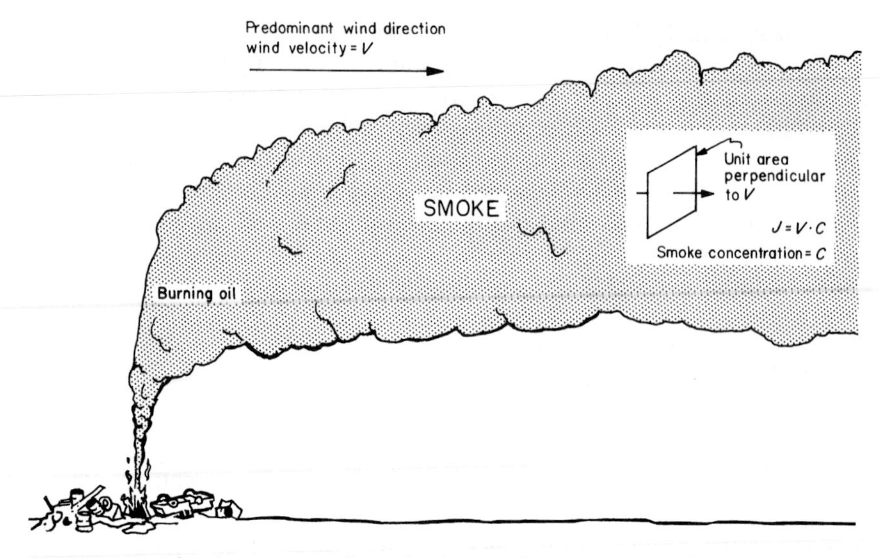

FIGURE 1-5 Advective transport of a smoke plume as shown in Fig. 1-4. The imaginary square frame is oriented perpendicular (⊥) to fluid flow and for convenience has an area of one (in whatever units we prefer—m^2, ft^2, etc.). The flux density of smoke, J, is the product of the wind velocity V and the concentration of smoke in the air, C.

fluctuations, and decisions must be made on how to average the data when reporting the wind or water speed or the associated chemical fluxes.

EXAMPLE 1-2

If the salt concentration in a river is 20 mg/liter and the average river velocity is 100 cm/sec, what is the average flux density J of salt in the downstream direction?

First, convert 20 mg/liter to units consistent with the velocity:

$$C = \frac{20 \text{ mg}}{\text{liter}} \times \frac{1 \text{ liter}}{1000 \text{ cm}^3} = \frac{0.02 \text{ mg}}{\text{cm}^3}.$$

Then use Eq. [1-2] to estimate the average flux density of salt:

$$J = 0.02 \text{ mg/cm}^3 \cdot 100 \text{ cm/sec} = 2 \text{ mg/cm}^2 \cdot \text{sec}.$$

1.4.2 Quantification of Fickian Transport

Turbulent Diffusion

Turbulent air and water motions contain constantly changing swirls of fluid, known as *eddies,* of many different sizes. One needs only to observe smoke rising from a factory smokestack or to experience gusty winds to appreciate the swirling and billowing that occur in air; in water, turbulence is evident in river rapids and breaking surf. These ubiquitous eddies give rise to another type of mass transport, known as *turbulent* (or *eddy*) *diffusion.* Turbulent diffusion, one of the mass transport processes commonly modeled as Fickian, arises from the random mixing of the air or water by these eddies. This type of mass transport neither augments nor impedes the downwind or downstream advective motion of a chemical. By mixing the chemical in the air or water, however, turbulent diffusion has the net effect of carrying mass in the direction of decreasing chemical concentration. The effects of turbulent diffusion on a mass of chemical are visible in many environmental situations: the spreading of a dye blob injected into a river, the expanding of a puff of smoke from fireworks, and the widening and blurring of condensation trails (contrails) of high altitude jets. Note in Fig. 1-4 that the oil smoke plumes become broader due to Fickian transport as they move downwind from their sources.

Fick's first law is typically used to describe the flux density of mass transport by turbulent diffusion,

$$J = -D(dC/dx) \qquad \text{(in one dimension)}, \qquad [1\text{-}3]$$

where J is the flux density [M/L^{2}T], D is the Fickian mass transport coefficient [L^2/T], C is the chemical concentration [M/L^3], and x is the distance over which a concentration change is being considered [L]. (In simple calculations the minus sign is often omitted if the direction of Fickian transport is clear.)

The parameter D is usually called a *turbulent* (or *eddy*) *diffusion coefficient* when it arises from fluid turbulence; its value varies enormously from one situation to another, depending on the intensity of turbulence and on whether the environmental medium is air or water. The diagram in Fig. 1-6 shows the Fickian mass flux arising from a concentration gradient in a smoke plume.

Fick's first law can also be expressed in three dimensions using vector notation,

$$\vec{J} = -D\nabla C \qquad \text{(in three dimensions)}, \qquad [1\text{-}4]$$

where ∇ is the gradient operator and D is assumed to be equal in all directions.

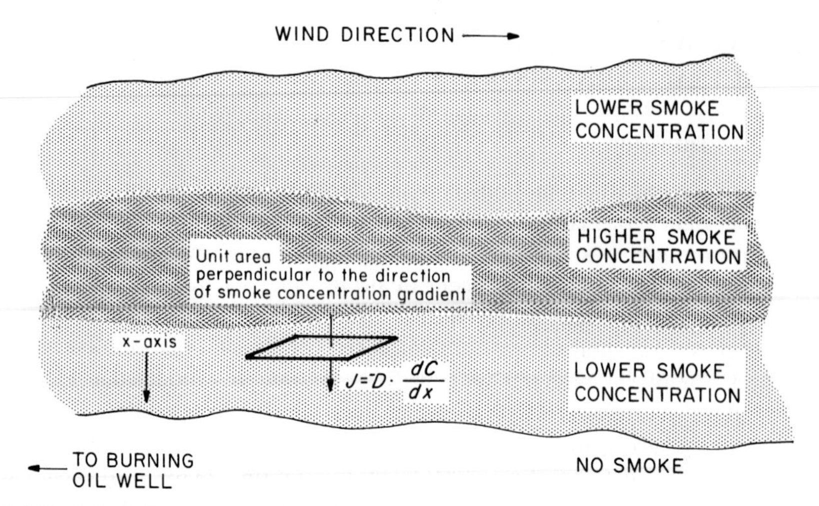

FIGURE 1-6 Fickian transport by turbulent diffusion in a smoke plume as shown in Figure 1-4. As in Figure 1-5, the square frame is of unit area, but in this case is oriented perpendicular to the direction of the concentration gradient (defined as the direction in which the concentration changes the most per unit distance.) In this case the x-axis is drawn in the direction of the gradient. The flux density, J, is equal to the concentration gradient, dC/dx, multiplied by the Fickian transport coefficient D. (In this situation, D is called a turbulent or eddy diffusion coefficient, because the major agent of Fickian transport is turbulence.)

In Eq. [1-4] the vector notation indicates that the direction of flux is in the direction of the steepest change in concentration with distance (the direction of the gradient vector), assuming that D is equal in all directions. For illustrative purposes, this book works mostly with the one-dimensional form of Fick's first law—Eq. [1-3]; in practice, many environmental situations also can be modeled in one dimension. Note that in the most general case, not only may D be *anisotropic* (i.e., not equal in all directions), but also D may vary with time and location.

Dispersion

Turbulent diffusion is an important mode of chemical transport in both surface water and air. In the subsurface environment, groundwater flow normally lacks the eddy effects that characterize surface water and air movements because typical groundwater velocities are so much lower. Nevertheless, groundwater must take myriad detours as it moves from one point to another, traveling over, under, and around soil particles, as shown in Fig. 1-7. These random detours cause mixing, thus the net transport of a chemical from

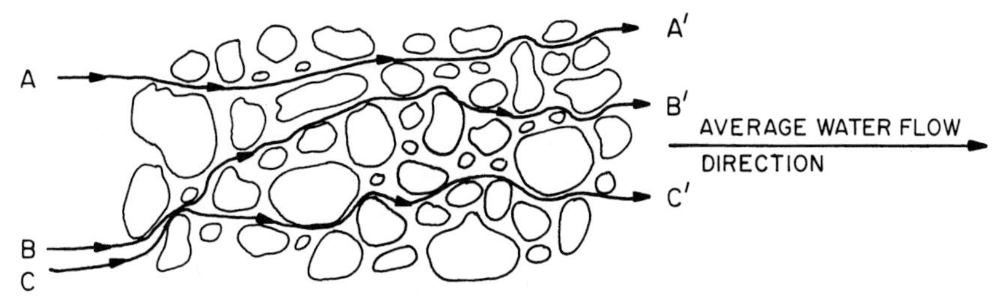

FIGURE 1-7 Fickian transport by dispersion as water flows through a porous medium such as a soil. Seemingly random variations in the velocity of different parcels of water are caused by the tortuous and variable routes water must follow. This situation contrasts with that of Fig. 1-6, in which turbulence is responsible for the random variability of fluid paths. In this case as well as in the previous one, Fickian mass transport is driven by the concentration gradient and can be described by Fick's first law. The mass transport effect arising from dispersion can be further visualized in Fig. 3-17. There, a mass initially present in a narrow slice in a column of porous media is transported by mechanical dispersion in such a way as to form a wider but less concentrated slice. At the same time, the center of mass also is transported longitudinally in the direction of water flow.

regions of higher concentration to regions of lower concentration. Despite the different physical mechanism causing the mixing, the net mass transport is entirely analogous to that of turbulent diffusion.

As in the case of turbulent diffusion, the chemical flux is often expressed by Fick's first law, as shown in Eqs. [1-3] and [1-4], but in this case D is called a *mechanical dispersion coefficient.* Dispersion also occurs at much larger scales than that of soil particles; for example, groundwater may detour around regions of relatively less permeable soil that are many cubic meters in volume. At this scale, the process is called *macrodispersion.*

Molecular Diffusion

The Fickian mass transport processes discussed so far involve parcels of fluid taking irregular paths, due to either turbulence or obstructions, in such a complex manner that the individual eddies and obstructions cannot be tracked. Even if a fluid is entirely quiescent and without obstructions, however, chemicals will still move from regions of higher concentration to regions of lower concentration, due to the ceaseless random movement (*thermal* motion) of molecules. This type of mixing is called *molecular diffusion* and is also described by Fick's first law, but for a given chemical gradient it usually results in lower flux densities than those of the other Fickian mass transport processes. In this case, D in Eqs. [1-3] and [1-4] is called a *molecular diffusion coefficient*. Unlike the coefficients for the previous two examples of Fickian transport, molecular diffusion coefficients can be estimated for a particular situation without much site-specific data, because they depend primarily on the size of the molecules that are diffusing. At environmental temperatures, most chemicals exhibit a molecular diffusion coefficient in air of about 0.2 cm^2/sec, and in water of about 10^{-5} cm^2/sec. Molecular diffusion increases in magnitude at higher temperatures and for smaller molecules or particles (which, at any given temperature, have higher average speeds than larger molecules or particles). Molecular diffusion sets the lower limit on the amount of Fickian mixing that can be expected in a fluid. The total Fickian transport coefficient equals the sum of the contributing Fickian coefficients due to molecular diffusion, turbulent diffusion, and mechanical dispersion.

EXAMPLE 1-3

Gasoline-contaminated groundwater has been transported under a residential dwelling from a nearby gasoline station. Two meters beneath the 100 m^2 dirt floor of the residential basement, the concentration of hydrocarbon vapors in the soil air is 25 ppm on a mass/mass basis. Estimate the flux density of gasoline vapor and the daily rate of vapor inflow transported into the basement by molecular diffusion. Assume an approximate diffusion coefficient of 10^{-2} cm^2/sec for gasoline vapor in the soil (this value includes a correction for the presence of soil grains, as discussed in Chapter 3). Also assume the basement is well ventilated, so that the gasoline vapor concentration in the basement is much less than 25 ppm. Air density is approximately 1.2 g/liter at 1 atm pressure and 20°C (Weast, 1990).

Diffusion calculations require that concentration be expressed as mass per unit volume. To express the concentration of vapor as mass per unit volume, consider that 25 ppm is the same as 25 g per million grams of air. Given that 1000 cm^3 of air has a mass of about 1.2 g at 1 atm pressure, the concentration of gasoline vapor 2 m below the dirt floor can be expressed as

$$C = 25 \text{ g}/10^6 \text{ g air} \cdot 1.2 \text{ g air}/1000 \text{ cm}^3 \text{ air} = 3 \times 10^{-8} \text{ g/cm}^3.$$

Treating this as a one-dimensional problem, the upward concentration gradient of vapor is approximately

$$dC/dz = (3 \times 10^{-8} \text{ g/cm}^3)/200 \text{ cm} = 1.5 \times 10^{-10} \text{ g/cm}^4.$$

The flux density, by Fick's first law in Eq. [1-3], is

$$J = -DdC/dz = (10^{-2}\ \text{cm}^2/\text{sec}) \cdot (1.5 \times 10^{-10}\ \text{g/cm}^4)$$
$$= 1.5 \times 10^{-12}\ \text{g/cm}^2\ \text{sec}.$$

(Note informal treatment of the minus sign.)
Then the daily rate of vapor flow into the house is

$$(1.5 \times 10^{-12}\ \text{g/cm}^2\ \text{sec}) \cdot (10^6\ \text{cm}^2) \cdot (3600\ \text{sec/hr}) \cdot (24\ \text{hr/day}) = 0.13\ \text{g/day}.$$

This flux is probably not enough to worry about from a flammability perspective because the house is well ventilated.

1.5 MASS BALANCE IN AN INFINITELY SMALL CONTROL VOLUME: THE ADVECTION–DISPERSION–REACTION EQUATION

So far, the concept of mass conservation has been applied to large, easily measurable control volumes such as lakes. Mass conservation also can be usefully expressed in an infinitesimal control volume, mathematically considered to be a point. Conservation of mass is expressed in such a volume with the *advection–dispersion–reaction equation.* This equation states that the rate of change of chemical storage at any point in space, dC/dt, equals the sum of both the rates of chemical input and output by physical means and the rate of net internal production (sources minus sinks). The inputs and outputs that occur by physical means (advection and Fickian transport) are expressed in terms of the fluid velocity (V), the diffusion/dispersion coefficient (D), and the chemical concentration gradient in the fluid (dC/dx). The input or output associated with internal sources or sinks of the chemical is represented by r. In one dimension, the equation for a fixed point is

$$\frac{dC}{dt} = -V \cdot \frac{dC}{dx} + \frac{d}{dx}\left(D \cdot \frac{dC}{dx}\right) + r. \qquad [1\text{-}5]$$

The only difference between Eqs. [1-5] and [1-1a] and [1-1b] is that, because the control volume is of an unspecified, arbitrarily small size, each term is expressed as mass per unit time *per unit volume.* Thus, the leftmost term, dC/dt, represents the rate at which a chemical's *concentration* (storage per unit time) changes at a fixed point in a flowing fluid. The concentration can change if there is a different concentration elsewhere in the flowing fluid and this different concentration is carried by advection to the fixed point of interest; this process corresponds to the term $V \cdot dC/dx$. The concentration can also change by Fickian transport if there is a spatially varying concentration gradient in the fluid; this process corresponds to the term $d/dx(D \cdot dC/dx)$. Changes in the concentration also can occur if a source or sink process, such as a chemical or biological reaction, is introducing or removing the compound of interest (r).

Equation [1-5] is pertinent to a one-dimensional system, such as a long, narrow tube full of water, where significant variations in concentration may be assumed to occur only along the length of the tube. In a three-dimensional situation, the advection–dispersion–reaction equation can be represented most succinctly using vector notation, where ∇ is the divergence operator:

$$\frac{dC}{dt} = -\vec{V} \cdot \nabla C + \nabla \cdot D(\nabla C) + r. \qquad [1\text{-}6]$$

Note that the transport terms (the second and third terms) in Eq. [1-6] are the three-dimensional counterparts of the corresponding terms in Eq. [1-5]. As in Eq. [1-4], D is assumed equal in all directions. In many cases, this assumption is an oversimplification; the value of D in the direction of flow can be very different than the value perpendicular to flow (i.e., D may be anisotropic). Furthermore, D may vary with location (i.e., be inhomogeneous), or vary with time. Often, a larger value of D may become applicable as the scale of the problem increases.

Although the forms of the mass conservation equation shown in Eqs. [1-5] and [1-6] may not appear to be directly applicable to large-scale environmental situations, they actually are very powerful tools. These equations can be integrated to yield mathematical solutions to chemical distributions in many physical systems. Given information on the inflow rates and chemical concentrations at the boundaries of a control volume, the chemical concentrations throughout the control volume may be determined by invoking solutions to Eqs. [1-5] and [1-6].

1.6 BASIC ENVIRONMENTAL CHEMISTRY

Now that the transport and storage terms of Eqs. [1-1a] and [1-1b] have been discussed, consider the source and sink terms. These correspond to the introduction or removal of a chemical due to either chemical reactions or transfer to another phase (e.g., volatilization of a chemical from water to air). Chemical reactions, which may occur spontaneously (abiotically) or be biologically mediated, result in the transformation of one chemical substance into another. Chemical substances can be uncharged atoms or molecules, ions (electrically charged atoms or aggregations of atoms), or free *radicals* (highly reactive atoms or aggregations of atoms having an unpaired valence electron). The degradation of pollutant chemicals is a subset of all possible chemical reactions occurring in the environment.

By definition, chemical reactions involve the formation or breakage of chemical bonds between atoms. Chemical bonds hold atoms together in a variety of ways. Bond types include *covalent bonds,* in which electrons are shared between atoms; *ionic bonds,* in which the bonding force arises from electric charges of opposite sign on adjacent atoms; and *hydrogen bonds,* in which the somewhat positively charged hydrogen at one end of a molecule loosely bonds with a somewhat negatively charged atom of another molecule. Other forces, such as *Van der Waals forces,* which cause weak mutual attractions between all molecules, can also contribute to bonding. A chemical bond may have characteristics of more than one idealized type of bonding. For further discussions on chemical reactions, the reader is referred to texts such as Petrucci (1989), Radel and Navidi (1990), and Shriver *et al.* (1994) for inorganic chemicals; and to Roberts and Caserio (1977), McMurry (1992), and Streitwieser *et al.* (1992) for organic chemicals.

1.6.1 Chemical Kinetics versus Chemical Equilibrium

Chemical reactions may be described from the standpoint of either *kinetics* or *equilibrium.* Kinetics describes the *rate* at which a reaction takes place and is significant when comparing a particular reaction rate with the rate at which some other process occurs. For example, if a degradable chemical is discharged into a stream by an industrial outfall, the rate of degradation and the travel time to a downstream municipal water supply intake can be used to calculate the chemical concentration at the intake. The rates (kinetics) at which degradation and advection occur are the significant parameters.

Equilibrium, by contrast, describes the *final expected chemical composition* in a control volume. In chemical parlance, a control volume with its chemical contents is often referred to as a *system.* Equilibrium is relevant in the case of reactions that are rapid compared with other environmental processes of interest. For example, if potassium hydroxide (KOH) is added to municipal drinking water to decrease its acidity, the reaction of the potassium hydroxide and the acids present in the drinking water may be assumed to be instantaneous compared with the time it takes to transport water to customers. A consideration of equilibrium chemistry also is very useful to determine the expected final composition toward which a system is proceeding, even if at a particular time it has not yet reached equilibrium. The context of the environmental problem of concern usually makes it clear whether it is the equilibrium chemical composition of a system or the rate at which a system proceeds toward equilibrium that is of particular interest. In this section, chemical reactions are discussed in terms of equilibrium before kinetics is addressed.

1.6.2 Free Energy

Regardless of the exact nature of the chemical bonds involved in a chemical reaction, a consideration of the *Gibbs free energy* of the chemical system provides information on both the direction in which a chemical reaction will proceed and the final equilibrium composition of the system. Gibbs free energy (G) is a function of both the *enthalpy*, which is the energy possessed by a mixture of chemicals in a system, and the *entropy*, or disorder, of the system. Gibbs free energy depends on the chemical composition, pressure, and temperature of the system and is quantitatively expressed as

$$G = H - TS, \qquad [1\text{-}7]$$

where G is the Gibbs free energy, H is the enthalpy, S is the entropy, and T is the absolute temperature.

Enthalpy refers to the energy in a system. Enthalpy includes energy associated with all intramolecular forces (due to bonds and attractions within molecules) and intermolecular forces (due to bonds and attractions between molecules). Entropy refers to the degree of disorganization, or randomness, of a system. At the molecular level, the entropy of a chemical system at any given state (with the state defined on the basis of macroscopic quantities such as pressure, volume, and temperature) can be shown to be a measure of the probability that the molecules within the system will occur with a distribution that creates that particular state. At the macroscopic level, the incremental change in entropy associated with any given reversible process is equal to the heat energy entering the system divided by the absolute temperature of the system; thus, entropy is expressed in units of energy per Kelvin. A Kelvin (K) is a measurement of absolute temperature; degrees Celsius are converted to Kelvins by the equation K = °C + 273.15. For a more thorough discussion of entropy, the reader is referred to a text on chemical thermodynamics, such as that of Wall (1974).

It takes work to organize chemical molecules; an increase in the level of organization is reflected in a decrease in the entropy term (more organization, less uniformity). Conversely, a disorganized, more random system of molecules has an increased entropy (less organization, more uniformity). Reactions in a chemical system proceed toward a final composition that minimizes the Gibbs free energy of the system, by a combination of maximizing disorder and of minimizing enthalpy. For any reaction, the actual pathway taken, known as the *course of the reaction,* does not affect the final equilibrium state.

The change in Gibbs free energy (ΔG), which occurs as a system proceeds toward equilibrium, can be expressed as the sum of two terms. The first term is the *standard free energy change* (ΔG°), which is fixed for any given reaction. ΔG° can be calculated from the *stoichiometry* of the reaction (i.e., how many moles of one compound react with how many moles of another compound) and the standard free energies of the chemicals involved. The second term contains the *reaction quotient* (Q), which depends on the concentrations of chemicals present. The fact that ΔG can be expressed in terms of the concentrations of all chemicals present in a system makes it possible to determine in which direction a chemical reaction will proceed and to predict its final composition when it reaches equilibrium.

To get a better feel for Gibbs free energy, consider a reversible reaction occurring in water, where lowercase letters represent *stoichiometric coefficients* and uppercase letters represent four chemical compounds, A, B, C, and D:

$$aA + bB \rightleftharpoons cC + dD \qquad [1\text{-}8]$$

As this system proceeds toward equilibrium, the change in Gibbs free energy per additional mole reacting is

$$\Delta G = \Delta G^\circ + RT \ln Q, \qquad [1\text{-}9]$$

where ΔG° is the standard free energy change (and is a constant for a given reaction), R is the *gas constant,* T is the absolute temperature, and

$$Q = \frac{[C]^c[D]^d}{[B]^b[A]^a}, \qquad [1\text{-}10]$$

with [A], [B], [C], and [D] referring to the *molar* concentrations of chemicals A, B, C, and D in water. Equivalent values of R are 0.0821 liter · atm/(mol · K), 8.31 J/(mol · K), and 1.99 cal/(mol · K), as shown in the Appendix.

If ΔG is negative, the reaction will proceed from left to right until ΔG becomes zero. (In mathematical terms, the only way for ΔG to become less negative is for $\ln Q$ to become less negative, which occurs as the product $[C]^c[D]^d$ in the numerator gets larger.) Conversely, if ΔG is positive, the reaction will proceed from right to left until ΔG becomes zero. For greater detail and derivations of these equations, the reader is referred to texts on chemical thermodynamics (Guggenheim, 1967; Wall, 1974; Denbigh, 1981; Castellan, 1983).

1.6.3 Chemical Equilibrium

As previously mentioned, the Gibbs free energy of a chemical system is minimized at equilibrium. At equilibrium, the forward and reverse reactions of a chemical reaction, such as given in Eq. [1-8], are necessarily occurring at the same rate, and the change in Gibbs free energy is zero:

$$\Delta G = \Delta G^\circ + RT \ln Q = 0 \qquad \text{(at equilibrium).} \qquad [1\text{-}11a]$$

Therefore,

$$\Delta G^\circ = -RT \ln Q \qquad \text{(at equilibrium).} \qquad [1\text{-}11b]$$

For the moment, consider only reactions involving chemicals dissolved in water. The preceding equations can be combined with the definition of the reaction quotient, Eq. [1-10], to define an *equilibrium constant*, K, that applies to the *final expected* chemical composition of the system:

$$K = \frac{[C]^c [D]^d}{[B]^b [A]^a} = e^{-\Delta G^\circ/RT}. \qquad [1\text{-}12]$$

Note that molar concentrations must be used in the expression for K and that K equals the reaction quotient Q *only* at equilibrium (Fig. 1-8). Equation [1-12], the expression of the equilibrium state of a reaction, is known as the *mass action law* and is a direct consequence of the minimization of Gibbs free

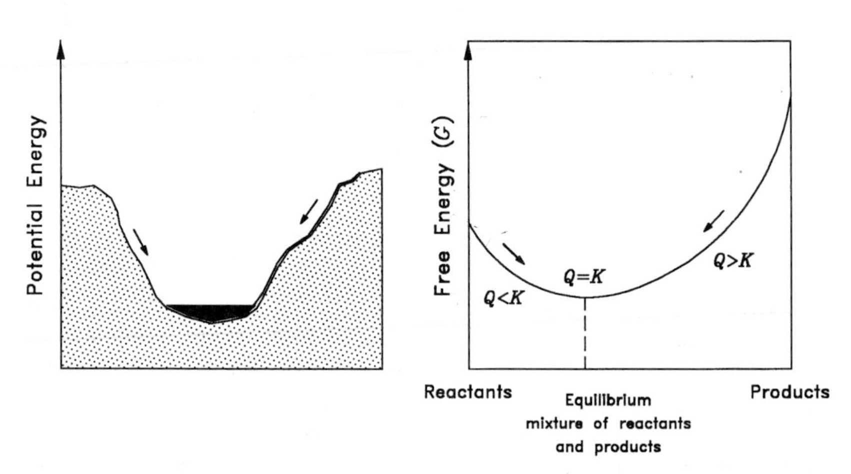

FIGURE 1-8 Free energy change in a chemical reaction. Analogous to water seeking its position of lowest gravitational energy on a hill slope, a chemical system will tend to move toward a composition having the lowest Gibbs free energy. Adapted from *Chemistry, 1st edition,* by S. Radel and M. Navidz. © 1990. Reprinted with permission of Brooks/Cole Publishing, a division of International Thomson Publishing. Fax 800-730-2215.

TABLE 1-1 Equilibrium Constants and Standard Free Energy Changes for Some Common Environmental Reactions[a]

					Approx. log K	$\Delta G°$ (kcal/mol)
			Acid–base reactions			
$H_2CO_3^*$ Carbonic acid	→	H^+ Hydrogen ion	+	HCO_3^- Bicarbonate ion	-6.3[b]	8.6
HCO_3^- Bicarbonate ion	→	H^+ Hydrogen ion	+	CO_3^{2-} Carbonate ion	-10.3[b]	14.1
CH_3COOH Acetic acid	→	H^+ Hydrogen ion	+	CH_3COO^- Acetate ion	-4.8[c]	6.5
$NH_3(g)$ Ammonia	→	NH_4^+ Ammonium ion	+	OH^- Hydroxide ion	-4.8[b]	6.5
			Precipitation–dissolution reactions			
$CaCO_3(s)$ Calcium carbonate (limestone)	→	Ca^{2+} Calcium ion	+	CO_3^{2-} Carbonate ion	-8.3[b]	11.6
$SiO_2(s)$ Quartz	+	$2H_2O$	→	H_4SiO_4 Silicic acid	-3.7[d]	5.0
$CaSO_4(s)$ Gypsum	→	Ca^{2+} Calcium ion	+	SO_4^{2-} Sulfate ion	-4.6[d]	6.3

[a] $\Delta G°$ values were calculated for a temperature of 25°C.
[b] Morel and Hering (1993).
[c] Weast (1990).
[d] Stumm and Morgan (1981).

energy. Table 1-1 presents standard free energy changes and equilibrium constants for some common environmental reactions.

EXAMPLE 1-4

Given the reaction between copper and hydroxide ions dissolved in water:

$$Cu^{2+} + OH^- \longrightarrow CuOH^+; \qquad \log K = 6.3,$$

what is the ratio of the Cu^{2+} (cupric) ion to the $CuOH^+$ ion in water if the hydroxide ion (OH^-) concentration is 10^{-4} mol/liter?

From mass action:

$$\frac{[CuOH^+]}{[OH^-][Cu^{2+}]} = 10^{6.3}.$$

Given that $[OH^-] = 10^{-4}$ mol/liter,

$$\frac{[CuOH^+]}{[Cu^{2+}]} = 10^{2.3}$$

or

$$\frac{[Cu^{2+}]}{[CuOH^+]} = 0.005.$$

Environmental reactions may involve not only chemicals dissolved in water but also chemicals in solid and gaseous forms. If a solid is part of a chemical reaction (i.e., if it is being formed or dissolved), its concentration is entered as one; if a gas is part of a reaction, its concentration is represented, by convention, as pressure, as discussed further in Section 1.8.1. Note that the letters (s) or (g) are often entered after a chemical formula to indicate solid or gas, respectively.

EXAMPLE 1-5

Limestone, $CaCO_3(s)$, is in equilibrium with water in which the carbonate ion, CO_3^{2-}, concentration is 10^{-5} *M*. What is the concentration of calcium ions, Ca^{2+}, in the water?

From Table 1-1, the relevant reaction and equilibrium constant can be obtained:

$$\frac{[Ca^{2+}][CO_3^{2-}]}{[CaCO_3]} = 10^{-8.3}$$

$$[Ca^{2+}] = 10^{-8.3} \cdot \frac{[CaCO_3]}{[CO_3{}^{2-}]}$$

$$= 10^{-8.3} \cdot \frac{1}{10^{-5}}$$

$$[Ca^{2+}] = 10^{-3.3}\ M.$$

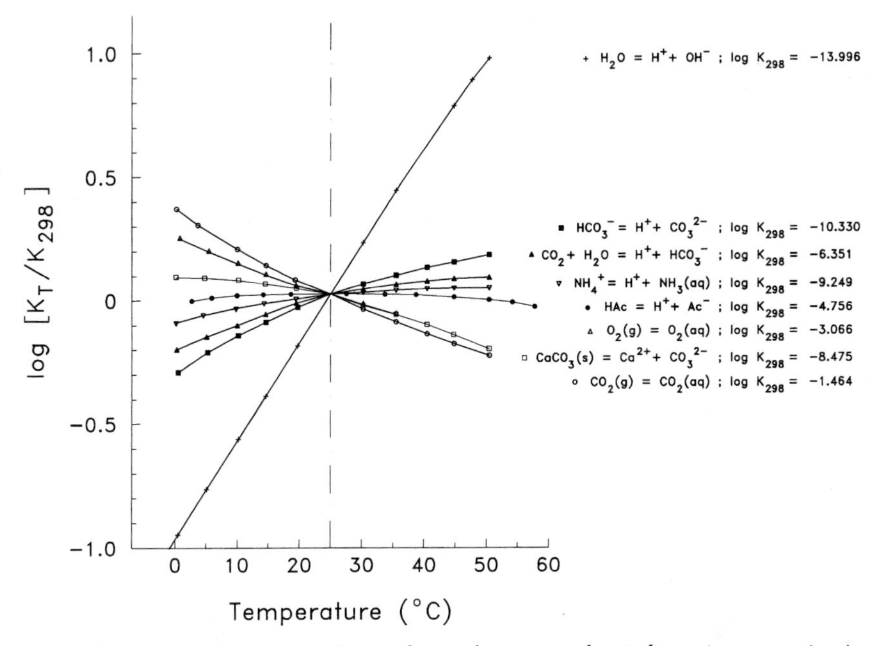

FIGURE 1-9 Temperature dependence of several common chemical reactions occurring in water. Although temperature may often be neglected in approximate calculations, for maximum accuracy, equilibrium constants must be corrected for the temperature of the chemical system of interest [adapted from *Principles and Applications of Aquatic Chemistry*, by F.M.M. Morel and J.G. Hering. Copyright © 1993, John Wiley & Sons, Inc. Reprinted by permission of John Wiley & Sons, Inc.].

The equilibrium constant K varies somewhat with changes in environmental temperature and pressure, as indicated in Fig. 1-9 for several common environmental reactions. Equilibrium constants are known for a wide variety of reactions and form the basis (when combined with mass balance and electroneutrality equations as described later) for both manual and computerized techniques for determining the equilibrium composition of many complex mixtures of chemicals in surface waters and groundwater (Stumm and Morgan, 1996; Morel and Hering, 1993).

One of the most basic equilibrium reactions of environmental concern is the ionization of water (H_2O) to form hydrogen ions (H^+) and hydroxide ions (OH^-):

$$H_2O \rightleftharpoons H^+ + OH^-. \qquad [1\text{-}13]$$

As written, H_2O is a *reactant* and H^+ and OH^- are *products*. The equilibrium constant K for Eq. [1-13] can be written as

$$K = \frac{[H^+][OH^-]}{[H_2O]} = 1.8 \times 10^{-16} \text{ mol/liter.} \qquad [1\text{-}14]$$

It is conventional, however, to omit the concentration of water from mass action expressions and instead to absorb its value into a revised equilibrium constant. The concentration of water is 55.4 *M*; thus, Eq. [1-14] can be rewritten as

$$K_w = [H^+][OH^-] = 10^{-14} \text{ mol}^2\text{/liter}^2. \qquad [1\text{-}15]$$

This convention is valid for most environmental situations because the concentration of water is essentially constant at 55.4 *M*, negligibly affected by the mass of chemicals dissolved in it or by its self-ionization.

The acidity of a given water or chemical solution is determined by the concentration of hydrogen ions present. This concentration is usually expressed by the pH. For now, consider pH to be approximately equal to $-\log[H^+]$ (strictly speaking, pH is the negative $\log_{10}$ of the hydrogen ion *activity*, as described further in Section 1.6.5). The pH of most natural waters lies between 4 and 9, although both higher and lower values occur in some locations. Note that a pH of 7 means a water is neutral, because there are equal concentrations of hydrogen and hydroxide ions. Waters containing more hydrogen ions than hydroxide ions have a pH less than 7 and are *acidic*; waters containing fewer hydrogen than hydroxide ions are *basic* (or *alkaline*) with a pH greater than 7.

EXAMPLE 1-6

A barrel of hard cider is bought too early for a party. Bacteria in the barrel begin to convert the ethanol (C_2H_5OH) present (at approximately 6% by weight) to acetic acid (CH_3COOH) according to the following reaction:

$$O_2 + C_2H_5OH \longrightarrow CH_3COOH + H_2O.$$

By the time it is discovered that the hard cider is going bad, bacteria have converted approximately one-third of the ethanol. Assuming the pH is determined primarily by the acetic acid concentration, what is the pH of the cider at this time? The relevant reaction and equilibrium constant are

$$CH_3COOH \rightleftharpoons H^+ + CH_3COO^-; \qquad K = 1.75 \times 10^{-5} \text{ mol/liter.}$$

First, estimate the molar concentration of ethanol that has been converted to acetic acid. If the density of cider is approximately that of water (1000 g/liter), the cider initially contains approximately 60 g/liter of ethanol. To convert one-third of that quantity, 20 g/liter, into a molar concentration, the molecular weight of ethanol needs to be calculated, using Table A-10 to obtain atomic weights:

$$2(12) + 6(1) + 1(16) = 46 \text{ g/mol}.$$

Therefore,

$$\frac{20 \text{ g}}{\text{liter}} \cdot \frac{1 \text{ mol}}{46 \text{ g}} = 0.43\ M.$$

Given that 1 mol of ethanol produces 1 mol of acetic acid, the cider will be a 0.43 *M* solution of acetic acid, once bacteria have converted one-third of the ethanol.

Second, estimate the pH due to the presence of acetic acid. The relevant mass action expression is

$$\frac{[H^+][CH_3COO^-]}{[CH_3COOH]} = 1.75 \times 10^{-5} \text{ mol/liter}.$$

Now make two simplifying assumptions: (1) all H^+ ions come from acetic acid, so $[H^+] = [CH_3COO^-]$ and (2) only a small amount of acetic acid actually reacts to form H^+. The first assumption is justifiable because the contribution of $[H^+]$ or $[OH^-]$ from the ionization of water is usually neglible when significant quantities (10^{-5} *M* or more) of acids or bases are added to water. The second assumption will be checked after $[CH_3COOH]$ is approximated by its initial concentration:

$$\frac{[H^+][H^+]}{0.43\ M} = 1.75 \times 10^{-5} \text{ mol/liter}$$

$$[H^+] = 2.8 \times 10^{-3}\ M$$

$$\text{pH} \approx -\log(2.8 \times 10^{-3}) = 2.6.$$

The concentration of H^+ is consistent with the second assumption; less than 1% of the acetic acid reacts to form H^+ and CH_3COO^-.

1.6.4 ELECTRONEUTRALITY

The principle of *electroneutrality* in aqueous chemical systems states that the sum of the concentrations of all positively charged ions (expressed in equivalents) equals the sum of the concentrations of all negatively charged ions, so that the overall charge of the solution is zero. (If this were not true, we would be constantly bombarded with electrical shocks!) When an equation based on the principle of electroneutrality is combined with equations provided by *conservation of mass,* and by the *mass action law,* Eq. [1-12], the equilibrium chemical composition of a system can be calculated.

EXAMPLE 1-7

A water sample is taken from a stream in Nevada receiving acid mine drainage. The stream passes through an area containing gypsum, $CaSO_4$. Laboratory analysis shows that the pH of the water is 4, the total concentration of sulfate (SO_4^{2-}) is 6×10^{-3} M, and the total concentration of chloride (Cl^-) is 3×10^{-4} M. (a) Assuming that the only other ionic species present is calcium (Ca^{2+}), what is the calcium concentration in the water? (b) Will the precipitation reaction,

$$Ca^{2+} + SO_4^{2-} \rightleftharpoons CaSO_4(s); \qquad K = 10^{4.62},$$

occur further downstream (i.e., will a solid form)?

(a) The electroneutrality equation for this water is

$$2[Ca^{2+}] + [H^+] = 2[SO_4^{2-}] + [OH^-] + [Cl^-]$$

$$2[Ca^{2+}] = -10^{-4} + 2(6 \times 10^{-3}) + 10^{-10} + (3 \times 10^{-4})$$

$$[Ca^{2+}] = 6.1 \times 10^{-3}\ M.$$

(b) From mass action:

$$\frac{[CaSO_4]}{[Ca^{2+}][SO_4^{2-}]} = \frac{1}{(6.1 \times 10^{-3})(6 \times 10^{-3})} = 10^{4.4}.$$

Because the reaction quotient $10^{4.4}$ is less than the equilibrium constant $10^{4.62}$, the reaction will proceed to the right and a solid should form; the extent to which a solid actually forms may be limited by kinetic considerations.

1.6.5 ACTIVITY

Although the mass action equation, Eq. [1-12], is written in terms of *concentration,* in fact the quantities in the brackets should be *activities,* which may be thought of as "corrected" concentrations that take into account nonideal effects in aqueous systems. These nonideal effects arise from electrostatic forces between ions dissolved in the water, and they increase as the total concentration of dissolved ions, measured as *ionic strength,* increases,

$$I = \frac{1}{2} \sum (z_i)^2 S_i, \quad [1\text{-}16]$$

where I is the ionic strength (mol/liter), z_i is the charge number on each ion, and S_i is the concentration of each ion (mol/liter).

Depending on the ionic strength, an activity for each ion can be calculated to account for nonideal effects,

$$\{x_i\} = \gamma_i [x_i], \quad [1\text{-}17]$$

where $\{x_i\}$ is the activity of an ion (mol/liter), γ_i is the activity coefficient, and $[x_i]$ is the concentration of the ion in the water (mol/liter).

Table 1-2 provides the negative $\log_{10}$ of activity coefficients for varying ionic strengths and ion charges. At low ionic strength, less than about 0.001 M, γ becomes very close to one. For ions in natural waters, γ is generally less than or equal to one, and for approximate calculations in many fresh waters, ionic strength effects may be neglected. In saline waters, such as ocean water, activity coefficients usually need to be used. For dissolved gases, γ tends to be greater than one.

TABLE 1-2 $-\text{Log}_{10}$ of Activity Coefficients for Specific Charge Numbers and Ionic Strengths[a]

z[b] I[c] (M)	1	2	3	4
0.1	0.11	0.44	0.99	1.76
0.3	0.13	0.52	1.17	2.08
0.5	0.15	0.60	1.35	2.40
1.0	0.14	0.56	1.26	2.24
2.0	0.11	0.44	0.99	1.76
3.0	0.07	0.28	0.63	1.12

[a] Morel and Hering (1993).
[b] z, charge number on ion.
[c] I, ionic strength.

EXAMPLE 1-8

A water sample with a pH of 8 is found to contain 10^{-1} *M* chloride (Cl^-), 2×10^{-1} *M* sodium (Na^+), and 5×10^{-2} *M* sulfate (SO_4^{2-}). The cupric ion Cu^{2+} is detected at 10^{-7} *M*. If the behavior of copper in the water is of interest, what copper activity should be used in equilibrium expressions involving copper and other chemical species?

The ionic strength of the solution is

$$I = \frac{1}{2}\Big[(-1)^2[Cl^-] + (1)^2[Na^+] + (-2)^2[SO_4^{2-}] + (2)^2[Cu^{2+}] + (1)^2[H^+] + (-1)^2[OH^-]\Big]$$

$$I = \frac{1}{2}\Big[(-1)^2(10^{-1}\,M) + (1)^2(2 \times 10^{-1}\,M) + (-2)^2(5 \times 10^{-2}\,M) + (2)^2(10^{-7}\,M) + (1)^2(10^{-8}\,M) + (-1)^2(10^{-6}\,M)\Big]$$

$$I \approx 0.25\,M.$$

By using Table 1-2, the negative $\log_{10}$ of the activity coefficient for copper should be approximately 0.50 (interpolating between 0.44 and 0.52).

Therefore, the activity coefficient is 0.32, and the activity of copper is

$$\{Cu^{2+}\} = 0.32\,(10^{-7}\,M) = 3.2 \times 10^{-8}\,M.$$

Note that this is only a third of the activity that would exist in water of low ionic strength.

1.6.6 Chemical Kinetics

Many environmental fate processes, such as the degradation of pollutant chemicals, are not usefully modeled as equilibrium chemistry problems because the *rate* of the reaction is more important to quantify than the final composition of the system. For example, even though it may be known that at equilibrium a certain chemical will be fully degraded, it is crucial to know whether degradation will take seconds, years, or perhaps centuries.

The rate at which a chemical reaction occurs may be limited by the frequency of collisions between the reacting atoms or molecules, as well as the likelihood that any particular collision will cause the reaction to proceed. Often several steps must occur in sequence to bring about a certain overall chemical process; the slowest is known as the *rate-limiting step*. Under other conditions, a molecule may react by itself without any collision to form another molecule. Under this condition, the number of molecules reacting in a time interval will simply be proportional to the number of molecules present. Then the rate at which the concentration of reacting molecules changes with time is described by *first-order kinetics*,

$$dC/dt = -k \cdot C, \qquad [1\text{-}18]$$

where C is the concentration of the *parent* compound [M/L^3], t is the time [T], and k is the rate constant [T^{-1}].

Such a chemical reaction, in which molecules are not colliding with other atoms or molecules, is called a *first-order reaction* because the rate at which chemical concentration changes at any instant in time is proportional to the concentration raised to the first power. Certain chemical processes, such as radioactive decay, are described by first-order kinetics. In the absence of any other sources of the chemical, first-order kinetics may lead to *exponential decay* or *first-order decay* of the chemical concentration (i.e., the concentration of the parent compound decreases exponentially with time):

$$C_t = C_0 e^{-kt,} \qquad [1\text{-}19]$$

where C_t is the concentration of the parent compound at some time t [M/L^3], C_0 is the initial concentration of the parent compound [M/L^3], k is the rate constant [T^{-1}], and t is the time [T].

Note that the integration of Eq. [1-18] leads to Eq. [1-19]. From Eq. [1-19] can be derived an expression, known as the *half-life* ($t_{1/2}$), which represents the amount of time it takes for the parent compound to decay to half its initial concentration:

$$\frac{C_t}{C_0} = 0.5 = e^{-kt_{1/2}}$$

$$\ln(0.5) = -kt_{1/2} \qquad [1\text{-}20]$$

$$t_{1/2} = \frac{0.693}{k}.$$

Note that half-lives convey exactly the same information as first-order decay rate constants, but may be more intuitive to some users. Half-lives for different decay processes may be easily compared to determine which decay mechanism is the most significant; they may also be compared with a transport time to determine whether a high or low chemical concentration will remain at a point downstream or downwind of a chemical source.

EXAMPLE 1-9

A sealed radioactive source used for physics demonstrations in 1940 contained 10 microcuries (μCi) of ^{60}Co (cobalt-60). Given a half-life of 1900 days for ^{60}Co, what would be the source strength in 1993?

The decay constant can be determined from the half-life using Eq. [1-20]:

$$k = \frac{0.693}{1900 \text{ days}} = 3.6 \times 10^{-4}/\text{day}.$$

Use Eq. [1-19] to calculate the concentration after 53 years:

$$C_t = 10e^{-(3.6\times10^{-4}/\text{day})(53 \text{ years})(365 \text{ days/year})}$$

$$C_t = 8.6 \times 10^{-3}\ \mu\text{Ci}.$$

Now consider the first type of reaction described in this section, in which more than one type of chemical molecule is involved. If two molecules A and B must collide to cause a reaction to occur, then the rates at which the concentrations of A and B decrease (in the absence of other sources) will be of the form:

$$dA/dt = -k \cdot A \cdot B \qquad [1\text{-}21a]$$

$$dB/dt = -k \cdot A \cdot B, \qquad [1\text{-}21b]$$

where A is the concentration of chemical A [M/L^3], B is the concentration of chemical B [M/L^3], k is the rate constant for the reaction [L^3/MT], and t is the time [T].

In this case, the rate at which a chemical disappears is proportional not only to its own concentration but also to the concentration of the other chemical with which it reacts. Increasing the concentration of either will increase the rate of reaction. Note that the units for the rate constant k are different from those used for the rate constant describing first-order kinetics in Eqs. [1-18] and [1-19]. Equations [1-21a] and [1-21b] are mathematically more complex than first-order kinetics. In some cases, however, the situation can be simplified because one of the chemicals involved in a reaction is approximately constant in concentration. For example, if a particular chemical is reacting with water, the concentration of water in an environmental water sample can be taken as nearly constant at 55.4 *M* (recall the treatment of water in Eqs. [1-13] through [1-15]). If a chemical is reacting with the hydrogen ion in a system where other chemical reactions keep the pH nearly constant, then the concentration of H^+ may be taken as constant. In such cases, it is convenient to multiply the rate constant k by the constant chemical concentration to obtain a new rate constant, k'. For example, if the concentration of chemical B is approximately constant, then the rate of reaction of chemical A could be described by

$$dA/dt = -k' \cdot A, \qquad [1\text{-}22]$$

where $k'[T^{-1}]$ is the product of $k[L^3/MT]$ and $B[M/L^3]$. As long as the concentration of one reacting chemical remains approximately constant, the simple mathematics of first-order kinetics—Eqs. [1-18] and [1-19]—can be used. This simplification is described as modeling the chemical reaction with *pseudo-first-order* kinetics.

Like equilibrium constants, rate constants also depend on environmental factors such as pressure and, especially, temperature. An increase in temperature usually gives rise to an increase in the chemical reaction rate, because molecules are moving faster and colliding more frequently with greater energy. If rate constants are known for two different temperatures, the rate constant for any other temperature can be calculated using the *Arrhenius rate law*,

$$k = A \cdot e^{-E_a/RT}, \qquad [1\text{-}23]$$

where k is the rate constant at a specific temperature, A is a constant (the *pre-exponential factor*), E_a is the Arrhenius activation energy, R is the universal gas constant, and T is the absolute temperature. When the rate constants for two different temperatures are available, the pre-exponential factor and Arrhenius activation energy can be estimated and then used to calculate the rate constant at a third temperature.

EXAMPLE 1-10

An ethylene *bis*-dithiocarbamate (EBDC) fungicide degrades in a storage tank to *N,N*-ethylene thiourea (ETU) at a rate of 0.046/day at 20°C; at 0°C, the reaction proceeds at a rate of 0.011/day. How fast will the reaction proceed at 15°C?

From Section 1.6.2, R is approximately 8.3 J/(mol · K). The two equations to be solved, based on the Arrhenius rate law in Eq. [1-23], are

$$0.046/\text{day} = A \cdot e^{-E_a/(8.3\ \text{J/mol}\cdot\text{K})(293\ \text{K})}$$

$$0.011/\text{day} = A \cdot e^{-E_a/(8.3\ \text{J/mol}\cdot\text{K})(273\ \text{K})}.$$

Solving for E_a,

$$E_a = 47.5\ \text{kJ/mol}.$$

Solving for A,

$$A = 0.046e^{(47{,}500)/(8.3)(293)} = 1.4 \times 10^7/\text{day}.$$

Therefore, at 15°C, the fungicide will degrade at a rate of

$$k = (1.4 \times 10^7/\text{day})(e^{-(47{,}500)/(8.3)(288)})$$

$$k = 0.033/\text{day}.$$

Note that rounding to an appropriate number of significant figures (two) is performed on the final result; for illustrative purposes, more figures are reported for the intermediate calculations. The next section discusses significant figures in greater detail.

1.7 ERROR IN MEASUREMENTS OF ENVIRONMENTAL QUANTITIES

The previous section discussed some of the basic concepts of environmental chemistry. Many of the chemical data required for the types of analyses presented in that section are based on experimental measurements made in the laboratory or the field. Any measurement of a chemical or physical quantity is always inaccurate to some degree because the exact, or *accurate,* value of a quantity can never be measured. The *error of observation* is the difference between the measured value of a quantity and the accurate value. Although the true value of the error of observation cannot be known, an estimate of the magnitude of the error is necessary to report meaningful uncertainty bounds for data. Uncertainty bounds allow a user to determine if the data are sufficiently accurate for a particular analysis. Sometimes the error of observation is negligible, but in other situations it is large enough to compromise an analysis or even render it completely meaningless.

Two types of errors of observation exist: systematic and random. The total error associated with a measurement is a function of the *systematic error,* commonly called *bias,* and the *random error.* Systematic errors in environmental measurements can be divided into three general categories: instrument errors (e.g., nonideal functioning of an instrument); interference from environmental factors (e.g., neglecting the effect of temperature on the instrument); and personal errors (e.g., operator bias in reading a measurement scale) (Velikanov, 1965). These systematic errors may be constant, or may vary over time in a predictable way. For a set of measurements for which total error is an important consideration, systematic error should be minimized. It is sufficient to make systematic error negligible compared with random error.

Even if all systematic error could be eliminated, the exact value of a chemical or physical quantity still would not be obtained through repeated measurements, due to the presence of random error (Barford, 1985). Random error refers to random differences between the measured value and the exact value; the magnitude of the random error is a reflection of the *precision* of the measuring device used in the analysis. Often, random errors are assumed to follow a *Gaussian,* or *normal, distribution,* and the precision of a measuring device is characterized by the sample standard deviation of the distribution of repeated measurements made by the device. [By contrast, systematic errors are *not* subject to any probability distribution law (Velikanov, 1965).] A brief review of the normal distribution is provided below to provide background for a discussion of the quantification of random error.

The Gaussian distribution has a symmetrical, bell shape and is sufficiently characterized by its *mean* (μ) and *variance* (σ^2). Its *standard deviation* (σ) is the square root of the variance. The symbols μ and σ refer to the mean and standard deviation *of a population* (i.e., the set of all possible measurements of a particular quantity). In practice, μ and σ often are not known because a population is too large to sample in its entirety. A subset of measurements of a particular quantity represents a *sample* of a population. If the sample is drawn from a normal population, the parameters characterizing the sample distribution are the sample mean of the measurements or *observations* ($\bar{y}$) and the *sample standard deviation* (s). Both of these parameters are calculated based on the values of the sample observations and the number of observations,

$$\bar{y} = \frac{\sum_{i=1}^{n} y_i}{n} \qquad [1\text{-}24]$$

$$s = \sqrt{\frac{\sum (y_i - \bar{y})^2}{n - 1}}, \qquad [1\text{-}25]$$

where $\bar{y}$ is the sample mean of the measurements, y_i is an observation, n is the number of observations, and s is the sample standard deviation. Note that the sample parameters $\bar{y}$ and s are estimates of the population parameters μ and σ, respectively, and become very close to μ and σ as n gets very large.

For greater detail on the Gaussian distribution and its statistical properties, the reader is referred to Dingman (1994), Davis (1986), or McCuen (1985).

Random error for repeated measurements of a particular quantity can be estimated as a multiplicative function of a t statistic, which reflects the probability that a value falls within a certain interval, and the *standard error of the mean,*

$$\delta = \pm t \cdot \mathrm{SE} = \pm t \cdot \frac{s}{\sqrt{n}}, \qquad [1\text{-}26]$$

where δ is the random error, t is the *t statistic for the sample mean,* SE is the standard error of the mean, s is the sample standard deviation, and n is the number of replications of the measurement. As in Eq. [1-26], the standard error of the mean depends on both the precision of the instrument and the number of replicate measurements. It is evident that as the number of replications increases, the standard error of the mean, and thus the random error, approaches zero; the total error then approaches the systematic error. If systematic error has been eliminated, then as n becomes infinitely large, the mean of the set of measurements of a particular quantity approaches the exact value of the quantity. The estimate of precision of the measuring device (s) does *not* decrease as n increases; it just approaches the true value of precision for the measuring device, which may be large or small.

Therefore, systematic errors affect the accuracy of the results, and random errors affect the precision. Accuracy describes the closeness of the measurements to the true value, whereas precision describes the closeness with which measurements of the same quantity agree with one another, independent of any systematic error. Sometimes the precision of a measuring device is known better than its accuracy. To significantly improve accuracy, it is always important to reduce the largest error first.

The overall uncertainty of an experimental result may be given as a function of the random and systematic errors. For methods of estimating individual bias components and propagated errors to predict the total systematic error, the reader is referred to Currie and DeVoe (1977) or Peters *et al.* (1974). Methods for estimating propagation of errors in products, quotients, sums, or differences are provided in Gans (1992), Kline (1985), and Topping (1962). Statistical methods for addressing measurements that appear to be *outliers* (i.e., extreme values that are not part of the population) are presented in McCuen (1992).

Thus, measurements (and the results of calculations that are based on measurements) are often described in terms of the best estimate of the population mean and an uncertainty range based on the total error. Often a *confidence interval* is constructed to give a range within which the population mean μ is expected to occur. If systematic error has been made negligible compared with random error, the best estimate of the population mean can be expressed as

$$\mu = \bar{y} \pm \delta, \qquad [1\text{-}27]$$

where $\bar{y}$ is the sample mean of the measurements and δ is the random error.

If n is seven or greater, the t statistic in Eq. [1-26], rounded to one significant figure, has a value of two for a 95% confidence interval. Thus, if a sample can be assumed to be drawn from a normal population distribution, Eq. [1-27] can be approximated as

$$\mu = \bar{y} \pm \frac{2s}{\sqrt{n}}. \qquad [1\text{-}28]$$

Equation [1-28] represents a 95% confidence interval and can be interpreted as saying that the true value of the mean has only a 5% chance of being outside the given range (i.e., a 2.5% chance of being below the given range and a 2.5% chance of being above the given range). If the sample size is six, the t statistic for a 95% confidence interval rounds to three. For small sample sizes, the actual values for the t statistic must be used because they increase rapidly as n decreases. (Note that the value of the t statistic is a function of the *degrees of freedom*, which in the case presented here, correspond to one less than the sample size n.) For values of the t statistic and further applications, the reader is referred to Berthouex and Brown (1994), Davis (1986), and McCuen (1985).

Even when confidence intervals or other types of error measurements are not specified, error estimates are implied by the number of significant figures used to present data. The number of significant figures suggests an error range; there should be uncertainty only for the least significant digit. Reporting a chemical concentration as 4.035 ppm implies that it is certainly less than 4.04 ppm and more than 4.03 ppm; otherwise, the last digit should be dropped. The number of significant figures used in reporting measurements depends on the precision of the measuring device and the number of measurements. As a rule of thumb, for a given parameter, the standard deviation (which is a good estimate of the error associated with the parameter) should typically have just one significant figure. When a calculation involving several measurements is being performed, the final result usually should have no more significant figures than the measurement with the fewest significant figures. Beware numerical clutter!

Note that in the examples provided throughout the rest of this book, all figures are carried throughout a calculation, and rounding to an appropriate number of significant figures is performed only for the final result. Intermediate results *appear* to be rounded only for presentation purposes.

In work with data, it is necessary to have an awareness of how much error can be tolerated. In environmental analyses, quantities are rarely known within a few percent of their true values, and useful work is sometimes performed with data that are known only within a factor of two. Other times errors as small as a few percent can render certain calculations not useful; the classic example is one in which a mass balance is being constructed and it is desired to estimate a missing term from the known values of all other terms in the mass balance. If the missing term is a relatively small difference between larger terms, error may make even the sign (+ or −) of the calculated term highly uncertain.

For an illustration of this concept, recall Example 1-1. If the average streamflow at the outlet of the lake were 2×10^5 m^3/day, then the estimated stream output rate of butanol would be 20 kg/day. Although there may still be a small internal sink of butanol in the lake, its magnitude cannot be distinguished from zero because it is masked by the uncertainty in the difference between two large numbers in the mass balance equation.

EXAMPLE 1-11

The state is assessing the performance of a particular wastewater treatment plant by determining the biochemical oxygen demand (BOD) of its effluent (see Section 2.5). An inspector collects seven replicate samples of the plant effluent and submits them to a water-testing laboratory. The laboratory reports the following results:

Sample	*BOD (mg/liter)*
1	9.7
2	10.2
3	10.5
4	10.0
5	9.9
6	10.2
7	10.4

Assuming negligible bias in the effluent testing, what is the best estimate of the effluent BOD concentration and what is the range in which the inspector can be 95% confident that the true BOD mean occurs?

First, calculate the mean BOD for the seven replicate water samples, using

Eq. [1-24]:

$$\bar{y} = 10.1.$$

The best estimate of the effluent BOD concentration is 10.1 mg/liter.

Next, calculate the sample standard deviation for the BOD using either Eq. [1-25] or a preprogrammed function on a calculator:

$$s = 0.281.$$

From Eq. [1-26], the standard error of the mean is

$$\text{SE} = \frac{s}{\sqrt{n}} = \frac{0.281}{\sqrt{7}} = 0.106.$$

Given that seven samples were taken and a 95% confidence interval is of interest, the approximation for the t statistic presented in Eq. [1-28] can be used:

$$\mu = 10.1 \pm (2 \cdot 0.106) = 10.1 \pm 0.2.$$

Thus, the inspector can be 95% confident that the true BOD mean lies within the range 10.1 ± 0.2 mg/liter. There is a 5% probability that the true BOD mean lies outside the range: a 2.5% chance that the true mean exceeds 10.3 mg/liter and a 2.5% chance that it is less than 9.9 mg/liter.

1.8 CHEMICAL DISTRIBUTION AMONG PHASES

Thus far the discussion of the transport and reactions of chemicals has predominantly focused on processes occurring in one of two fluids: air or water. Of course, the natural environment contains more than these two media; furthermore, one medium often contains different phases. For example, the atmosphere contains not only air, but also water and solids (particulate matter) in small, varying amounts; surface waters often contain solid particles and gas bubbles. The subsurface medium contains not only solids, but also a substantial volume of water and air. Therefore, a consideration of the principles that determine how a chemical becomes distributed among air, water, and solid phases is necessary, not only to understand chemical movement between media (atmosphere to surface water, soil to atmosphere, etc.), but also to understand the behavior of a chemical within a single environmental medium.

Pure air is an example of a *gas phase;* pure water is an example of the *aqueous phase. Solid phases* include soil grains, solid particles suspended in water or air, and pure solid chemicals. In addition, an immiscible liquid (i.e., a liquid that does not mix freely with water) can form its own *nonaqueous phase liquid* (NAPL, pronounced “napple”). An oil slick or a pool of industrial solvent floating on a water surface is an example of a NAPL. This section presents how to estimate the relative amounts of chemicals expected in different phases that are at equilibrium with one another.

In the discussion of chemical distribution among phases, it is assumed that chemicals are not transformed (i.e., no chemical bonds are formed or broken). For example, when liquid gasoline evaporates and enters the air in a partially empty gas tank, the bonds within individual molecules of the chemicals that compose gasoline are not being disrupted; the molecules are simply moving from a nonaqueous liquid phase to the gas phase without changing their identities. The *rate* at which this chemical movement occurs from one phase to another, relative to the timescale of interest, determines whether the problem is an equilibrium problem or a kinetics problem. Examples of both types abound in the environment; this section, however, refers only to the principles that govern *equilibrium*.

1.8.1 Solubility and Vapor Pressure

Aqueous solubility is a fundamental, chemical-specific property. It is defined as the concentration of a chemical dissolved in water when that water is both in contact and at equilibrium with the pure chemical. (For the moment, consider the chemical to be in either liquid or solid form; solubility of gases is discussed in Section 1.8.2.) Although aqueous solubility is temperature dependent, it does not vary greatly for a given chemical over the typical range of temperatures encountered in the environment.

As an example of solubility, pure liquid trichloroethene (C_2Cl_3H, commonly abbreviated TCE) dissolves into water until an aqueous concentration of approximately 1000 mg/liter is reached. See Table 1-3 for solubility values of several common chemicals.

Vapor pressure, another chemical-specific property, is defined as the *partial pressure* of a chemical in a gas phase that is in equilibrium with the pure liquid or solid chemical. For example, if at 20°C a bottle contains both air and pure liquid TCE, the partial pressure of TCE vapor in the air-filled neck of the bottle (the *headspace*) will be approximately 0.08 atm (61 mm Hg), which corresponds to 0.0033 mol/liter (440 mg/liter). The *ideal gas law* is used to convert the vapor pressure into the corresponding moles of vapor per unit volume:

$$\frac{n}{V} = \frac{P}{RT}, \qquad [1\text{-}29]$$

where P is the vapor pressure, V is the volume, n is the number of moles of the chemical, R is the gas constant, and T is the absolute temperature. Note

TABLE 1-3 Some Properties of Various Chemicals

Chemical	Molecular weight (g/mol)[a]	Density (g/cm^3)[a]	Solubility (mg/liter)[b]	Vapor pressure (atm)[b]	Henry's law constant (atm · m^3/mol)[b]	Henry's law constant (dimensionless)[b]	log K_{ow}[c]	Comments
Acetic acid	60.05	1.05	∞					
Aroclor 1254	325.06[d]	1.50[d]	1.2×10^{-2}	1×10^{-7}	2.7×10^{-3}	1.2×10^{-1}	6.5[e]	Polychlorinated biphenyl mixture (PCB)
Aroclor 1260	371.22[d]	1.57[d]	2.7×10^{-3}	5.3×10^{-8}	7.1×10^{-3}	3.0×10^{-1}	6.7[e]	Polychlorinated biphenyl mixture (PCB)
Atrazine	215.68[d]		33[b]	4×10^{-10}[f]	3×10^{-9}[f]	1×10^{-7}[f]	2.68[b]	Herbicide
Benzene	78.11	0.88	1780	1.25×10^{-1}	5.5×10^{-3}	2.4×10^{-1}	2.13	Gasoline constituent
Benz[*a*]anthracene	228.29		2.5×10^{-4}[g]	6.3×10^{-9}[g]	5.75×10^{-6}[g]	2.4×10^{-4}[g]	5.91	Polycyclic aromatic hydrocarbon (PAH)
Benzo(*a*)pyrene	252.32		4.9×10^{-5}[g]	2.3×10^{-10}[g]	1.20×10^{-6}[g]	4.9×10^{-5}[g]	6.50	PAH
Carbon tetrachloride	153.82	1.59	800	0.12	2.3×10^{-2}	9.7×10^{-1}	2.83[b]	
Chlorobenzene	112.56	1.11	472	1.6×10^{-2}	3.7×10^{-3}	1.65×10^{-1}	2.92	
Chloroform	119.38	1.48	8000	0.32	4.8×10^{-3}	2.0×10^{-1}	1.97[b]	
m-Cresol	108.14		2780[g]				1.96	
Cyclohexane	84.16	0.78	60[g]	0.13[g]	0.18[g]	7.3[g]	3.44	
1,1-Dichloroethane	98.96	1.18	4960[g]	3.0×10^{-1}[g]	6×10^{-3}[g]	2.4×10^{-1}[g]	1.79	
1,2-Dichloroethane	98.96	1.24	8426[g]	9.1×10^{-2}[g]	10^{-3}[g]	4.1×10^{-2}[g]	1.47	

(*continues*)

TABLE 1-3 (*continued*)

Chemical	Molecular weight (g/mol)[a]	Density (g/cm^3)[a]	Solubility (mg/liter)[b]	Vapor pressure (atm)[b]	Henry's law constant ($atm \cdot m^3/mol$)[b]	Henry's law constant (dimensionless)[b]	log K_{ow}[c]	Comments
cis-1,2-Dichloroethene	96.94	1.28	3500[h]	0.26[h]	3.4×10^{-3}[h]	0.25[h]	1.86[h]	
trans-1,2-Dichloroethene	96.94	1.26	6300[h]	0.45[h]	6.7×10^{-3}[h]	0.23[h]	2.06[h]	
Ethane	30.07		2.4×10^{-3}[g]	39.8[g]	4.9×10^{-1}[g]	20[g]		Gas
Ethanol	46.07	0.79	∞	7.8×10^{-2}[h]	6.3×10^{-6}[h]		−0.31[h]	Booze
Ethylbenzene	106.17	0.87	152	1.25×10^{-2}	8.7×10^{-3}	3.7×10^{-1}		
Lindane	290.9		7.3	1.2×10^{-8}	4.8×10^{-7}	2.2×10^{-5}		Pesticide
Methane	16.04				0.66[g]	27[g]		Natural gas
Methylene chloride	84.93	1.33	1.3×10^{4}	0.46	3×10^{-3}	1.3×10^{-1}	1.15	Also called dichloromethane
Naphthalene	128.17	1.03	33	3×10^{-4}	1.15×10^{-3}	4.9×10^{-2}	3.36	PAH
Nitrogen	28.01							Atmospheric gas
n-Octane	114.23	0.70	0.72[g]	0.019[g]	2.95[g]	121[g]	4.00[b]	Alkane
Oxygen	32.00							Atmospheric gas
Pentachlorophenol	266.34	1.98	14	1.8×10^{-7}	3.4×10^{-6}	1.5×10^{-4}		
n-Pentane	72.15	0.63	40.6[g]	0.69[g]	1.23[g]	50.3[g]	3.62	
Perchloroethene	165.83	1.62	400	2×10^{-2}	8.3×10^{-3}	3.4×10^{-1}	2.88	Commonly used in dry cleaning; tetrachloroethene
Phenanthrene	178.23	0.98	6.2[g]	8.9×10^{-7}[g]	3.5×10^{-5}[g]	1.5×10^{-3}[g]	4.57	PAH
Styrene	104.15	0.91					2.95[b]	

(*continues*)

TABLE 1-3 *(continued)*

Chemical	Molecular weight (g/mol)[a]	Density (g/cm^3)[a]	Solubility (mg/liter)[b]	Vapor pressure (atm)[b]	Henry's law constant ($atm \cdot m^3/mol$)[b]	Henry's law constant (dimensionless)[b]	log K_{ow}[c]	Comments
Toluene	92.14	0.87	515	3.7×10^{-2}	6.6×10^{-3}	2.8×10^{-1}	2.69	A common solvent
1,1,1-Trichloroethane (TCA)	133.40	1.34	950	0.13	1.8×10^{-2}	7.7×10^{-1}	2.48	A common solvent
Trichloroethene (TCE)	131.39	1.46	1000	8×10^{-2}	1×10^{-2}	4.2×10^{-1}	2.42	A common solvent
o-Xylene	106.17	0.88	175	8.7×10^{-3}	5.1×10^{-3}	2.2×10^{-1}	3.12	1,2-Dimethyl benzene
Vinyl chloride	62.50	0.91	2790[g]	3.4	2.4	99	0.60	Degradation product of TCE

[a] Values from Weast (1990), unless otherwise noted. Densities measured between 15.5 and 22°C, except for *o*-xylene at 10°C and phenanthrene at 4°C.
[b] Lyman *et al.* (1990). Solubility, vapor pressure, and Henry's law constants are for 20°C, unless otherwise noted.
[c] Schwarzenbach *et al.* (1993). Note that K_{ow} values are for 25°C.
[d] Values from Budavari (1989). Average number of chlorines per molecule for Aroclor 1254 and Aroclor 1260 is 4.96 and 6.30, respectively.
[e] Estimated from values in Anderson and Parker (1990).
[f] Riederer (1990).
[g] Schwarzenbach *et al.* (1993). Solubility, vapor pressure, and Henry's law constants are for 25°C.
[h] Howard (1990). Vapor pressure for *cis*-1,2-dichloroethene is for 35°C. Solubility and vapor pressure for *trans*-1,2-dichloroethene are for 25°C. Vapor pressure for ethanol is for 25°C.

that the quotient, n/V, is the concentration of a chemical in the gas phase (typically expressed in moles per liter).

Vapor pressures are quite temperature dependent, and can vary appreciably over as little as 5 or 10°C. For example, the vapor pressure of trichloroethene increases by approximately 27% over less than 5°C: the vapor pressure at 25.50°C is approximately 9.546 kilopascals (kPa), compared with 7.506 kPa at 20.99°C (Boublík *et al.*, 1984). Therefore, it is extremely important to note the temperature at which a vapor pressure was measured or estimated. (See the Appendix for conversions between kilopascal and other units of pressure.) Over narrow temperature ranges, the *Antoine equation* is commonly used to predict vapor pressure of a liquid at a particular temperature,

$$\ln VP = \frac{-B}{T + C} + A, \qquad [1\text{-}30]$$

where VP is the vapor pressure of the chemical at a particular temperature, T is the absolute temperature (Kelvin) of interest, and A, B, and C are constants based on a regression equation fitted to vapor pressure data measured at many temperatures. In the CRC handbook (Weast, 1990), a very similar regression equation for $\log VP$ is provided; the third parameter C is not used. The CRC handbook contains tabulated values for A and B for both inorganic and organic chemicals. See Table 1-3 for vapor pressures of several common chemicals.

EXAMPLE 1-12

An automobile fuel tank has a filler pipe 2 ft in length with a diameter of 1.5 in. Estimate the amount of fuel lost (a) by molecular diffusion (if the gas cap is left off for a day) and (b) by advective "pumping" through a tank vent, when atmospheric pressure decreases from 30.0 to 29.5 in. Hg. Use an approximate diffusion coefficient of 0.1 cm²/sec. Assume the fuel is octane (C_8H_{18}, with a molecular weight of 114) having a vapor pressure of 0.015 atm at the ambient temperature of 70°F (21°C). The 70-liter tank is half full.

(a) Fick's first law can be used to estimate the rate of fuel loss by molecular diffusion. First, estimate the octane concentration in the tank headspace. Assume that the partial pressure of octane is equal to its vapor pressure inside the tank, because the air in the tank is in close contact with the fuel. Concentration of octane in the vapor phase is then given by the ideal gas law;

see Eq. [1-29]:

$$\frac{n}{V} = \frac{P}{RT} = \frac{0.015 \text{ atm}}{0.082 \text{ liter} \cdot \text{atm/(mol} \cdot \text{K)} \cdot (21 + 273)\text{K}} = 6.2 \times 10^{-4} \, M.$$

Given octane's molecular weight of 114, the preceding concentration corresponds to

$$6.2 \times 10^{-4} \frac{\text{mol}}{\text{liter}} \cdot 114 \frac{\text{g}}{\text{mol}} = 7.1 \times 10^{-2} \text{ g/liter},$$

or 7.1×10^{-5} g/cm^3.

The concentration gradient along the filler pipe can be estimated as

$$\frac{dC}{dx} = \frac{(7.1 \times 10^{-5} \text{ g/cm}^3 - 0 \text{ g/cm}^3)}{2 \text{ ft} \cdot 30.48 \text{ cm/ft}} = 1.2 \times 10^{-6} \frac{\text{g}}{\text{cm}^4}.$$

Fick's first law, Eq. [1-3], can then be used to estimate the flux density of octane:

$$J = -D \cdot \frac{dC}{dx} = 0.1 \frac{\text{cm}^2}{\text{sec}} \cdot 1.2 \times 10^{-6} \frac{\text{g}}{\text{cm}^4}$$

$$= 1.2 \times 10^{-7} \frac{\text{g}}{\text{cm}^2 \cdot \text{sec}}.$$

The rate of fuel loss is the flux density multiplied by the cross-sectional area:

$$1.2 \times 10^{-7} \frac{\text{g}}{\text{cm}^2 \cdot \text{sec}} \cdot \pi \cdot \left(\frac{1.5 \text{ in.}}{2} \cdot \frac{2.54 \text{ cm}}{\text{in.}}\right)^2 = 1.3 \times 10^{-6} \text{ g/sec}$$

$$= 0.11 \text{ g/day}.$$

(b) To estimate the amount of fuel lost by advective "pumping," consider that the amount of air leaving the tank can be estimated from the fact that the mass of air in the tank is proportional to pressure. Initially, air in the tank contains

$$35 \text{ liter} \cdot 7.1 \times 10^{-2} \text{ g/liter} = 2.5 \text{ g octane}.$$

When pressure drops from 30 to 29.5 in. Hg, the mass of air decreases to

$$\frac{29.5}{30} \cdot 100\% = 98.3\% \text{ of original mass}.$$

Therefore, approximately 1.7% of the air—and thus approximately 1.7% of the 2.5 g of octane in the air—leaves the tank:

$$1.7\% \cdot 2.5 \text{ g} = 0.041 \text{ g}.$$

Note that temperature changes, as well as atmospheric pressure changes, can cause fuel loss by pumping air into and out of the tank. Loss via pumping is proportional to the empty volume in the tank, and can be minimized by keeping the tank full. Loss via diffusion can be inhibited by putting the gas cap back on; diffusive loss is not greatly affected by the quantity of fuel in the tank. Note also that diffusive loss may be much greater than predicted if wind creates turbulence in the filler pipe, thereby increasing the Fickian transport coefficient.

1.8.2 Henry's Law Constants

Now that solubility and vapor pressure have been defined, consider how a volatile chemical *partitions,* or distributes itself, between water and air phases at equilibrium. In general, a *partition coefficient* is the ratio of the concentrations of a chemical in two different phases, such as water and air, under equilibrium conditions. The *Henry's law constant,* H (or K_H), is a partition coefficient usually defined as the ratio of a chemical's concentration in air to its concentration in water at equilibrium. [Occasionally, a Henry's law constant is interpreted in an inverse fashion, as the ratio of a chemical's concentration in water to its concentration in air; see, e.g., Stumm and Morgan (1981, p. 179). Note that in that table, K_H is equivalent to $1/H$ as H is defined above.] Values of Henry's law constants are tabulated in a variety of sources (Lyman *et al.*, 1990; Howard, 1989, 1991; Mackay and Shiu, 1981; Hine and Mookerjee, 1975); Table 1-3 lists constants for some common environmental chemicals. When H is not tabulated directly, it can be estimated by dividing the vapor pressure of a chemical at a particular temperature by its aqueous solubility at that temperature. (Think about the simultaneous equilibrium among phases that would occur for a pure chemical in contact with both aqueous and gas phases.) Henry's law constants generally increase with increased temperature, primarily due to the significant temperature dependency of chemical vapor pressures; as previously mentioned, solubility is much less affected by the changes in temperature normally found in the environment.

Confusion about Henry's law constants often occurs because H can be expressed either in a dimensionless form or with units. The dimensionless form is obtained by using the same units for the chemical concentrations in both the air and the water phases, for example, (mol chemical/liter air) divided by (mol chemical/liter water). If the temperature at which H was measured or estimated is known, the dimensionless form can be readily converted to a form with units by using the ideal gas law—Eq. [1-21]—such that the molar air concentration is converted to units of partial pressure:

$$H = \frac{\text{mol chemical/liter air}}{\text{mol chemical/liter water}} \cdot \text{R}\left(\frac{\text{liter air} \cdot \text{atm chemical}}{\text{mol chemical} \cdot \text{K}}\right) \cdot \text{T (K)} \qquad [1\text{-}31]$$

$$= \frac{\text{atm chemical}}{\text{mol chemical/liter water}} = \frac{\text{atm} \cdot \text{liter}}{\text{mol}}.$$

Another common dimensional form of H has units of atm · m^3/mol. If the Henry's law constant has been measured or estimated at 25°C, it can be converted to a dimensionless form by multiplying by 40.9 mol/m^3 · atm:

$$H = \frac{\text{atm chemical}}{\text{mol/m}^3} \cdot \frac{40.9 \text{ mol/m}^3}{\text{atm}} = \text{dimensionless form of } H. \qquad [1\text{-}32]$$

Note that the value of 40.9 in Eq. [1-32] is obtained by multiplying the gas constant (0.0821 liter · atm/mol · K) by the absolute temperature (298 K) and by dividing by 1000 to convert liters into cubic meters. Similar factors can be readily derived for other temperatures.

EXAMPLE 1-13

Consider an unsaturated soil (i.e., a soil that contains both air and water in the pores between soil grains). Suppose the concentration of dissolved oxygen in soil water at equilibrium with soil air is 100 μmol/liter (μM). Given a Henry's law constant of 26 (dimensionless) for oxygen at 20°C, what is the corresponding oxygen concentration in soil air? What is the Henry's law constant in units of atm · m^3/mol at 20°C?

Given equilibrium between soil water and air, the oxygen concentration in soil air would be

100 μmol/liter (water) × 26 (Henry's law constant) = 2600 μmol/liter (air).

Use Eq. [1-22], with an additional conversion factor for liters to cubic meters, to convert the dimensionless Henry's law constant to a constant with units of atm · m^3/mol:

$$H = \frac{26 \text{ mol/liter (air)}}{1 \text{ mol/liter (water)}} \cdot RT \cdot 10^{-3}\text{m}^3/\text{liter}$$

$$H = \frac{26 \text{ mol/liter (air)}}{1 \text{ mol/liter (water)}} \cdot \frac{0.082 \text{ liter} \cdot \text{atm}}{\text{mol} \cdot \text{K}} \cdot \frac{(20°\text{C} + 273)\text{ K}}{1} \cdot \frac{1 \text{ m}^3}{1000 \text{ liter}}$$

$$H = \frac{0.63 \text{ atm} \cdot \text{m}^3}{\text{mol}}.$$

1.8.3 Chemical Partitioning to Solids

Chemical partitioning also occurs between water and solid phases and between air and solid phases, in a process most generally termed *sorption*. Types of sorption include *adsorption*, in which a chemical sticks to the two-dimensional surface of a solid, and *absorption*, in which a chemical diffuses into a three-dimensional solid. Chemical sorption in the environment is much more difficult to predict than is chemical partitioning between air and water, partly because the types of sorptive solid phases (*sorbents*) vary enormously, and partly because there are many different mechanisms by which sorption can occur. In this section, only partitioning between water and solid phases is considered.

Solids capable of sorbing chemicals (*sorbates*) include minerals, such as clays and metal oxides; natural organic material; and plastic, for example, polyvinyl chloride (PVC), commonly used in groundwater monitoring wells. The mechanisms by which sorption can occur include absorption into natural organic matter; adsorption to mineral surfaces via van der Waals, dipole–dipole, and other weak physical intermolecular forces; adsorption through electrostatic attractions to oppositely charged surface sites on the solids; and adsorption through covalent bonding to surface groups on the solids. The symbol K_p is frequently used to represent a solid–water partition coefficient; K_d, symbolizing a *distribution coefficient*, is an equivalent notation.

Use of a partition coefficient without reference to the conditions under which it was measured suggests that the coefficient is a constant. A constant coefficient implies a linear relationship between the amount of dissolved chemical and the amount sorbed. In actuality, the relationship between dissolved and sorbed chemical concentrations is often nonlinear and may be expressed as a *sorption isotherm*. (Use of the term isotherm indicates that sorption measurements are being made at a constant temperature.) Laboratory measurements of sorption sometimes fit a relationship known as the *Freundlich isotherm*,

$$C_{\text{sorb}} = K_f \cdot (C_w)^n, \qquad [1\text{-}33]$$

where C_{sorb} is the concentration of sorbed chemical [M/M], K_f is the Freundlich constant, C_w is the concentration of dissolved chemical [M/L^3], and n reflects nonlinearity—if n equals one, Eq. [1-33] reduces to a linear partition coefficient. If the value of n is less than one and the dissolved chemical concentration increases, the additional molecules sorb in a smaller proportion. If the value of n is greater than one, proportionally more sorption occurs if the dissolved concentration increases (Fig. 1-10).

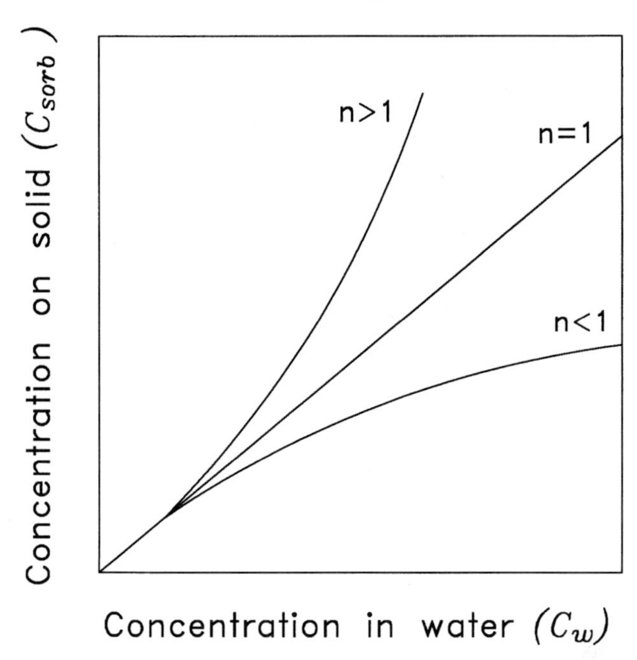

FIGURE 1-10 Freundlich isotherms having exponents less than, equal to, and greater than one. An exponent of one corresponds to a linear isotherm, in which case the relationship between the aqueous concentration of a chemical and the chemical's concentration on a solid phase at equilibrium with the water can be described by a single partition coefficient. If the exponent is less than one, the ratio of sorbed concentration to aqueous concentration decreases as the chemical concentration in the system increases. Such a decrease would happen, for example, if the solid contained a finite number of sites that became filled (saturated) at higher aqueous chemical concentrations. An exponent greater than one might occur if the sorbed chemical modified the solid phase to favor further sorption.

Unlike a Henry's law constant, whose value in a given environmental situation is predominantly dependent on only temperature, a solid–water partition coefficient can also be dependent on other factors, including water pH, type of solid, and ionic strength. For inorganic chemicals, it is not appropriate to apply a K_d value for a given pollutant in a particular environmental situation to the same chemical under other conditions. Furthermore, there is considerable uncertainty in predicting *a priori* a K_d value for inorganic chemicals, as well as for charged organic chemicals which sorb through chemical reactions with solid surfaces. For these types of chemicals, K_d often must be measured in the laboratory. However, for many neutral (uncharged)

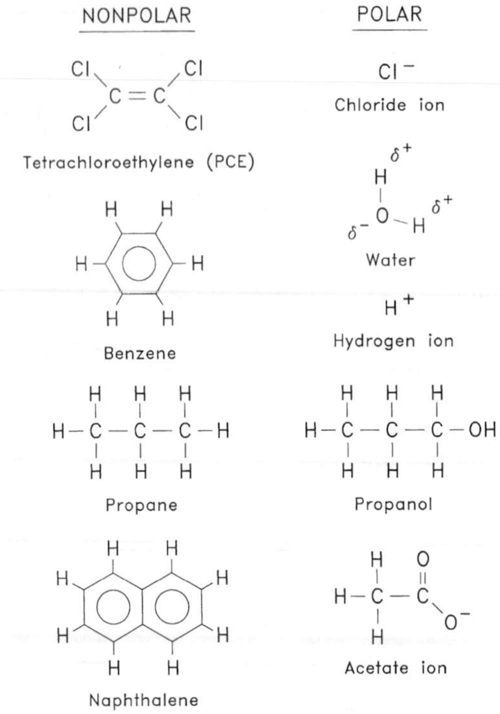

NONPOLAR

Tetrachloroethylene (PCE)

Benzene

Propane

Naphthalene

POLAR

Chloride ion

Water

Hydrogen ion

Propanol

Acetate ion

FIGURE 1-11 Some examples of polar and nonpolar chemical species. Note that unbalanced electrical charge, asymmetry, and presence of oxygen all tend to make chemicals more polar.

organic chemicals, a robust method for approximating K_d values exists. This method depends on both sorbate properties and sorbent properties (i.e., the concentration of natural organic matter in the solid), and assumes that absorption into natural organic matter is the primary sorption mechanism. The sorbate properties necessary for predicting K_d values are described below; the necessary sorbent properties and further estimation techniques are presented in Chapter 3.

Many neutral organic chemicals with low water solubility tend to absorb into natural organic matter because they are *nonpolar*. *Polarity* refers to the extent to which charge is unevenly distributed within a chemical molecule or substance (Fig. 1-11). Water itself is very polar, with an excess of negative charge associated with the oxygen atom and an equal excess of positive charge associated with the hydrogen atoms. Most natural minerals are also polar. Because of water's polarity, it readily dissolves other polar *(hydrophilic)* chemicals. However, the nonpolar *(hydrophobic)* chemicals tend to avoid the less favorable state (from a free energy point of view) of being dissolved in water by absorbing into natural organic matter.

The polarity of a chemical has a strong inverse correlation with the chemical's K_{ow}, the *octanol–water partition coefficient.* K_{ow} is the ratio of a chemical's concentration in octanol ($C_7H_{15}CH_2OH$) to its concentration in water at equilibrium. The concept of K_{ow} was developed by the pharmaceutical industry as a useful index of a drug's behavior in the body, because partitioning between water and octanol roughly mimics partitioning between water and body fat. In general, smaller molecules and more polar molecules dissolve more readily in water, have lower K_{ow} values, and have less tendency to sorb to solids. Larger molecules and less polar molecules are less soluble, have higher K_{ow} values, and are more likely to sorb to solids. K_{ow} is a very valuable index of the partitioning behavior of many organic compounds in the environment. Methods for using K_{ow} values in conjunction with sorbent parameters to predict K_d values are presented in Chapter 3.

K_{ow} has been measured and tabulated for many chemicals and also lends itself to estimation based on the chemical's structure (Lyman *et al.*, 1990; Howard, 1989–1991; Schwarzenbach *et al.*, 1993; Yalkowsky and Valvani, 1979; Syracuse Research Corporation's Environmental Fate Data Base; and Hansch and Leo, 1985). See Table 1-3 for K_{ow} values for some chemicals commonly found in the environment.

EXAMPLE 1-14

In a toxicity test, small fish are placed in aquaria containing various concentrations of toluene. For partitioning of toluene, these particular fish can be modeled as bags containing, by volume, 5% fatty tissue similar to octanol in its affinity for toluene, 3% air in a swim bladder, and 85% water. Treat the missing 7% as being nonsorptive for toluene. If equilibrium is assumed, what fractions of toluene in the fish will reside in fatty tissue and in the air bladder?

To solve this problem, first obtain the relevant partition coefficients, and then set up an expression in which all masses of toluene are written as functions of the volume of fish (V_{fish}) and the toluene concentration in the aquaria water (C_w). From Table 1-3, $H = 0.28$ and $\log K_{ow} = 2.69$ for toluene ($K_{ow} = 490$).

$$\text{Mass of toluene in water of fish} = (V_{fish})(0.85)C_w.$$

$$\text{Mass of toluene in swim bladder} = (V_{fish})(0.03)(H_{toluene} \cdot C_w).$$

$$\text{Mass of toluene in fatty tissue} = (V_{fish})(0.05)(K_{ow} \cdot C_w).$$

Then the fraction of toluene in fatty tissue can be written:

$$\frac{(V_{\text{fish}})(0.05)(K_{\text{ow}} \cdot C_{\text{w}})}{(V_{\text{fish}})(0.05)(K_{\text{ow}} \cdot C_{\text{w}}) + (V_{\text{fish}})(0.03)(H_{\text{toluene}} \cdot C_{\text{w}}) + (V_{\text{fish}})(0.85)C_{\text{w}}}$$

$$= \frac{0.05\, K_{\text{ow}}}{0.05\, K_{\text{ow}} + 0.03\, H_{\text{toluene}} + 0.85}$$

$$= \frac{0.05(490)}{0.05(490) + 0.03(0.28) + (0.85)}$$

$$= \frac{25}{25 + 0.008 + 0.85} = 97\%.$$

(Note that this is dimensionless—all units have cancelled out.) Similarly, only 0.03% is in the air bladder (and approximately 3% resides in the water fraction).

Although this model for an aquatic organism may seem very crude, such models are useful in many situations, as discussed further in Chapter 2.

1.8.4 Equilibrium Partitioning among All Phases: Fugacity

From the preceding three sections, it is evident that the relative concentrations of a chemical in air, water, and soil phases at equilibrium can be predicted from a knowledge of the chemical's partition coefficients (i.e., vapor pressure, Henry's law constant, and distribution coefficient). Each of the chemical's partition coefficients describes a behavior that may also be thought of in terms of chemical potential; when equilibrium partitioning among phases is attained, the chemical potentials in all phases are equal. A convenient measure of chemical potential is *fugacity*, literally, the "tendency to flee." The fugacity of a chemical in a given phase is equal to the vapor pressure the chemical would have in a gas volume in equilibrium with the phase. The fugacity concept can provide a useful framework within which previously described partition coefficients may be used to compute the partitioning of chemicals among the phases of an ecosystem.

Fugacity has units of pressure, and can be related to the concentration of a chemical in a system through a *fugacity capacity constant*, commonly with units of (mol/atm $\cdot$ m^3). Thus the chemical concentration in a given phase is

$$C_i = Z_i \cdot f, \qquad [1\text{-}34]$$

where C_i is the chemical concentration in phase i [M/L^3], Z_i is the fugacity capacity [T^2/L^2], and f is the fugacity [M/LT^2]. At equilibrium, the fugacity for the entire system can be calculated as

$$f = \frac{M_{\text{tot}}}{\sum_i (Z_i \cdot V_i)}, \qquad [1\text{-}35]$$

where M_{tot} is the total moles of chemical in the system and V_i is the volume of phase i with which the chemical is associated [L^3].

In the air phase, under pressures normally found in the environment, fugacity equals the pressure exerted by the chemical's vapor. (At higher pressures, vapors do not exactly obey the ideal gas law, and a correction must be applied; this is small enough to ignore for practical purposes of fate and transport modeling in the environment.) By combining the ideal gas law (Eq. [1-29]) and Eq. [1-34], it is evident that the fugacity capacity for air is $1/RT$, for all chemicals:

$$P = \frac{n}{V} \cdot RT = f.$$

$$C_i = \frac{n}{V} \qquad [1\text{-}36]$$

$$\therefore Z_i = \frac{1}{RT}.$$

The fugacity capacity for other phases is a function of both the chemical's partition coefficient between that phase and water and the chemical's Henry's law constant. For water, the fugacity capacity is

$$Z_{\text{water}} = \frac{1}{H}, \qquad [1\text{-}37]$$

where H is the Henry's law constant [L^2/T^2].

For sediment, the fugacity capacity for a chemical can be expressed as

$$Z_{\text{sediment}} = \frac{\rho_s \cdot K_d}{H}, \qquad [1\text{-}38]$$

where ρ_s is the density of sediment [M/L^3] and K_d is the soil–water partition coefficient [L^3/M].

For fish, the fugacity capacity for a chemical can be described by

$$Z_{\text{fish}} = \frac{\rho_{\text{fish}} \cdot BCF}{H}, \qquad [1\text{-}39]$$

where ρ_{fish} is the density of fish [M/L^3] and BCF is the partition coefficient between fish and water (discussed in Chapter 2) [L^3/M].

Once the fugacity capacity for each phase has been calculated, the moles of chemical in each phase are given by

$$M_i = f \cdot V_i \cdot Z_i, \qquad [1\text{-}40]$$

where M_i is the moles of chemical in phase i.

Fugacity modeling does not allow any new calculations to be made that cannot already be made with the partition coefficients described in the previous three sections. However, a comparison of the fugacity capacity of a chemical in different phases permits a direct assessment of which phase will have the highest chemical concentration at equilibrium. For further details, the reader is referred to Mackay and Paterson (1981) and Schwarzenbach et al. (1993).

EXAMPLE 1-15

Consider a simplified ecosystem consisting of 10^{10} m^3 of air, 7×10^6 m^3 of water, and 3.5 m^3 of fish. Released into the water is 10 kg of methylene chloride. Predict the equilibrium partitioning of methylene chloride into each phase using the fugacity concept. Assume a BCF of 4.4 liter/kg, a fish density of 1 g/cm^3, and a temperature of 25°C.

First, gather the necessary data from Table 1-3: vapor pressure is 0.46 atm; H is 3×10^{-3} atm · m^3/mol; molecular weight is 84.93 g/mol.

Then convert the mass of methylene chloride into moles:

$$10 \text{ kg} \cdot \frac{1000 \text{ g}}{1 \text{ kg}} \cdot \frac{1 \text{ mol}}{84.93 \text{ g}} = 118 \text{ mol}.$$

Next, calculate the fugacity capacity for each phase. For the air phase, use Eq. [1-36]:

$$Z_{\text{air}} = \frac{1}{RT} = \frac{1}{0.0821 \text{ liter} \cdot \text{atm/(mol} \cdot \text{K)} \cdot 298 \text{ K}} \cdot \frac{1000 \text{ liter}}{1 \text{ m}^3}$$

$$Z_{\text{air}} = 40.9 \text{ mol/(atm} \cdot \text{m}^3).$$

For the water phase, use Eq. [1-37]:

$$Z_{\text{water}} = \frac{1}{3 \times 10^{-3}\ \text{atm} \cdot \text{m}^3/\text{mol}} = 333\ \text{mol/(atm} \cdot \text{m}^3).$$

For the fish phase, use Eq. [1-39]:

$$Z_{\text{fish}} = \frac{1000\ \text{g/m}^3 \cdot 4.4\ \text{liter/kg}}{3 \times 10^{-3}\ \text{atm} \cdot \text{m}^3/\text{mol}} \cdot \frac{1\ \text{kg}}{1000\ \text{g}} \cdot \frac{1\ \text{m}^3}{1000\ \text{liter}} = 1.5\ \text{mol/(atm} \cdot \text{m}^3).$$

Then calculate the fugacity for the entire system, from Eq. [1-35]:

$$f = \frac{118\ \text{mol}}{40.9 \cdot 10^{10} + 333 \cdot 7 \times 10^{6} + 1.5 \cdot 3.5} = 2.9 \times 10^{-10}\ \text{atm}.$$

Finally, the moles of methylene chloride in each phase can be calculated from Eq. [1-40]:

$$M_{\text{air}} = 2.9 \times 10^{-10} \cdot 10^{10} \cdot 40.9 = 117\ \text{mol}$$

$$M_{\text{water}} = 2.9 \times 10^{-10} \cdot 7 \times 10^{6} \cdot 333 = 0.7\ \text{mol}$$

$$M_{\text{fish}} = 2.9 \times 10^{-10} \cdot 3.5 \cdot 1.5 = 1.5 \times 10^{-9}\ \text{mol}.$$

Therefore, at equilibrium, the mass of methylene chloride will be overwhelmingly in the air as compared with the other two phases. However, the highest concentration of methylene chloride is in the water (1×10^{-7} mol/m^3), the phase with the highest fugacity capacity.

1.9 CONCLUSION

In this chapter the basic concepts of physical transport and environmental chemistry have been introduced. Next the three principal environmental media, surface waters, the subsurface environment, and the atmosphere, are examined in detail. In the following chapters, the physical, chemical, and biological structures of these media are discussed. By applying physical and chemical principles, one can interpret, explain, and make predictions about chemical behavior in each medium.

It should be kept in mind that an extensive body of literature exists, within which the basic concepts of this and the following chapters are further developed. The reader is encouraged to explore in greater detail the concepts presented in this book by consulting the literature; the references cited in each chapter provide a starting point.

EXERCISES

1. An analytical chemist determines that an estuarine water sample contains 1.5 g/liter of sulfate ion (SO_4^{2-}). What is the concentration in terms of
   a. grams per liter of sulfur (S)?
   b. molar concentration of sulfate?
   c. normality?
   d. parts per million of sulfate?
2. A particular air sample at 1 atm pressure has a density of approximately 1.3 g/liter and contains SO_2 at a concentration of 25 $\mu g/m^3$. What is this concentration in terms of
   a. parts per million (mass SO_2 per 10^6 units mass of air)?
   b. moles SO_2 per 10^6 mol of air?
3. Fuel oil is pumped into a leaking tank at a rate of 1 liter/min. The hole at the bottom has an effective area of 0.1 cm^2. What is the maximum depth that the fuel oil can reach if the tank is initially empty? (Note that fuel oil spurts out through the hole with a velocity of $\sqrt{2\,gh}$, where h is the depth of oil and g is the acceleration due to gravity, 981 cm/sec^2.)
4. Natural dissolved organic material (DOM) concentration in the streams entering a 2×10^5 m^3 water reservoir averages 7 mg/liter; total annual inflow is 10^5 m^3. Due to evaporation, annual outflow of liquid water (via a dam and spillway, and a municipal water intake) is only 9×10^4 m^3, and DOM concentration in the outflow is 6.5 mg/liter. What is the sink strength for DOM, expressed per cubic meter of water per day?
5. a. Anaerobic bacteria living at the bottom of a shallow inlet to a salt marsh are generating hydrogen sulfide (H_2S) as a by-product of their metabolism. Although this compound (which is responsible for the "rotten egg" smell characteristic of salt marshes and is extremely toxic) is ordinarily found as a gas, it is reasonably soluble in water. What is the maximum concentration of dissolved hydrogen sulfide species (including H_2S, HS^-, and S^{2-}) that could theoretically accumulate in the waters of the inlet if the pH is 6.0? (This absolute limit would be reached when the pressure of H_2S reached 1 atm and bubbles of pure H_2S formed and escaped. In practice, H_2S concentrations would typically be lower.) Some relevant equilibria include

$$H_2S \rightleftharpoons H_2S(g) \qquad K_H = 10^{0.99}\ \text{atm} \cdot \text{liter/mol}$$

$$H_2S \rightleftharpoons H^+ + HS^- \qquad K = 10^{-7.02}\ M$$

$$HS^- \rightleftharpoons H^+ + S^{2-} \qquad K = 10^{-13.9}\ M.$$

b. A small amount of dibromomethane (CH_2Br_2) has been spilled into the described inlet. Assume that the partial pressure of H_2S is 0.1 atm. If the principal process that will affect the fate of this species is its reaction with HS^-, for which the relevant rate constant is 5.25×10^{-5}/*M*/sec (note that other dissolved hydrogen sulfide species do not react at measurable rates), how many days will be required for the CH_2Br_2 concentration to decrease to 10% of its initial value?

6. What is the flux density of
   a. organic nitrogen (org-N) in a wastewater infiltration basin, if the org-N concentration is 10 mg/liter and water seeps into the soil at a rate of 2 cm/hr?
   b. salt in a horizontal tube 10 cm in length connecting a tank of seawater (salinity = 30 g/liter) and a tank of freshwater (salinity ≈ 0), assuming no advection occurs?
   c. CO_2 in an automobile exhaust pipe, where gas velocity is 30 cm/sec and CO_2 concentration is 0.05 g/liter?
7. Is a kinetic or an equilibrium model more likely to be useful in each of the following situations? Qualify, if necessary.
   a. Toxic components of an antifouling ship paint are leaching from a yacht into a river.
   b. Deicing salt that has been sprinkled on the sidewalk is dissolving into a puddle.
   c. There is a concentration of volatile fuel oil constituents in the head-space of a tank of heating oil.
8. In manufacturing a fruit drink, citric acid is added to water in the amount of 0.1 mol/liter. What is the pH of the resulting solution? Use an acid dissociation constant of 8.4×10^{-4} for the ionization of citric acid (assume only the first deprotonation reaction occurs).
9. The chemistry laboratory returns to you the following water analysis:

$$\mathrm{pH} = 6.3$$

$$\mathrm{Na^+} = 2\ \mathrm{mg/liter}$$

$$\mathrm{K^+} = 0.5\ \mathrm{mg/liter}$$

$$\mathrm{Mg^{2+}} = 0.3\ \mathrm{mg/liter}$$

$$\mathrm{Cl^-} = 1.5\ \mathrm{mg/liter}$$

$$\mathrm{SO_4^{2-}} = 2\ \mathrm{mg/liter}$$

$$\mathrm{NO_3^-} = 0.5\ \mathrm{mg/liter}$$

$$\mathrm{HCO_3^-} = 0.9\ \mathrm{mg/liter.}$$

Do you believe this analysis is both complete and accurate? Explain why or why not.

10. A body of water contains 10^{-5} mol/liter of carbonic acid, $H_2CO_3^*$.
    a. What is the concentration of bicarbonate, HCO_3^-, at pH values of 4, 7, and 10?
    b. Use the same pH conditions in (a) but in seawater having an ionic strength of 0.5 *M*. (Because pH is properly defined on the basis of hydrogen ion *activity*, ionic strength corrections need only be made for HCO_3^-.)
11. Methyl dichloroacetate ($Cl_2CHCO_2CH_3$) decays into methanol (CH_3OH) and dichloroacetic acid (Cl_2CHCO_2H) on reaction with water. Given a rate constant of 2.7×10^{-4}/sec and an initial concentration of 1 ppm methyl dichloroacetate in the water, how much methanol will be present in the water after 30 min? Assume that no methanol is initially present in the water and neglect the possibility that methanol volatilizes or undergoes further decay after being formed.
12. Tritium, a radioactive isotope of hydrogen, has a half-life of approximately 12 years.
    a. What is its decay constant?
    b. If a sample of water containing only tritium is sealed up for 25 years, what will be the molar ratio of tritium to its decay product, helium-3, at the end of this time period?
13. If 10 Ci of cesium-137, having a half-life of 1.1×10^4 days, is blown into the upper atmosphere by a nuclear test, approximately how much cesium-137 may eventually return to the earth, if its typical residence time in the atmosphere is 2 years?
14. Small fish in an *m*-cresol contaminated lake acquire an *m*-cresol concentration of 0.1 ppm in their fatty tissue. If their fatty tissue behaves similarly to octanol with respect to *m*-cresol partitioning, what is the probable approximate concentration of *m*-cresol in the lake water?
15. A stoppered flask at 25°C contains 250 ml of water, 200 ml of octanol, and 50 ml of air. An unknown amount of *o*-xylene is added to the flask and allowed to partition among the phases. After equilibrium has been established, 5.0 mg of *o*-xylene is measured in the water. What is the total mass of *o*-xylene present in the flask?
16. The following sketch represents the concentration of a colored dye at two different times, t_1 and t_2, in a water-filled tube. The concentration gradient at each point is equal to the slope of the curve; $D \approx 10^{-5}$ cm²/sec.
    a. Estimate the flux of dye at points 1.5 and 2 cm along the tube at time t_1 and the later time t_2.
    b. How is the answer to (a) for time t_1 consistent with the development of the t_2 profile?

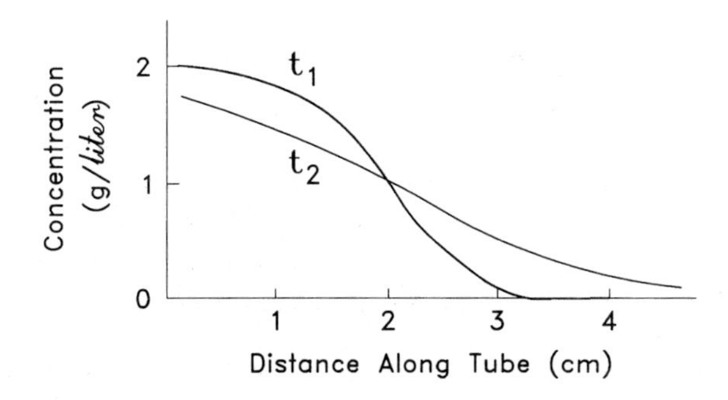

17. A printed circuit board manufacturing plant is discharging the chemical 1,1,1-trichloroethane (TCA) into a river that subsequently flows into a lake, the source of drinking water for a small town. The levels of TCA in this lake currently fail to meet drinking water objectives of 10 ppb. As an environmental specialist with the state department of environmental protection, you have been asked to help determine discharge requirements for the manufacturing plant. By assuming that the lake is a well-mixed system, that concentrations are at steady state, and that the only processes acting on this chemical in the lake are volatilization and biodegradation, how high can the influent concentration of TCA (C_{in}) be at the point where the river enters the lake? Some pertinent data are

TCA (ignore effect of concentration on these rates)	
Volatilization rate	3×10^{-4} μg/cm$^2 \cdot$ sec
Biodegradation rate	6×10^{-3} μM/day
River inflow to lake	200 gal/min
River outflow from lake	200 gal/min
Lake volume	1.5×10^6 gal
Average lake depth	10ft

18. Nitrous acid (HNO_2) is an essentially nonvolatile acid that ionizes in water with an equilibrium constant of 4×10^{-4} mol/liter; 10 mmol of nitrous acid are added to a liter of water that is open to the atmosphere.
    a. List all chemical species that you expect to occur in the water. For which of these do you know the concentration *a priori?* (You may assume the water has reached equilibrium with the air, and therefore $H_2CO_3^*$ concentration is 10 μM, per Henry's law.)
    b. Write every equation that constrains the composition of the system at equilibrium, identifying each as a mass conservation, mass action, or electroneutrality constraint.
19. A permeation tube is a device for preparing gas standards having a known concentration of some trace constituent of interest. For example, one

could be made for carbon tetrachloride, CCl_4, by sealing some liquid in a plastic tube of small but finite permeability for the chemical. When the tube is placed in a flowing gas stream, CCl_4 vapor slowly diffuses through the plastic into the gas stream.

a. What rate of CCl_4 release do you expect to occur if the tube has a wall area of 8 cm^2 and a thickness of 1 mm? (The density of CCl_4 is 1.6 g/cm^3. Use a diffusion coefficient of 10^{-10} cm^2/sec.)
b. Besides assuming that the permeability of the tube is well known and that the inside of the tube is fully in contact with the liquid, what other conditions must be met to ensure that the concentration of CCl_4 thus created in the gas stream is constant? (There are at least three additional conditions that must be provided if accurate results are desired.)

20. 200 Ci of cesium-137 are used for the gamma ray source of an experimental food irradiation unit. What is the amount of activity remaining after 5 years? The half-life of cesium-137 is 11,000 days.
21. Calculate the approximate pH of a 0.08 *M* solution of formic acid (K_a = 1.8×10^{-4} mol/liter), assuming the pH is entirely controlled by the ionization of this acid to yield H^+ and formate ion, $HCOO^-$.
22. What is the flux density of
    a. chloride in a river whose average velocity is 10 cm/sec and whose chloride concentration is 220 μmol/liter?
    b. chloride across the freshwater–saltwater interface in an estuary, where the chloride concentration changes from 20 to 2 ppt (parts per thousand) in a distance of 5 m?
    c. octane vapor in a tank 4 m tall that is open at the top and contains a thin layer of liquid octane at the bottom? (Neglect any possible turbulence, and use a molecular weight of 114 g/mol and a vapor pressure of 0.019 atm for the octane.)
23. A flask contains a liter of water to which 0.01 g of the salt $MgCl_2$ and 0.1 g of acetic acid (HAc = CH_3COOH) have been added. The latter ionizes according to $CH_3COOH = H^+ + CH_3COO^-$, with an equilibrium constant *K* of 10^{-4} mol/liter.
    a. What chemical species are present? Which have concentrations that are known *a priori*? Which are unknown?
    b. Write all the equations that constrain this system.
24. If 10^{-3} mol of baking soda (sodium bicarbonate, $NaHCO_3$) is dissolved in a liter of pure water:
    a. The pH of the solution is 8.3. What is $[H_2CO_3^*]$?
    b. Strong acid is added until the pH is 3.5, at which point you may consider that essentially all of the carbonate system is present as $H_2CO_3^*$. If the water is then equilibrated with a *very* small gas volume,

what is PCO_2, the partial pressure of carbon dioxide in the gas volume?

25. If 10 mg of naphthalene is added to 1 liter of water in a 20-liter sealed bottle (the solubility of naphthalene is 2.6×10^{-4} mol/liter, and the vapor pressure is 3×10^{-4} atm at 20°C):
    a. What is the *dimensionless* Henry's law constant?
    b. What percentage of the total naphthalene ends up in the air-filled volume of the bottle at equilibrium?
26. A large tank of water containing 500 ppm of salt is connected to a large tank of distilled water with a narrow tube 10 cm long. Pressures are adjusted so that there is no bulk fluid flow. What is the approximate flux density of the salt in the tube due to molecular diffusion when the salt gradient in the tube has become steady?
27. 0.01 mol each of acetic acid and ammonia are added to a liter of water. They can undergo reactions as follows:

$$CH_3\,COOH \rightarrow CH_3\,COO^- + H^+ \quad \log K = -4$$

$$NH_3 \rightarrow NH_4^+ + OH^- \quad \log K = -4.8$$

    a. List the chemical species that can occur at equilibrium (assume carbonate species are excluded).
    b. Give the mass action constraints.
    c. Give the mass conservation constraints.
    d. Cite any other constraints on the composition of the system.
    e. Are there enough constraints to determine the final pH of the system?

REFERENCES

Anderson, M. A. and Parker, J. C. (1990). "Sensitivity of Organic Contaminant Transport and Persistence Models to Henry's Law Constants: Case of Polychlorinated Biphenyls." *Water, Air, Soil Poll.* 50, 1–18.

Barford, N. C. (1985). *Experimental Measurements: Precision, Error and Truth.* 2nd ed. Wiley, Chichester.

Berthouex, P. M. and Brown, L. C. (1994). *Statistics for Environmental Engineers.* Lewis Publishers, Boca Raton, FL.

Boublík, T., Fried, V., and Hála, E. (1984). *The Vapour Pressures of Pure Substances.* 2nd revised ed. Elsevier, Amsterdam.

Budavari, S. (Ed). (1989). *The Merck Index,* 11th ed. Merck, Rahway, NJ.

Castellan, G. W. (1983). *Physical Chemistry.* 3rd ed. Benjamin/Cummings, Menlo Park, CA.

Currie, L. A. and DeVoe, J. R. (1977). "Systematic Error in Chemical Analysis." *In: Validation of the Measurement Process* (J. R. DeVoe, Ed.), ACS Symposium Series 63, American Chemical Society, Washington, DC.

Davis, J. C. (1986). *Statistics and Data Analysis in Geology.* 2nd ed. Wiley, New York.

Denbigh, K. G. (1981). *The Principles of Chemical Equilibrium: With Applications in Chemistry and Chemical Engineering*, 4th ed. Cambridge University Press, Cambridge.

Dingman, S. L. (1994). *Physical Hydrology*. Prentice-Hall, Englewood Cliffs, NJ.

Gans, P. (1992). *Data Fitting in the Chemical Sciences by the Method of Least Squares*. Wiley, Chichester.

Guggenheim, E. A. (1967). *Thermodynamics: An Advanced Treatment for Chemists and Physicists*, 5th ed. North-Holland Publishing Co., Amsterdam.

Hansch, C. and Leo, A. J. (1985). MEDCHEM Project. Issue No. 26. Pomona College, Claremont, CA.

Haynes, W. (1954). *American Chemical Industry—A History*, Vols. I–VI. Van Nostrand, New York.

Hine, J. and Mookerjee, P. K. (1975). "The Intrinsic Hydrophilic Character of Organic Compounds. Correlations in Terms of Structural Contributions." *J. Org. Chem.* **40**(3), 292–298.

Howard, P. H. (Ed.). (1989, 1990, 1991). *Handbook of Environmental Fate and Exposure Data for Organic Chemicals*, Vols. I–III. Lewis, Chelsea, MI.

Kline, S. J. (1985). "The Purpose of Uncertainty Analysis." *J. Fluids Eng.* **107**, 153–160.

Lyman, W. J., Reehl, W. F., and Rosenblatt, D. H. (1990). *Handbook of Chemical Property Estimation Methods*, 2nd printing. American Chemical Society, Washington, DC.

Mackay, D. and Paterson, S. (1981). "Calculating Fugacity." *Environ. Sci. Technol.* **15**(9), 1006–1014.

Mackay, D. and Shiu, W. Y. (1981). "A Critical Review of Henry's Law Constants for Chemicals of Environmental Interest." *J. Phys. Chem. Ref. Data* **10**(4), 1175–1199.

McCuen, R. H. (1985). *Statistical Methods for Engineers*. Prentice-Hall, Englewood Cliffs, NJ.

McCuen, R. H. (1992). *Microcomputer Applications in Statistical Hydrology*. Prentice-Hall, Englewood Cliffs, NJ.

McMurry, J. (1992). *Organic Chemistry*. 3rd ed. Brooks/Cole, Pacific Grove, CA.

Morel, F. M. M. and Hering, J. G. (1993). *Principles and Applications of Aquatic Chemistry*. Wiley, New York.

Morrison, R. T. and Boyd, R. N. (1973). *Organic Chemistry*, 3rd ed. Allyn & Bacon, Boston.

Peters, D. G., Hayes, J. M., and Hieftje, G. M. (1974). *Chemical Separations and Measurements: Theory and Practice of Analytical Chemistry*. Saunders Golden Sunburst Series, Philadelphia, PA.

Petrucci, R. H. (1989). *General Chemistry: Principles and Modern Applications*, 5th ed. Macmillan, New York.

Radel, S. R. and Navidi, M. H. (1990). *Chemistry*. West Publishing, St. Paul, MN.

Riederer, M. (1990). "Estimating Partitioning and Transport of Organic Chemicals in the Foliage/Atmosphere System: Discussion of a Fugacity-Based Model." *Environ. Sci. Technol.* **24**, 829–837.

Roberts, J. D. and Caserio, M. C. (1977). *Basic Principles of Organic Chemistry*. 2nd ed. W. A. Benjamin, Menlo Park, CA.

Schwarzenbach, R. P., Gschwend, P. M., and Imboden, D. M. (1993). *Environmental Organic Chemistry*. Wiley, New York.

Shriver, D. F., Atkins, P., and Langford, C. H. (1994). *Inorganic Chemistry*. 2nd ed. W. H. Freeman, New York.

Streitwieser, A., Jr., Heathcock, C. H., and Kosower, E. M. (1992). *Introduction to Organic Chemistry*, 4th ed. Macmillan, New York.

Stumm, W. and Morgan, J. J. (1981). *Aquatic Chemistry: An Introduction Emphasizing Chemical Equilibria in Natural Waters*. Wiley, New York.

Stumm, W. and Morgan, J. J. (1996). *Aquatic Chemistry: Chemical Equilibria and Rates in Natural Waters*. 3rd ed. Wiley-Interscience, New York.

Syracuse Research Corporation. (Continually updated). Environmental Fate Data Base. Syracuse, NY.

Topping, J. (1962). *Errors of Observation and their Treatment.* 3rd ed. Chapman & Hall, London.

Velikanov, M. A. (1965). *Measurement Errors and Empirical Relations* (Trans. from Russian), Israel Program for Scientific Translations, Jerusalem.

Wall, F. T. (1974). *Chemical Thermodynamics; A Course of Study,* 3rd ed. Freeman, San Francisco, CA.

Weast, R. C. (Ed.). (1990). *CRC Handbook of Chemistry and Physics,* 70th ed., 2nd printing. CRC Press, Boca Raton, FL.

Yalkowsky, S. H. and Valvani, S. C. (1979). "Solubilities and Partitioning. 2. Relationships between Aqueous Solubilities, Partition Coefficients, and Molecular Surface Areas of Rigid Aromatic Hydrocarbons." *J. Chem. Eng. Data* **24**, 127–129.

CHAPTER 2

Surface Waters

2.1 INTRODUCTION

2.1.1 NATURE OF SURFACE WATERS

Surface waters, including rivers, streams, lakes, and estuaries, were among the first environmental media to receive widespread attention for chemical pollution problems. This attention was due in part to the high visibility and extensive public usage of surface waters, as well as to their historical use as waste receptors. Fish kills, odiferous industrial discharges, sewage, floating refuse, and other obvious signs of pollution mobilized scientific and regulatory communities to study and regulate sources and sinks of pollutants. A major goal of such work was to determine "acceptable" levels of waste loadings to surface waters. For example, estuaries, which are the transitional zones between rivers and the ocean, were at one time regarded as potential pollutant filters. However, research on the very high productivity of these waters, as well as the dependence on them of many sensitive life cycle stages of organisms, has alerted people to the need to maintain these waters in as pristine a condition as possible.

FIGURE 2-1 Bass Brook, a small fast-moving stream in New Britain, Connecticut. Upstream of this location the stream goes over a steep rapids; further downstream, it travels through a floodplain, a broad and flat area of bordering land that becomes part of the river during high water. Downslope flow of water due to gravity distinguishes streams and rivers from lakes. (Photo by H. Hemond.)

Before discussing in detail any of the fate and transport processes occurring in surface waters, the major characteristics of surface waters must be defined. As illustrated in Fig. 2-1, rivers and streams are relatively long, shallow, narrow water bodies characterized by a pronounced horizontal movement of water in the downstream direction. Often the water flow is sufficiently turbulent to erode the stream channel and carry sediment for considerable distances. Due to this movement of sediment, some river channels are constantly shifting in geometry. Compared with rivers, lakes tend to be deeper and wider and are not dominated by a persistent downstream current (Fig. 2-2). Lakes are often vertically *stratified* for part of the year, with two distinct layers of water whose temperatures and chemistries are markedly different. Estuaries (Fig. 2-3), the interfaces between rivers and the ocean, also are often vertically stratified, due to the denser saline seawater sinking beneath the freshwater discharged from the river. Estuaries have tides due to their connection to the ocean, and they tend to be rich in nutrients.

FIGURE 2-2 Bickford Reservoir, a lake in central Massachusetts. Lakes such as this typically stratify during the summer season, but become fully mixed during the spring and fall. Water currents in a lake are mostly wind-driven and vary in velocity. On large lakes, wave action also becomes an important transport factor. (Photo by H. Hemond.)

After a brief description of pollutant sources to surface waters, the physics, biology, and chemistry of surface waters are discussed. The major fate processes for chemicals in surface waters are then presented, including the physical processes of volatilization and sedimentation; the chemical processes of reduction–oxidation reactions, hydrolysis, and photodegradation; and the biological processes of biodegradation and accumulation in aquatic organisms.

2.1.2 Sources of Pollutant Chemicals to Surface Waters

Sources of pollutants are commonly divided into two categories: point sources and nonpoint sources. *Point sources* refer to discrete, localized, and often readily measurable discharges of chemicals. Examples of point sources are industrial outfall pipes, untreated storm water discharge pipes, and treated

FIGURE 2-3 The mouth of the Poquonock River in Groton, Connecticut, where it enters the coastal waters of Fishers Island Sound (photo left). Here, a barrier beach (the prominent curving spit of sand) separates the Sound from the estuary; water enters and exits, driven by tides and by river inflow, through the relatively narrow inlet west (photo top) of the barrier. Not all estuaries have such a barrier, but all have a region where saltwater and freshwater come together and mix. This estuary is geometrically complex, with a quilt work of open water, marsh, natural upland, and a region of fill on which the airport (photo top right) is built. (Photo by H. Hemond.)

sewage outfalls. A spill of chemicals, due to an accident on or near a surface water body, can also be regarded as a point source of pollution because the amount and location of the discharge are often well characterized.

Nonpoint sources of pollution are more difficult to measure because they often cover large areas or are a composite of numerous point sources. Examples of nonpoint sources include pesticide and fertilizer runoff from agricultural fields, and urban runoff contaminated with pollutants from automobile emissions. Nonpoint sources may not be directly located next to a surface water body; pollutants may be transported to surface waters by runoff from the land, by groundwater inflow, or by atmospheric transport.

2.2 PHYSICAL TRANSPORT IN SURFACE WATERS

2.2.1 Rivers

Gravity-Driven Advection

In rivers, water flows downstream by gravity. The velocity of a river is usually measured directly because rivers are generally accessible and satisfactory current-measuring devices exist. Nevertheless, a great deal is known about the factors that control river flow; if river geometry is known, it is possible to determine river velocity without going near the water. One governing principle of river flow is that the gravitational energy gained by the water flowing downslope must either go into frictional energy loss (dissipated as heat) or kinetic energy (energy associated with the water velocity). Two equations are widely used to model the velocity of water in uniform river or stream channels: the Chezy equation and the Manning equation. Both equations relate water velocity to the channel's *hydraulic radius* and slope. The hydraulic radius is the ratio of the cross-sectional area of flowing water to the wetted perimeter (for a rectangular channel, width plus twice the depth). The Chezy equation is

$$V = C\sqrt{RS}, \qquad [2\text{-}1]$$

where V is the velocity [L/T], C is the Chezy friction coefficient [$L^{1/2}$/T], R is the hydraulic radius [L], and S is the slope of the water surface (dimensionless). The Manning equation is

$$V = \frac{1.49\ R^{2/3}S^{1/2}}{n}, \qquad [2\text{-}2]$$

TABLE 2-1 **Manning Roughness Coefficients (n)[a,b]**

Channel characteristics	Value
Smooth concrete	0.012
Ordinary concrete lining	0.013
Vitrified clay	0.015
Straight unlined earth canals in good condition	0.020
Winding natural streams and canals in poor condition—considerable moss growth	0.035
Mountain streams with rocky beds, and rivers with variable sections and some vegetation along banks	0.040–0.050

[a] Dunne and Leopold (1978).
[b] For use in Eq. [2-2]. Note that in Eq. [2-2], the hydraulic radius must be expressed in units of feet; the resultant velocity has units of feet per second.

where n is the Manning roughness coefficient, which describes river channel roughness. As commonly tabulated, values of n require that R be expressed in feet; the resultant velocity has units of feet per second.

The coefficients C or n are determined experimentally for different types of channel linings. A tabulation of Manning roughness coefficients is shown in Table 2-1. For further discussion of these two equations, the reader is referred to Henderson (1966), Dunne and Leopold (1978), or Linsley *et al.* (1982).

EXAMPLE 2-1

A 2-m wide rectangular culvert made of ordinary concrete is constructed to carry storm water flow away from a new housing development. The slope of the culvert is 0.001. After a heavy rainstorm, an 8-in. deep flow of water is measured in the culvert. Assume uniform, steady flow and estimate the velocity of this storm water.

Eq. [2-2] can be used to estimate the water velocity in the culvert. First, estimate the hydraulic radius, in units of feet:

$$R = \frac{\text{width} \cdot \text{depth}}{\text{width} + 2 \cdot \text{depth}}$$

$$= \frac{(2\text{ m} \cdot 3.281\text{ ft/m}) \cdot (8\text{ in.} \cdot 1\text{ ft/ }12\text{ in.})}{(2\text{ m} \cdot 3.281\text{ ft/m}) + 2 \cdot (8\text{ in.} \cdot 1\text{ ft/}12\text{ in.})}$$

$$= 0.55\text{ ft.}$$

Then use Table 2-1 to obtain a value of 0.013 for the Manning roughness coefficient for ordinary concrete lining in a channel.

The velocity of the storm water flow can then be estimated from Eq. [2-2] as

$$V(\text{ft/sec}) = \frac{1.49 \cdot (0.55)^{2/3} \cdot (0.001)^{1/2}}{0.013} = 2.4 \text{ ft/sec.}$$

When a mass of a chemical is released at a point in a river, the center of the chemical's mass moves downstream at the average velocity of the river (Fig. 2-4, upper panel). The average amount of time it takes a chemical to travel from an upstream point to a downstream point in a river (i.e., to traverse the length of a given segment, or *reach,* of river) is called the *travel time,* τ, and is expressed as

$$\tau = L/V, \qquad [2\text{-}3a]$$

where τ is the travel time [T], L is the length of river reach [L], and V is the average velocity of the river [L/T]. If the river velocity is not uniform along the river, then the travel time must be expressed as an integral,

$$\tau = \int_{x_1}^{x_2} \frac{1}{V(x)}\, dx, \qquad [2\text{-}3b]$$

where x is the distance along the river [L], points x_1 and x_2 are the endpoints of the reach, and $V(x)$ is the magnitude of the velocity [L/T] of the river at any given point x.

An estimate of travel time is important in many situations. For example, a municipal water supply operator needs to know how long it will take a chemical spilled upriver to reach downstream water intake pipes so that the valves can be closed before the spilled chemical arrives. An estimate of travel time is also necessary when calculating whether processes such as loss to the air (*volatilization*) or bacterial degradation will significantly decrease a chemical concentration along a reach of river.

The total volume of water passing any given point in the river per unit time is called *discharge.* The relationship between discharge and velocity is

$$Q = A \cdot V, \qquad [2\text{-}4]$$

where Q is discharge [L^3/T], A is the cross-sectional area of the river [L^2], and V is the average water velocity [L/T]. Discharge may be measured using a *weir* or a *flume;* these are structures that are built in a river channel and have

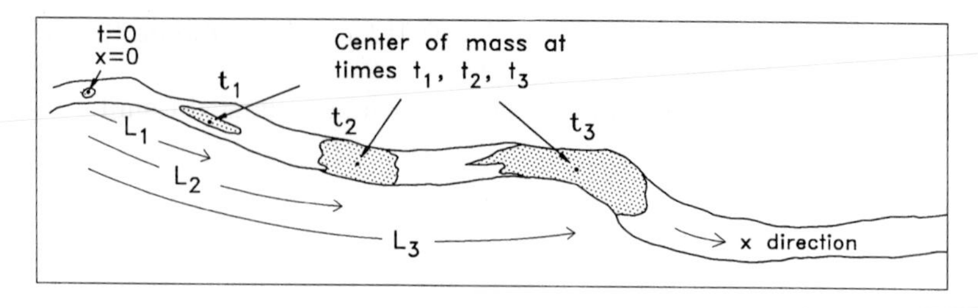

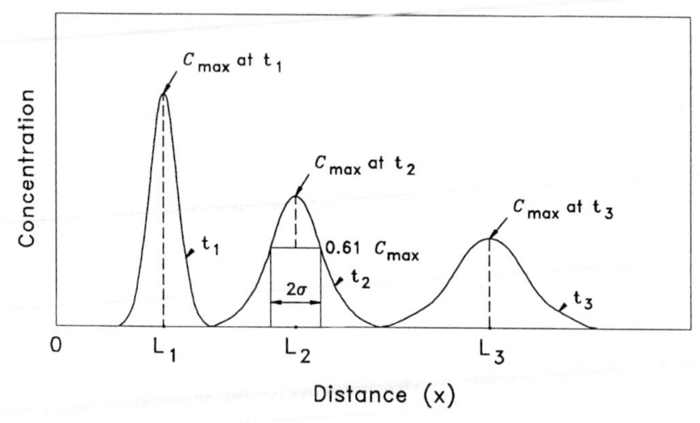

FIGURE 2-4 Transport of a chemical in a river. At time zero, a pulse injection is made at a location defined as distance zero in the river. As shown in the upper panel, at successive times t_1, t_2, and t_3, the chemical has moved farther downstream by advection, and also has spread out lengthwise in the river by mixing processes, which include turbulent diffusion and the dispersion associated with nonuniform velocity across the river cross section. Travel time between two points in the river is defined as the time required for the center of mass of chemical to move from one point to the other. Chemical concentration at any time and distance may be calculated according to Eq. [2-10]. As shown in the lower panel, C_{max}, the peak concentration in the river at any time t, is the maximum value of Eq. [2-10] anywhere in the river at that time. The longitudinal dispersion coefficient may be calculated from the standard deviation of the concentration versus distance plot, Eq. [2-7].

known depth–discharge relationships. Alternatively, discharge may be calculated from stream geometry and velocity measurements made with a current meter. The mass of chemical transported by a river past a given point per unit time is

$$J_{tot} = Q \cdot C, \qquad [2\text{-}5]$$

where J_{tot} is the total flux of chemical [M/T] and C is the average chemical concentration [M/L^3].

Fickian Mixing Processes

A mass of chemical released in a river will spread out as it moves downstream. This spreading is due to dispersion, caused by *velocity shear* within the river, and turbulent diffusion. Velocity shear in a river occurs because different parcels of water have different downstream velocities, depending on their position in a river cross section. Typically, water velocity in a river increases with distance from the river bottom and sides, reaching a maximum near the river center and usually somewhat below the water surface, as shown in Fig. 2-5. Chemical mass dissolved in the midchannel near-surface water will travel downstream more rapidly than mass dissolved in deeper water or in water near the channel sides. The distribution of the chemical mass thus elongates in the direction of flow. Turbulent diffusion is caused by the random motions of water associated with channel irregularities and with the eddies that are caused by velocity shear. A Fickian mixing approach can be used to describe the sum of the mixing due to dispersion and due to turbulent diffusion. The greater the spatial variability in water velocity and the greater the turbulence, the greater will be the mixing.

Mixing in a stream or river can be quantitatively illustrated by instantaneously releasing a mass of chemical uniformly throughout the cross section of a channel; this *pulse injection* is best expressed as mass per cross section of river, $[M/L^2]$. Ideally, one chooses a *conservative tracer* (i.e., a chemical that does not undergo degradation in the river and is not absorbed to the river channel or suspended particles). The lower panel of Fig. 2-4 shows a chemical concentration in a stream at several different times after a pulse injection. At any instant, the plot of concentration versus distance is "bell-shaped"; ideally, if the mixing is truly Fickian, the curve has the shape of a Gaussian, or *normal*, curve,

$$\phi(x) = \frac{1}{\sigma\sqrt{2\pi}} e^{-x^2/2\sigma^2}, \qquad [2\text{-}6]$$

where x is the abscissa (x coordinate) of each point on the curve, $\phi(x)$ is the ordinate (y coordinate) of each point on the curve, and σ is the standard deviation of $\phi(x)$ about the origin.

The equation that is used to describe the chemical concentration at various locations has the same mathematical form as Eq. [2-6] because the randomness assumed in Fickian mixing is similar to the randomness that gives rise to the normal curve. For a pulse injection, there is a close relationship between a Fickian mixing, or transport, coefficient D in a given direction, and the standard deviation of the chemical distribution in that direction. D can be

a

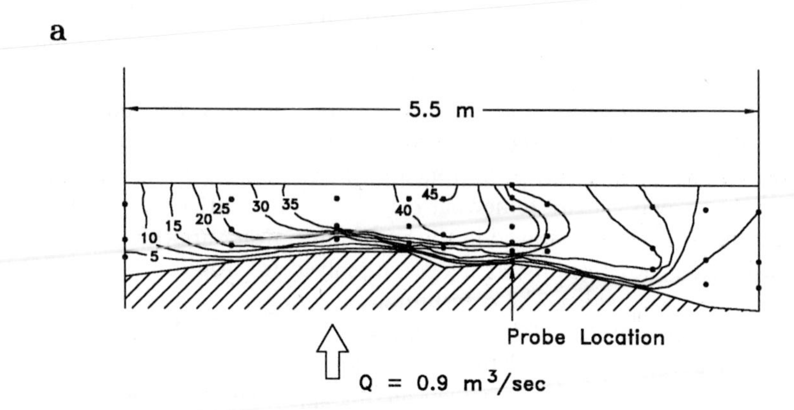

b

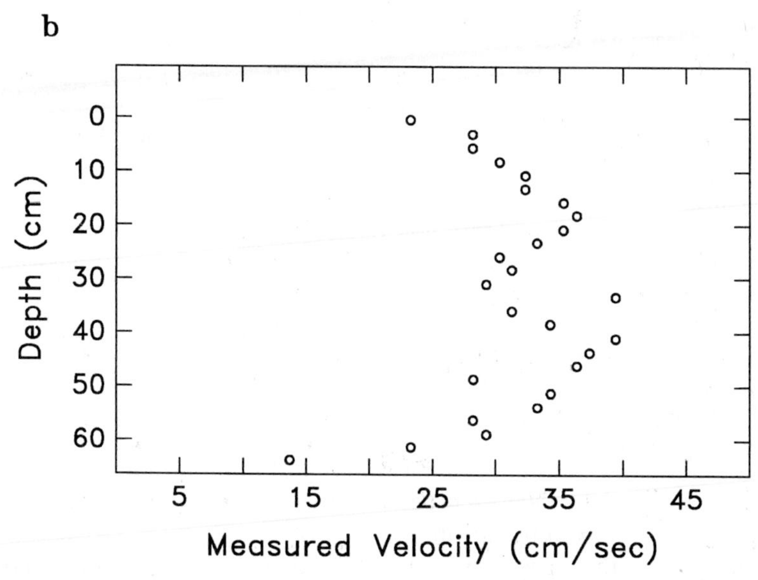

FIGURE 2-5 (a) Velocity distribution across a channel cross section in the Aberjona River in Woburn, Massachusetts. A nonuniform velocity distribution results in longitudinal dispersion of chemicals carried by the river. (b) Vertical velocity profile in the Aberjona River at the probe location. Velocity is highest at an intermediate depth, lower at the surface due to air resistance, and zero at the channel bottom [data from Solo-Gabriele (1995)].

calculated from the following equation,

$$D = \sigma^2/2\,t, \qquad [2\text{-}7]$$

where σ^2 is the spatial variance (the square of the standard deviation) of the chemical distribution [L^2], and t is the time since the pulse injection of the chemical was made. σ increases with time, but the tracer maintains the shape of a Gaussian distribution about its center of mass.

When Eq. [2-7] is solved for σ in the direction of river flow (the longitudinal direction), and then σ is substituted into Eq. [2-6], the resulting equation still describes a Gaussian curve,

$$\phi(x,t) = \frac{1}{\sqrt{2D_L t}\,\sqrt{2\pi}}\, e^{-x^2/2 \cdot 2D_L t}, \qquad [2\text{-}8]$$

where D_L is the longitudinal dispersion coefficient [L^2/T]. Given that the center of mass of the cloud of tracer is moving downstream at velocity V, the distance x must be replaced by $(x - Vt)$ in Eq. [2.8]:

$$\phi(x,t) = \frac{1}{\sqrt{2D_L t}\,\sqrt{2\pi}}\, e^{-(x-Vt)^2/2 \cdot 2\,D_L t}. \qquad [2\text{-}9]$$

To obtain an expression for chemical concentration in the river, the chemical mass M injected into the river must be taken into account. In Eq. [2-9], the area under the curve is unity, whereas for a conservative tracer, the area under a curve of concentration versus distance must equal M, the initial chemical mass injected into the river per unit area of cross section. Thus, Eq. [2-9], when multiplied by M, yields Eq. [2-10], which gives the concentration of a conservative tracer at any time t after injection and any distance x downstream:

$$C(x,t) = \frac{M}{\sqrt{4\pi D_L t}}\, e^{-(x-Vt)^2/(4\,D_L t)}, \qquad [2\text{-}10]$$

where C is the concentration of chemical [M/L^3], M is the mass of chemical injected per cross-sectional area of river [M/L^2], x is the distance downstream of injection location [L], V is the river velocity [L/T], t is the time elapsed since injection [T], and D_L is the longitudinal Fickian mixing coefficient [L^2/T]. Eq. [2-10] is also a solution, given this particular set of conditions, to the advection–dispersion–reaction equation shown in Eq. [1-5].

If the chemical of interest undergoes first-order decay during transport downstream, the righthand side of Eq. [2-10] can be multiplied by the factor e^{-kt}, where k is a first-order rate constant for chemical transformation and removal processes [T^{-1}]:

$$C(x, t) = \frac{M}{\sqrt{4\pi D_L t}} e^{-(x-Vt)^2/(4D_L t)} \cdot e^{-kt}. \qquad [2\text{-}11]$$

At any given time t, the maximum concentration of the chemical (C_{max}) is found at a distance downstream of the injection point (x) equal to the product of the time elapsed since injection (t) and the average river velocity (V). At this location, the quantity ($x - Vt$) in Eq. [2-10] equals zero, and thus

$$C_{max} = \frac{M}{\sqrt{4\pi D_L t}} e^{-kt}. \qquad [2\text{-}12]$$

It follows from the properties of the normal distribution that the portion of the river lying within one standard deviation (σ) upstream or downstream on either side of the point of maximum chemical concentration includes 68% of the chemical mass in the river. It also follows that the chemical concentration one standard deviation from the point of maximum concentration equals the maximum concentration multiplied by 0.61 (see Fig. 2-4, lower panel).

The preceding discussion assumed that a chemical is injected into a river or stream uniformly across a river cross section. In fact, spills and other inputs are rarely introduced uniformly across a channel, and a certain distance must be traveled before a chemical concentration becomes uniform across the channel. For a chemical released at a river bank, the length L of this *transverse mixing zone* can be roughly estimated by equating the lateral standard deviation (σ_t) of the chemical's concentration distribution to the width of the river,

$$\sigma_t = \sqrt{2D_t t} \approx w, \qquad [2\text{-}13]$$

where D_t is the transverse Fickian mixing coefficient [L^2/T], t is the time since the chemical was released, and w is the width of the river [L]. Combining Eq. [2-3a] with Eq. [2-13] results in the estimate of the length of the transverse mixing zone,

$$L \approx \frac{w^2 V}{2D_t}, \qquad [2\text{-}14]$$

where V is the average velocity [L/T].

Estimation of Fickian Transport Coefficients from Tracer Experiments

In the case of mixing primarily due to turbulent diffusion and dispersion, the Fickian transport coefficients are essentially independent of the chemical, so that the values of D determined from tracer experiments can be applied to other chemicals of interest in the same river. Two types of commonly used tracers are salts, such as sodium chloride (NaCl), and fluorescent dyes, such as rhodamine, which can be measured at very low concentrations.

For example, to estimate D_L, a pulse of tracer is injected into a river and the longitudinal distribution of the tracer is measured as the river carries it past a downstream location. The spatial standard deviation of tracer and the travel time are determined from tracer concentration data, and D_L is computed using Eq. [2-7].

Travel time and the longitudinal Fickian transport coefficient can also be evaluated from a *continuous injection* experiment, in which injection of tracer is initiated at time $t = 0$ and continues until steady-state conditions are reached downstream. This type of experiment is discussed for groundwater in Section 3.2.5; the equation describing concentrations resulting from a continuous injection in a river is identical to Eq. [3-18]. The equivalent equation presented in Fig. 3-19 could also be used, but with 100% porosity. In fact, equations such as Eq. [2-10] that describe advective and Fickian transport are surprisingly general. Although applied here to rivers and streams, they can be used to describe transport and evaluate Fickian mixing coefficients in many environmental situations, including groundwater and air.

EXAMPLE 2-2

The t_2 profile of Fig. 2-4 was measured 5 hr after a pulse injection of dye. What is the average river velocity if the maximum concentration is occurring 1025 m down the river from the pulse injection at this time? Estimate the longitudinal dispersion coefficient for this river if the standard deviation in the longitudinal direction, σ_L, is approximately 350 m when the chemical has traveled a distance of 1975 m to L_3.

The average velocity is

$$V = \frac{L_2}{t_2} = \frac{1025 \text{ m}}{5 \text{ hr}} = 205 \text{ m/hr}.$$

To estimate the dispersion coefficient, consider the concentration profile at time t_3; the peak of the profile (C_{max}) occurs at approximately 1975 m, and the standard deviation is roughly 350 m. Assuming the average river velocity is 205 m/hr, the travel time to L_3, using Eq. [2-3a], is

$$\tau_3 = \frac{L_3}{V} = \frac{1975 \text{ m}}{205 \text{ m/hr}} = 9.6 \text{ hr}.$$

The longitudinal dispersion coefficient D_L can then be estimated from Eq. [2-7]:

$$D_L = \sigma_L{}^2/2\tau = (350 \text{ m})^2/(2 \cdot 9.6 \text{ hr}) \simeq 6400 \text{ m}^2\text{/hr}.$$

Estimation of Fickian Transport Coefficients from Flow Data

In the absence of experimental tracer data, it is also possible to estimate Fickian transport coefficients from stream channel geometry and stream discharge (Fischer *et al.*, 1979). A substantial amount of work has been done on the problem of estimating both longitudinal dispersion coefficients and transverse diffusion coefficients (commonly called transverse dispersion coefficients) for rivers. The distinction between transverse (or lateral) mixing and longitudinal mixing is that transverse mixing occurs only by turbulence (there is by definition no lateral advection in a river), whereas longitudinal mixing is caused in part by turbulence, but is usually dominated by dispersion. The result is that different formulations have arisen for estimating Fickian transport in each direction.

As previously discussed, turbulence is caused in part by velocity shear due to a nonuniform velocity profile. In a river, a *shear velocity*, which is related to the shear force per unit area exerted by the water flow on the river channel, can be estimated (Fischer *et al.*, 1979) as

$$u^* = \sqrt{gdS}, \quad [2\text{-}15]$$

where u^* is the shear velocity [L/T], g is the acceleration due to gravity [L/T^2], d is the stream depth [L], and S is the channel slope (dimensionless). (Strictly speaking, the hydraulic radius R should be used instead of d, but most rivers are wide compared with their depth, so that the hydraulic radius is approximately equal to the depth.)

A fairly good correlation (within a factor of two) has been reported between u^* and the transverse dispersion coefficient, D_t:

$$D_t \approx 0.15 \cdot d \cdot u^* \text{ for straight channels,} \quad [2\text{-}16a]$$

$$D_t \approx 0.6 \cdot d \cdot u^* \text{ for typical natural channels.} \quad [2\text{-}16b]$$

Table 2-2 shows a range of reported D_t for straight, sinuous, and meandering natural channels.

In the case of longitudinal dispersion coefficients, D_L, the stream velocity and width become important predictors; the following equation typically predicts D_L within a factor of four (Fischer *et al.*, 1979),

$$D_L = \frac{0.011 \cdot V^2 \cdot w^2}{d \cdot u^*}, \quad [2\text{-}17]$$

where D_L is the longitudinal dispersion coefficient [L^2/T], V is the average velocity [L/T], w is the width of the channel [L], d is the stream depth [L],

TABLE 2-2 Reported Transverse Dispersion Coefficients[a]

River type/river	Transverse dispersion coefficients (m^2/sec)	Discharge during dispersion measurement (m^3/sec)
Straight channels		
Atrisco	0.010	7.4
South	0.0047	1.5
Athabasca	0.093	776
Bends		
Missouri	1.1	1900[b]
Beaver	0.043	20.5
Mississippi	0.1	92–120
Meandering		
Missouri	0.12	966
Danube	0.038	1030
Rea	0.0014	0.30
Orinoco	3.1	47,000
MacKenzie	0.67	15,000[b]

[a] Rutherford (1994).
[b] Estimated based on height, width, and velocity.

and u^* is the shear velocity [L/T]. Typical values of D_L range from 0.05 to 0.3 m^2/sec for small streams (Genereux, 1991) to greater than 1000 m^2/sec for large rivers such as the Rhine (Wanner *et al.*, 1989). Table 2-3 presents a range of reported D_L at particular locations and times for several rivers.

2.2.2 Lakes

Although lakes are distinguished from rivers in part by the relative absence of a pronounced downstream flow of water, the waters of a lake are by no means stationary. Water currents, typically driven by wind instead of gravity, are a major feature of these water bodies. Water currents not only provide advective transport of chemicals in lakes but also cause transport by eddy diffusion because the water currents are almost always turbulent. In a lake, the average amount of time that water remains in the lake is called the *hydraulic residence time,* which can be estimated by the ratio of the lake volume to the rate at which water is lost through all processes (e.g., outflow, seepage, and evaporation).

TABLE 2-3 Reported Longitudinal Dispersion Coefficients[a]

River	Depth (m)	Width (m)	Velocity (m/sec)	Longitudinal dispersion coefficient (m^2/sec)
Irrigation canal	0.14	1.5	0.33	1.9
Monocacy	0.32	35	0.21	4.7
Monocacy	0.45	37	0.32	13.9
Monocacy	0.88	48	0.44	37.2
Yadkin	2.33	70	0.43	111
Yadkin	3.85	72	0.76	260
Susquehanna	1.35	203	0.39	92.9
Sabine	2.04	104	0.58	316
Sabine	4.75	128	0.64	670
Missouri	2.70	200	1.55	1500

[a] Rutherford (1994).

Wind-Driven Advection

Figure 2-6 a shows the simplest pattern of water movement in a lake, caused by wind exerting a force on the water at the lake surface. The downwind surface current is called *wind drift* and typically moves at a rate of 2 to 3% of the average wind speed. Clearly, the water moving downwind cannot pile up indefinitely at the end of the lake; instead, it flows back upwind, typically at

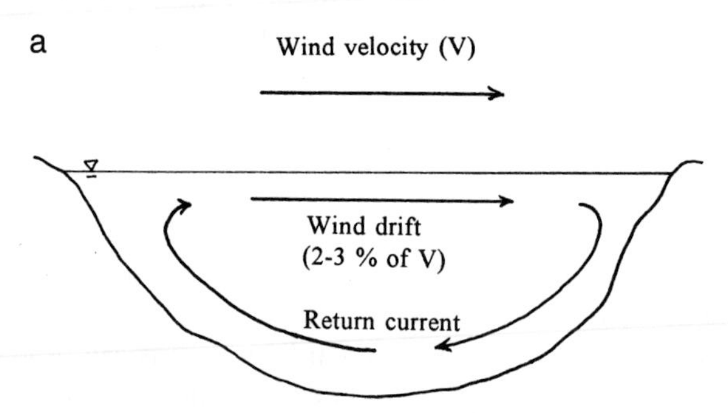

FIGURE 2-6 Wind-driven currents in a lake. (a) Circulation in a small lake of simple geometry. The surface water current, or *wind drift*, averages 2 to 3% of wind velocity. Deeper in the lake, a *return current* is established, returning water to the upwind end of the lake. (b) In a large lake such as Lake Michigan, variability of winds, complex lake geometry, and other forces (such as the Coriolis effect) lead to complex patterns of water movement [Ayers *et al.* (1958)] *(Figure continues).*

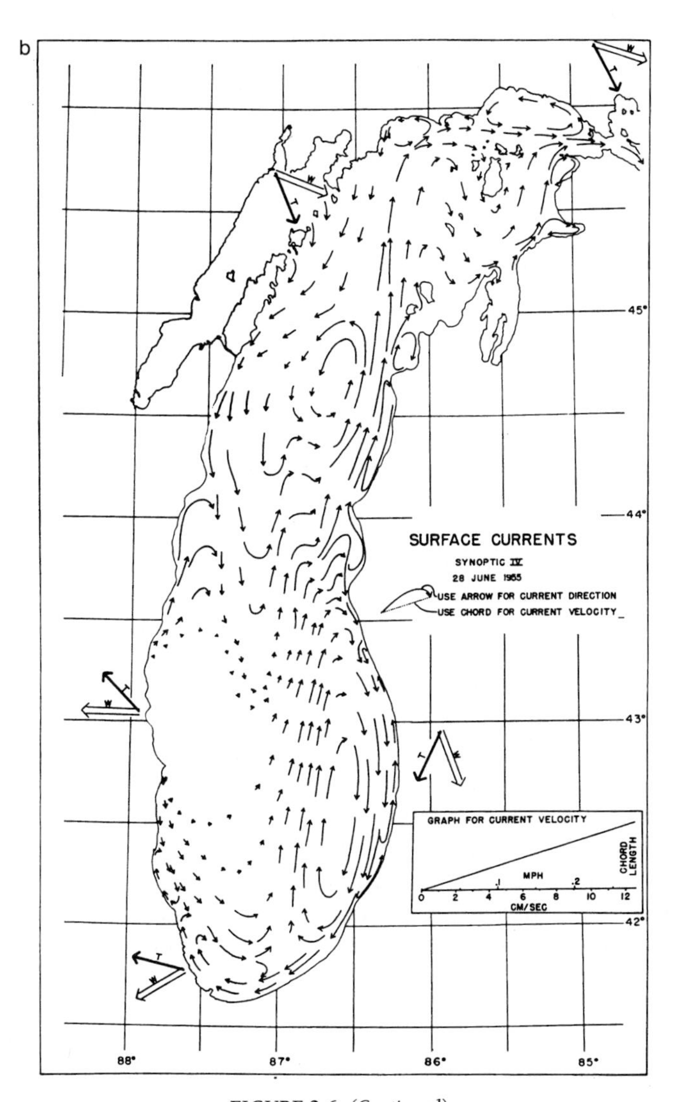

FIGURE 2-6 *(Continued)*

a greater depth, as a *return current*. Chemicals dissolved or suspended in the water are thereby transported within the lake.

In many lakes, the pattern of water movement is much more complicated, because the water movement also is affected by the shape of the lake basin, by variations in water density, by inflowing streams, and, especially in large lakes, by the Coriolis force (see Section 4.3.1). In large lakes of complex shape, the actual water currents can be exceedingly complicated, as shown in Fig. 2-6 b.

Fickian Mixing

A mass of tracer chemical injected into a lake not only will move by advection associated with water currents, but also will spread out into an ever-larger volume of water. Given enough time (perhaps a few days for a small lake), a tracer tends to become completely mixed throughout a lake; concentrations become essentially homogeneous (the same everywhere), and therefore concentration gradients become zero. Such mixing is due primarily to turbulence; eddies carry chemicals away from regions of relatively high concentration toward regions of lower concentration. The situation is comparable to that in a river, except that a one-dimensional description of Fickian mixing is rarely an adequate approximation in a lake. Lateral dispersion must almost always be taken into account because a lake's width is a significant fraction of its length, and a simplifying assumption of complete mixing across the width is not justified. If the lake is much longer and wider than it is deep (as most lakes are), and if one is mostly interested in horizontal chemical transport, it may be sufficient to consider Fickian mixing coefficients along two horizontal directions, while assuming the chemical of interest is well mixed vertically within the lake or within a stratified layer of the lake. For small time intervals after a release or in locations where the water is quite deep, however, the rate of vertical mixing must be evaluated as well. Note that Fickian mixing coefficients are generally different in each direction, just as they are for a river.

A useful two-dimensional expression for concentration of a chemical introduced as a pulse over the depth of a vertically mixed layer of water is given by Eq. [2-18]. (The equation includes the effect of first-order decay.) This equation is a solution to Eq. [1-6] under conditions of an instantaneous injection of mass into an infinite two-dimensional body of water,

$$C(x, y, t) = \frac{M}{4\pi t\sqrt{D_x D_y}} e^{-((x-V_x t)^2/4D_x t + (y-V_y t)^2/4D_y t)} \cdot e^{-kt}, \qquad [2\text{-}18]$$

where C is the concentration of tracer chemical [M/L^3], M is the mass of tracer chemical per depth of water [M/L], x and y are the distances from injection location along the x and y axes [L], t is the time elapsed since injection [T], V_x is the average velocity in the x direction [L/T], V_y is the average velocity in the y direction [L/T], D_x is the Fickian transport coefficient in the x direction [L^2/T], D_y is the Fickian transport coefficient in the y direction [L^2/T], and k is the first-order decay rate constant [T^{-1}]. Note that the depth of water could correspond to the total depth of a vertically well-mixed lake or to the depth of a particular layer in a stratified lake. As in Eqs. [2-10] and [2-11], simplification may be made for ideal tracer chemicals; in the absence of decay, the factor e^{-kt} becomes one.

Estimating D_x and D_y in a lake is more complicated than estimating Fickian mixing coefficients in a river, in part because of the larger areal extent of a lake, which leads to a scale issue in conducting tracer experiments. The eddies that exist in a lake may become larger than those in a river; the relatively smaller width of a river tends to limit the size of the largest eddies, while in a lake the largest eddies may be a sizable fraction of the lake basin size. The practical result is that a dispersion coefficient that is determined for a small plume of spreading chemical may not be appropriate to accurately model chemical dispersion at a later time, when the plume has spread and larger eddies are contributing to its mixing. Often the values of D are approximately proportional to the distance a plume has traveled raised to the 4/3 power. For more information on the estimation of dispersion coefficients and the use of lake models in which D_x and D_y increase with the areal extent of a chemical plume, the reader is referred to Fischer *et al.*, 1979.

Hydrostatic Pressure

In a surface water body, water pressure can be very closely approximated by the expression,

$$P = \rho g z, \qquad [2\text{-}19]$$

where P is the water pressure [M/LT^2], ρ is the density of water [M/L^3], g is acceleration due to gravity [L/T^2], and z is the depth below the water surface [L]. The proportionality between pressure and depth is a direct result of the fact that water is nearly incompressible, and hence has a nearly constant density. As will be seen in Section 4.1.1, the compressibility of air results in a nonlinear pressure–height relationship in the atmosphere.

Stratification

Although water density does not vary greatly from 1.0 g/cm^3, the small changes in density that do occur due to variations in water temperature and solute content can have profound effects on mixing processes in a surface

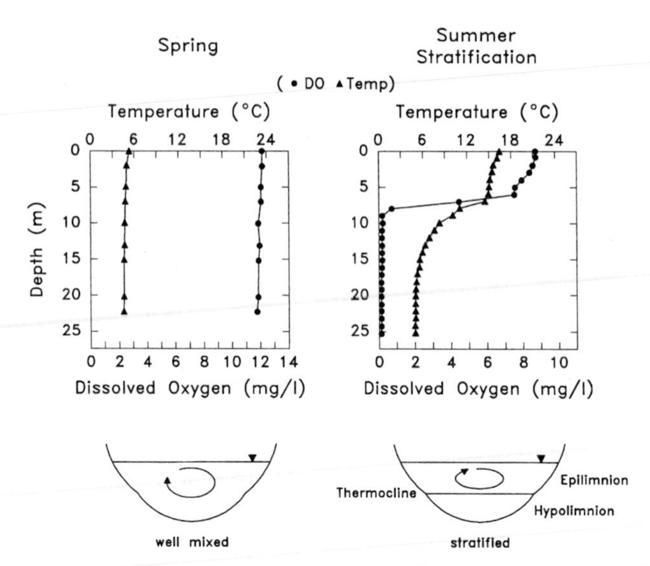

FIGURE 2-7 Measured temperature and oxygen profiles from the Upper Mystic Lake in eastern Massachusetts, on April 1, 1991 and September 30, 1991. (Left) the lake is unstratified and well mixed during turnover, which occurs in spring and fall. (Right) during summer, this eutrophic (productive) lake becomes depleted in oxygen in the lower layer of water (the hypolimnion), while its upper layer (epilimnion) remains well mixed by the wind and oxygenated by photosynthesis and by contact with the atmosphere. An oligotrophic (unproductive) lake may retain its high springtime concentration of oxygen in the hypolimnion throughout the summer [data from Aurilio (1992)].

water body. As alluded to earlier, *stratification* divides lakes into different layers by inhibiting vertical mixing between the layers. Stratification occurs when water at the bottom of a lake is denser than the surface water, and water currents (usually wind-driven) fail to generate eddies strong enough to penetrate the boundary between the water layers. Wind-driven circulation and turbulent mixing in such a lake are thereby restricted to the upper water layer (Fig. 2-7); the lower layer, isolated from wind effects, may be quite quiescent. Such a density difference is usually due to temperature differences between upper and lower water masses; the lake is then called *thermally* stratified. The upper layer, which is typically well mixed, is called the *epilimnion;* the lower layer is the *hypolimnion;* and the region separating them is the *thermocline.*

Thermal stratification is common in lakes located in climates with distinct warm and cold seasons. Although many variations are possible, the classic temperate zone lake begins a period of summer stratification when heat from solar radiation preferentially warms the uppermost water, decreasing its den-

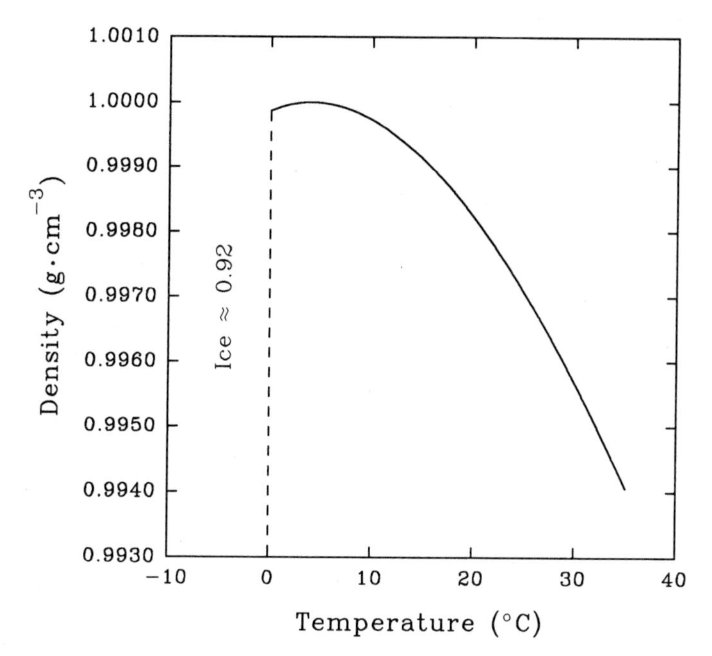

FIGURE 2-8 Density versus temperature curve for water. Maximum density occurs at 4°C; thus, stratification in a lake can occur in winter with bottom waters near 4°C and less dense surface waters closer to 0°C. In summer, if stratification occurs, the warmer water will be at the surface. Note that a given spread in water temperature conveys a larger density contrast between the waters (and hence a more stable stratification) at higher temperatures than at lower temperatures. The density of ice is *much* less than the density of liquid water (note the broken scale for ice density).

sity relative to deeper water. Usually this occurs in the spring during a few days of bright sunlight and low winds. Once started, the thermal stratification usually persists until the autumn; typical wind forces usually do not remix a lake once two layers of significantly differing densities have been created. The epilimnion does, however, thicken and become warmer throughout the summer as more solar radiation is absorbed. Remarkably little exchange of heat or chemicals occurs between the epilimnion and the hypolimnion; the lake does not fully mix again until diminished solar radiation in the autumn causes the epilimnion to cool to the approximate temperature of the hypolimnion. At this time, when the lake is nearly *isothermal,* there can be sufficient energy from the wind to mix the lake thoroughly. In the winter, *reverse stratification,* in which the deeper layer is *warmer* than the surface layer, may occur because water has a density maximum at 4°C (Fig. 2-8). As a lake cools in winter, the water temperature throughout becomes 4°C; then, as the surface water cools below 4°C, it becomes less dense than the deeper water. This stratification fosters the onset of ice, which prevents further wind mixing of the lake until the ice melts in the spring.

The effects of stratification on the lake environment are profound. During the warm season in a temperate zone lake, it is not unusual to have surface water temperatures between 15 and 25°C, while the lake remains only a few degrees above freezing near the bottom. Chemical reactions generally proceed more rapidly in the epilimnion due to warmer temperatures. Cold-water fish may be able to live in a particular lake even though its surface water temperatures are too high for their survival if the concentration of dissolved oxygen remains high enough in the hypolimnion (approximately 7 to 10 mg/liter, depending on species). Sufficiently high oxygen levels are commonly not maintained, however, especially in *eutrophic* (productive) lakes, in which large amounts of organic matter are produced in the epilimnion and settle into the hypolimnion. The isolation of the bottom waters from the atmosphere by stratification prevents the renewal of oxygen as it is consumed by organisms, and therefore the water may become *anoxic* or *anaerobic*. These two terms, anoxic and anaerobic, are often used interchangeably to imply an absence of molecular oxygen, O_2. Chemicals in an anoxic hypolimnion may undergo a set of chemical and biological transformations that are very different from those occurring in the epilimnion (see Section 2.4.3).

Vertical stratification due to water salinity differences also can occur in a lake; for example, in the winter, runoff water containing road salt can flow into a lake and travel along the bottom due to the greater density of the salty water, eventually accumulating in a low point of the lake and creating permanent stratification. Alternatively, natural processes, such as inflow of saline water from submerged springs, or accumulation of salts from chemical decomposition, may create permanent stratification. A lake in which this occurs is called a *meromictic* lake. Lack of mixing in meromictic lakes leads to a continuous depletion of oxygen in the bottom water (or *monimolimnion*).

2.2.3 Estuaries

Water Flow

Water flow in estuaries is more complicated than in rivers and in lakes; it is influenced by the inflow of freshwater from rivers and streams, by the tides of the ocean, and by the large salinity—and hence density—difference between fresh- and ocean water. The density difference tends to create a strong stratification, while the back-and-forth movement of water driven by the tides

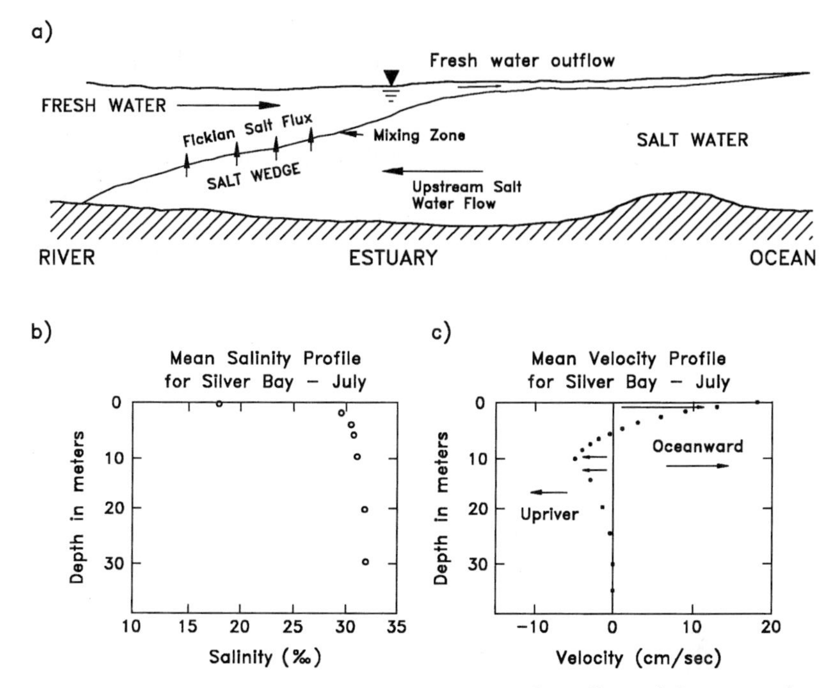

FIGURE 2-9 (a) An idealized estuary in cross section. In this well-stratified estuary, a distinct *salt wedge* extends upstream beneath fresher water at the surface. The freshwater/saltwater interface moves upriver at high tide and seaward at low tide. Data from Silver Bay, Alaska show (b) a steep average salinity gradient in parts per thousand (‰) and (c) the net upstream advection of saltwater at depth. More strongly mixed estuaries exhibit weaker vertical stratification [data are from Rattray (1967)].

enhances dispersion and mixing. In any estuary, *average* water movement is downstream (toward the ocean), driven by river flow; at any moment, however, the water can be flowing either way, depending on depth and whether the tide is rising or falling.

When a well-developed vertical stratification condition is established, an estuary is sometimes said to have a *salt wedge,* in reference to the underlying saltwater layer whose thickness tapers from the depth of the downstream saltwater body to zero in the upstream freshwater direction (Fig. 2-9). Interestingly, the average water velocity in a salt wedge is *upstream!* This can be shown by performing a mass balance for salt; upstream advection of salt by the saline layer must, on average, balance the outflow of salt in the freshwater layer. Salt is present in the freshwater layer due to upward diffusion from the salt wedge (i.e., Fickian transport).

Simple transport models have significant limitations in a complex estuarine setting; commonly, sophisticated numerical models are employed to predict transport in estuaries. In long, narrow estuaries, however, a simple one-dimensional model, such as is used in rivers, that incorporates a longitudinal dispersion coefficient and a time-averaged seaward water velocity can be useful. The results of such a model must be averaged over the tidal cycle; concentrations at each point in the estuary may be expected to vary significantly with the state of the tide. See Fischer *et al.* (1979) for a more complete discussion of transport in estuaries.

Stratification

Stratification in estuaries is in some respects similar to stratification in lakes, although in estuaries the density difference is primarily due to the difference in salinity between freshwater and ocean water, instead of being primarily due to temperature differences, as in most lakes. Freshwater has a density of approximately 1.00 g/cm^3, whereas ocean water has a density of approximately 1.03 g/cm^3 due to dissolved salts [primarily sodium (Na^+), chloride (Cl^-), calcium (Ca^{2+}), and sulfate (SO_4^{2-})]. This is a much larger density difference than that which occurs due to temperature differences in surface waters; hence, the stratification may be very strong. Whatever its cause, stratification always inhibits the vertical transfer of dissolved chemicals from layer to layer.

The salinity gradient in an estuary also has other effects on chemical fate and transport. As salinity increases in the region where freshwaters and saltwaters meet, particles brought in by the freshwaters tend to stick together (*flocculate*) and thus settle to the bottom more rapidly. The mechanism for increased flocculation is electrostatic. The rate of flocculation is given by the product of the frequency with which particles collide and the percentage of collisions that result in sticking. Particles of like charge repel each other, thereby decreasing the number of collisions that contribute to flocculation. A measure of the range at which repulsion occurs is the thickness of a "diffuse layer" of water, surrounding the particle, within which the charge of the particle is counterbalanced by a localized net excess of dissolved ions of opposite charge. As salinity increases, thereby increasing the ionic strength of the water, this diffuse layer decreases in thickness; this allows particles to approach each other more closely, resulting in more flocculation. Rising salinity also affects the activity of dissolved ionic chemical species due to the increasing ionic strength, thereby changing chemical equilibria in the water (see Section 1.6.5). Oxidation–reduction reactions are also affected because oxygen is less soluble in saline water (Table 2-4).

TABLE 2-4 Solubility of Oxygen (mg/liter) in Water Exposed to Water-Saturated Air at a Total Pressure of 760 mm Hg[a]

	Chloride concentration in water (mg/liter)				
Temperature (°C)	0	5,000	10,000	15,000	20,000
0	14.6	13.8	13.0	12.1	11.3
1	14.2	13.4	12.6	11.8	11.0
2	13.8	13.1	12.3	11.5	10.8
3	13.5	12.7	12.0	11.2	10.5
4	13.1	12.4	11.7	11.0	10.3
5	12.8	12.1	11.4	10.7	10.0
6	12.5	11.8	11.1	10.5	9.8
7	12.2	11.5	10.9	10.2	9.6
8	11.9	11.2	10.6	10.0	9.4
9	11.6	11.0	10.4	9.8	9.2
10	11.3	10.7	10.1	9.6	9.0
11	11.1	10.5	9.9	9.4	8.8
12	10.8	10.3	9.7	9.2	8.6
13	10.6	10.1	9.5	9.0	8.5
14	10.4	9.9	9.3	8.8	8.3
15	10.2	9.7	9.1	8.6	8.1
16	10.0	9.5	9.0	8.5	8.0
17	9.7	9.3	8.8	8.3	7.8
18	9.5	9.1	8.6	8.2	7.7
19	9.4	8.9	8.5	8.0	7.6
20	9.2	8.7	8.3	7.9	7.4
21	9.0	8.6	8.1	7.7	7.3
22	8.8	8.4	8.0	7.6	7.1
23	8.7	8.3	7.9	7.4	7.0
24	8.5	8.1	7.7	7.3	6.9
25	8.4	8.0	7.6	7.2	6.7

[a] American Public Health Association (1960).

2.2.4 Wetlands

Although wetlands are not strictly surface waters, they share enough attributes with surface water ecosystems to warrant being mentioned here. Wetlands contain soils that are saturated or nearly saturated with water and have a high organic content. Portions of the soil may be submerged beneath shallow water. Wetlands often border rivers and lakes; saline wetlands may border estuaries (see Fig. 2-3). The nomenclature of wetlands is diverse and inconsistent from region to region. Generally, *marshes* are wetlands vegetated by herbaceous plants such as grasses and sedges; *swamps* are wetlands vegetated by woody plants (trees and shrubs). *Bogs* and *fens* are typically northern wetlands; they may form deep deposits of *peat,* which is partially decomposed plant material. Bogs receive water predominantly from precipitation, whereas fens are partially recharged by more mineral-rich groundwater.

Chemical behavior in wetlands is strongly influenced by the organic content of the soil. Microbial decomposition of the organic matter in the soil (see Section 2.4.3) results in rapid oxygen consumption. Given that oxygen diffuses slowly through water-saturated soil relative to its transport through turbulent open water, wetland soils are usually strongly depleted in oxygen, and in many respects behave in a manner similar to bottom sediments (as discussed later). Wetland soils differ from bottom sediments, however, in that they are usually heavily vegetated and often are in contact with the atmosphere. The reader is referred to Mitsch and Gosselink (1993) for a more complete discussion of wetlands.

2.2.5 Particles in Surface Waters

In the foregoing discussions it has been implicit that advection and the Fickian mixing processes of diffusion and dispersion are responsible for the transport of *dissolved* chemicals. It is not necessary, however, for a chemical to be dissolved to be transported by these fluid processes; chemicals that are adsorbed onto the surfaces of particles or absorbed into particles can also be readily transported by these processes.

Types of Particles

Particles can be of mineral or organic origin. Mineral particles are derived from geologic materials, such as bedrock or glacial outwash, by two primary methods: the flow of water, ice, and wind, which mechanically erodes rock and sediment; and the chemical weathering of rocks. Oxyhydroxide particles, such as those of iron [$Fe(OH)_3$] and manganese [$Mn(OH)_4$], and clay particles consisting of aluminosilicates, are quite common. The density of many mineral particles ranges between 2 and 3 g/cm^3 and is often approximated as 2.6 g/cm^3. Organic particles are derived from plant material, dead bacterial or algal cells, and decaying aquatic organisms. Organic particles usually have a density only slightly greater than that of water and contain a high fraction of organic carbon, which is an excellent sorbent for many pollutants. Anthropogenic sources of mineral and organic particles to surface waters include industrial effluent and sewage outfalls as well as emissions of fugitive dusts, which are initially released to the atmosphere but subsequently settle into lakes and streams.

No matter what their origin, particles affect the transport of pollutants that are sorbed. Most of the discussion about advection and dispersion of dissolved chemicals in surface waters can also be applied to chemicals sorbed to suspended particles, provided that the time required for particles to settle out of the water is much longer than the time required for advection or mixing.

Suspended Sediment Load

The transport of suspended sediment is most prominent in rivers and in streams, which have higher water velocities than those of most lakes. The concentration of suspended sediment varies widely from river to river. The suspended sediment load also varies with discharge, often increasing as discharge rises because the greater turbulence at high flow allows a greater load of sediment to be held in suspension. Of course, the higher sediment-carrying capacity of a stream at a higher flow rate will lead to an increased suspended sediment load only if a supply of additional sediment exists. The supply of suspended sediment can be increased during times of high discharge by erosion from the land surface and by the *resuspension* of particles that previously had settled to the bottom of the water body. Consequently, the total advective flux of river sediment, which is the product of discharge and concentration, typically increases at a faster rate than discharge, as shown in Fig. 2-10. For further information on the relationships among stream channel geometry, discharge, and sediment flux, the reader is referred to Leopold and Maddock (1953).

Bed Load

The bed load of a river consists of particles that spend the majority of the time on the river bottom, but are periodically *entrained* into the turbulent water flow and carried a short distance downstream before settling again. Bed load consists mostly of particles in the size category of 1 mm in diameter; particles with diameters less than 0.1 mm are likely to be classified as suspended material, while particles with diameters larger than 10 mm move little at average flows. Bed load at times of high flow, however, can include surprisingly large particles, including rocks many centimeters in diameter. It has been proposed that bed load is approximately proportional to the mechanical *power* (work per time) being dissipated in a river at high flows.

The movement of particles, both as suspended sediment and as bed load, is of great importance to the evolution of river channels. The *meandering* of rivers (see Fig. 2-1) is an example of the physical effects of particle transport (Henderson, 1966; Reid and Wood, 1976).

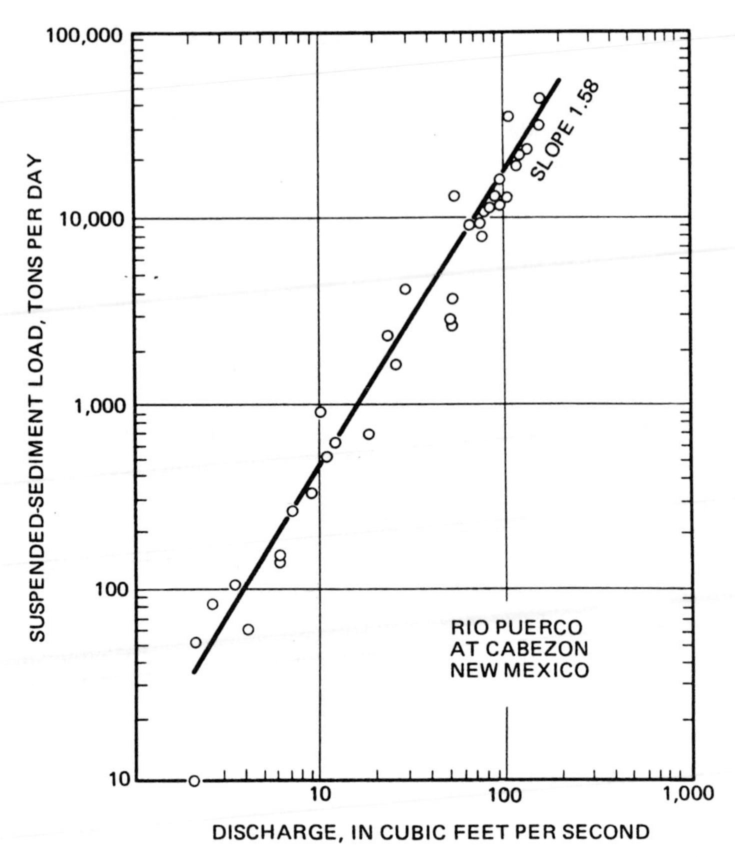

FIGURE 2-10 Suspended sediment load in the Rio Puerco, near Cabezon, New Mexico. Note the wide range of discharge (streamflow), which is typical of most rivers, and the even larger range of suspended sediment load (total flux of sediment), which varies over a factor of 1000. Sediment flux is the product of discharge and the concentration of suspended sediment in the river. Given that the slope of sediment flux versus discharge is greater than 1 (1.58), the suspended sediment concentration in the river must be increasing at higher flows [from *Ecology of Inland Waters and Estuaries, 2nd edition,* by G. K. Reid and R. D. Wood. © 1976. Reprinted with permission of Brooks/Cole Publishing, a division of International Thomson Publishing. Fax 800-730-2215.].

Particle Settling

Suspended particles in surface waters are eventually either transported out of the water body (e.g., carried to the ocean by a river) or deposited on the bottom of the water body by *settling*. Settling is especially important in relatively quiescent waters, such as lakes, which typically have a relatively short life span (as measured in geologic terms) because they tend to fill in with sediment. The average settling velocity of a particle can be approximated by *Stokes' law*, which estimates the settling velocity of a small sphere in a viscous fluid,

$$\omega_f = \frac{(2/9) \cdot g \cdot (\rho_s/\rho_f - 1) \cdot r^2}{\eta_f}, \qquad [2\text{-}20]$$

where ω_f is the settling velocity [L/T], g is the acceleration due to gravity [L/T^2], ρ_s is the density of the spherical particle [M/L^3], ρ_f is the density of the fluid [M/L^3], r is the radius of the particle [L], and η_f is the kinematic viscosity of the fluid [L^2/T]. *Kinematic viscosity* is the ratio of the dynamic viscosity of a fluid to the density of the fluid. Eq. [2-20] assumes that the particle is spherical, but is applicable to nonspherical particles if r is taken to be an empirical hydrodynamic radius.

Figure 2-11 presents data, including sizes and settling velocities, for many particles. Of particular note is the tremendous range of settling velocities among environmental particles. Given a finite settling velocity, it might seem that all particles in a surface water body should eventually settle out of the water. As particles settle toward the bottom, however, an upward concentration gradient is created, and upward Fickian transport begins to counteract the downward transport by settling. Even in very still water, particles have a diffusion coefficient, analogous to a molecular diffusion coefficient, that arises from their random *Brownian motion* (Fig. 2-11).

To illustrate particle diffusion, consider a tall water-filled volume in which particles have settled until a steady-state vertical concentration profile has been attained. Under this steady-state condition, the downward flux density of particles must equal the upward flux density of particles at every depth. The downward flux density can be expressed as

$$J_{\text{Stokes}} = C \cdot \omega_f, \qquad [2\text{-}21]$$

where J_{Stokes} is the downward flux density due to particle settling [M/L^{2}T], C is the particle concentration [M/L^3], and ω_f is the settling velocity [L/T]. The upward flux density is given by Fick's first law,

$$J_{\text{Fickian}} = D \cdot dC/dx, \qquad [2\text{-}22]$$

where J_{Fickian} is the upward flux density [M/L^{2}T], D is the particle diffusion coefficient [L^2/T], and x is the distance above the bottom of the water column [L]. Adding Eqs. [2-21] and [2-22] and equating their sum to zero gives

$$dC/dx = -(\omega_f/D) \cdot C = -(\text{constant}) \cdot C. \qquad [2\text{-}23]$$

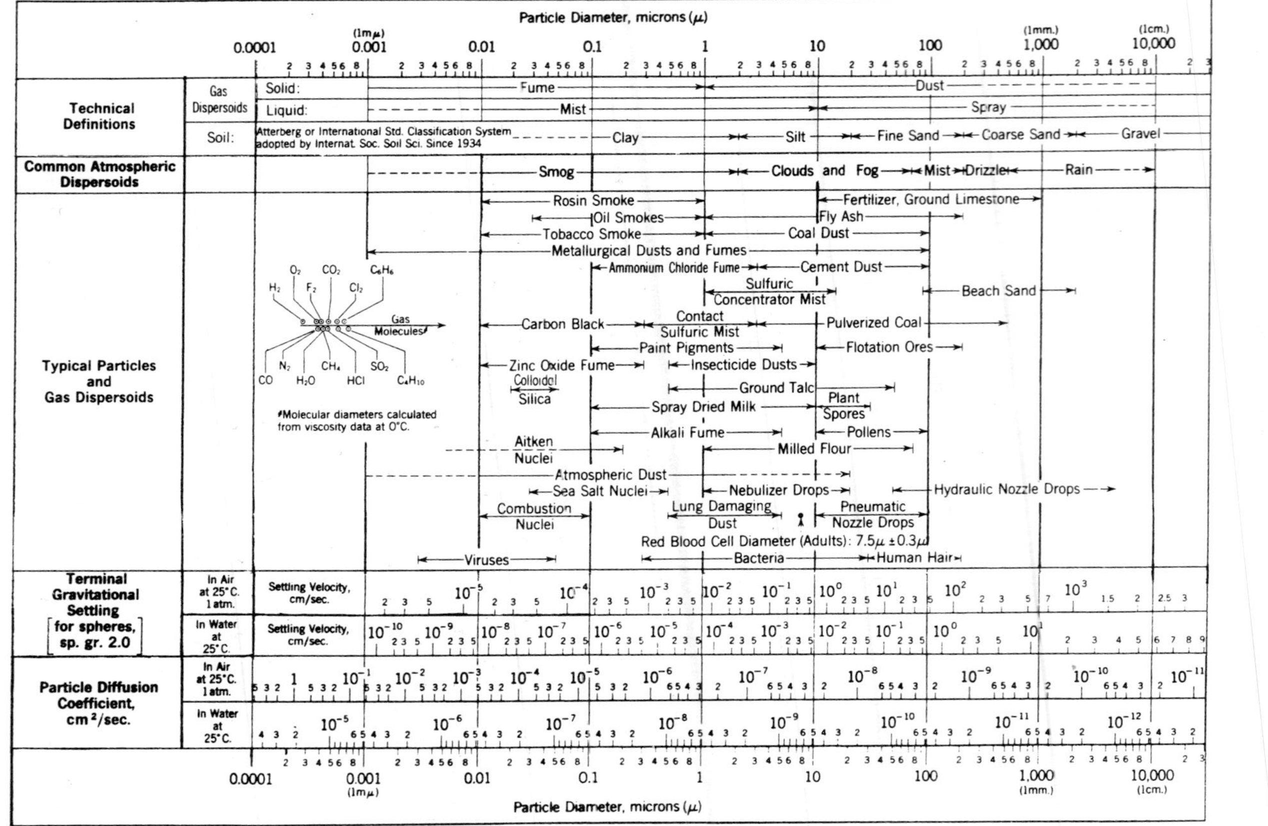

FIGURE 2-11 Some characteristics of environmental particles in both air and water. Note the enormous range of particle sizes, and the even larger range of settling velocities. Settling velocities and diffusion coefficients are calculated for spherical particles of density 2.0 g/cm³ (at the lower end of the density range for materials of mineral origin). In water, a particle of lesser density, such as a particle of organic material, whose density is commonly only slightly higher than the density of water, may settle much more slowly than shown

The solution to the differential equation shown in Eq. [2-23] should be familiar through comparison with Eqs. [1-18] and [1-19],

$$C = C_0 \cdot e^{-(\text{constant}) \cdot x}, \qquad [2\text{-}24]$$

where C_0 is the concentration at the bottom of the water column, where x is zero. Note, however, that "decay" is occurring over depth, not time, so the "decay constant" has units of [1/L], not [1/T], as in Eqs. [1-18] and [1-19]. The vertical concentration profile of particles at steady state can therefore be written as

$$C = C_0 \cdot e^{-(\omega_f/D) \cdot x}. \qquad [2\text{-}25]$$

Thus, particle concentration in the water column follows an exponential decay upward from the bottom. If the decay constant, ω_f/D, is small enough, the particle concentration might be nearly uniform over a substantial distance above the bottom of the water column. This could occur for extremely small particles in quiescent waters, or for larger particles if D is much larger than the value inferred from Brownian motion. Because surface waters are rarely quiescent, D can become many orders of magnitude larger in the presence of even mild turbulence. Under these conditions, Fickian transport may be sufficient to keep particles several micrometers in diameter suspended in the water indefinitely.

EXAMPLE 2-3

Describe the steady-state distribution of 1-μm diameter clay particles and 0.01-μm diameter clay particles in still water. For each particle size, calculate the depth above the bottom of the water column at which the particle concentration is one-half the particle concentration at the bottom. Assume a kinematic viscosity of water of 0.013 cm^2/sec at 50°F and a solid particle density of 2.6 g/cm^3.

First, calculate the settling velocity of 1-μm diameter particles from Eq. [2-20]:

$$\omega_f = \frac{(2/9) \cdot 981 \text{ cm/sec}^2 \cdot ((2.6 \text{ g/cm}^3)/(1 \text{ g/cm}^3) - 1) \cdot (5 \times 10^{-5} \text{ cm})^2}{0.013 \text{ cm}^2\text{/sec}}$$

$$= 6.7 \times 10^{-5} \text{ cm/sec}.$$

Next, estimate the first-order decay constant. From Fig. 2-11, the particle diffusion coefficient in water for a 1-μm particle is approximately 5×10^{-9}

cm^2/sec:

$$\omega_f/D = \frac{6.7 \times 10^{-5}\ \text{cm/sec}}{5 \times 10^{-9}\ \text{cm}^2/\text{sec}} = 1.3 \times 10^4/\text{cm}.$$

Then use Eq. [2-25] to estimate the distance at which the concentration is halved:

$$\frac{1}{2} = e^{-(1.3\times10^4/\text{cm})\cdot x}$$

$$x = 5 \times 10^{-5}\ \text{cm or } 0.5\ \mu\text{m}.$$

For a particle of 0.01-μm diameter, the diffusion coefficient in water from Fig. 2-11 is approximately 5×10^{-7} cm^2/sec. By repeating the preceding calculations, ω_f is approximately 6.7×10^{-9} cm/sec, the decay constant is approximately 0.013/cm, and therefore the distance at which the concentration is halved is approximately 50 cm.

Although small particles may remain separated from one another and travel long distances, small particles may also aggregate into larger particles. This process of flocculation creates particles that settle much faster than the original smaller particles, often leading to the deposition of the particles (and their sorbed pollutant loads) into bottom sediment, as will be discussed next. As previously mentioned, flocculation is a significant process in estuaries where the increasing salinity of water enhances the tendency of particles to stick to each other. Flocculation is also important in some wastewater treatment facilities, where chemicals such as polymers and ferric chloride ($FeCl_3$) are added to wastewater to enhance particle aggregation and settling.

EXAMPLE 2-4

A river of 2-m depth moving at an average velocity of 0.2 m/sec receives particles of 200-μm diameter from a storm drain emptying at the river surface. Assume that the particles are of mineral origin with a density of approximately 2.6 g/cm^3 and that the kinematic viscosity of water is 1.3×10^{-2} cm^2/sec (at 10°C). What is the minimum distance the particles will travel before settling to the river bottom?

First, use Eq. [2-20] to estimate the settling velocity:

$$\omega_f = \frac{(2/9)(981\ \text{cm/sec}^2)(2.6/1 - 1)(10^{-2}\ \text{cm})^2}{1.3 \times 10^{-2}\ \text{cm}^2/\text{sec}}$$

$$\omega_f = 2.7\ \text{cm/sec}.$$

The time required to settle 2 m to the river bottom can then be estimated as

$$\frac{200\ \text{cm}}{2.7\ \text{cm/sec}} = 75\ \text{sec}.$$

In 75 sec, the water will travel on average

$$75\ \text{sec} \cdot 0.2\ \text{m/sec} = 15\ \text{m}.$$

Therefore, the particles will travel at least 15 m before settling. They may travel further, depending on the degree of turbulence acting to keep material in suspension.

Bottom Sediment

Many chemicals in surface waters are sorbed onto suspended particles, which ultimately settle to the bottom of the water body. The settling of particles to the bottom of a water body is a mechanism for chemical removal from the water column. The magnitude of the settling flux of a chemical is equal to the product of the rate of sediment deposition and the chemical concentration associated with the settling particles.

Particle settling can represent a significant chemical flux, especially where flocculation or low turbulence promotes the settling process. Sometimes the sediment becomes a long-term repository for the chemical; alternatively, a chemical may be degraded within the sediments, or eventually be returned to the water by a variety of *remobilization* processes.

The chemical and biological conditions that develop in bottom sediment are similar to those that occur in wetlands. The physical presence of the sediment inhibits turbulent diffusion and thus inhibits the transport of dissolved oxygen into the *pore waters* (water between the solid particles) of the sediment. At the same time, the organic matter that composes a significant fraction of most bottom sediments promotes the growth of oxygen-consuming microorganisms. The usual result is that oxygen is consumed more rapidly than it can be replenished by the relatively slow process of molecular diffusion, and anoxic conditions result. Most animal life is inhibited, except for species that obtain oxygen from above the sediment [e.g., various worms that pump water through their burrows *(irrigation)* or extend gill structures above the bottom (good for bottom-feeding fish!)]. In addition, a whole suite of anaerobic microbial processes, some of which facilitate the remobilization of certain chemicals, occur. These processes are discussed further in Section 2.4.3; see also Berner (1980).

The Sedimentary Record

In a lake, as bottom sediment is deposited, it preserves a historical record of chemical and biological conditions in the lake basin. Remains of aquatic organisms, pollen from the adjoining land, and chemicals sorbed to sediment particles all reflect past conditions. The sediment record is sequential; in the absence of events that mix the sediment, deeper sediments are older. If it is possible to determine the date at which a particular layer of the sediment was deposited, the chemical and biological information contained in that layer can be assigned to a particular time in history.

An example of a sediment record from the Aberjona watershed in eastern Massachusetts is shown in Fig. 2-12. Figure 2-12a presents the arsenic concentration in bottom sediments of the Upper Mystic Lake as a function of depth below the sediment–water interface. These data reveal several features about the history of arsenic input into the Upper Mystic Lake from its watershed. The low concentrations of arsenic below approximately 80 cm show that natural, preindustrial levels of arsenic in this lake were low, as is typical of most unpolluted lakes. High concentrations around 60 cm indicate the release of large amounts of arsenic into the watershed, from the manufacture of sulfuric acid and arsenic-based pesticides. Although the arsenic concentrations are lower in sediments deposited during the period of decline of these industries (at depths of 50 to 40 cm), it is clear that arsenic pollution from the watershed continued long after active manufacturing stopped. Furthermore, there is a second major episode of arsenic release into the lake as indicated by the high concentration of arsenic at 30-cm depth; this episode probably occurred due to the disturbance associated with earthmoving at the former industrial sites where arsenic first was released.

Figure 2-12b shows the lead concentration as a function of depth in the same lake; in this instance the large industrial source revealed by the high concentrations near a 60-cm depth in the core is superimposed on another source that rises gradually until approximately a 20-cm depth, and then declines steadily. This other source is believed to be lead deposition from leaded gasoline, whose consumption increased steadily from its introduction in the 1930s until it was phased out in the United States during the 1970s and 1980s.

Thus, as shown in Fig. 2-12, the sedimentary record can reveal much about the magnitude and temporal sequence of chemical events occurring in a lake

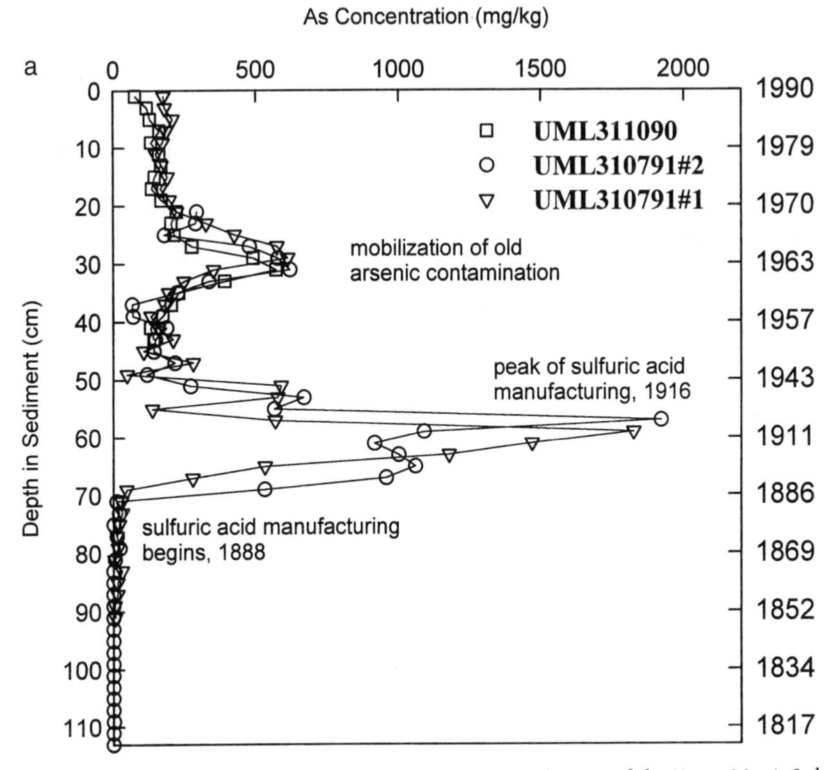

FIGURE 2-12 Arsenic and lead concentrations in bottom sediments of the Upper Mystic Lake on the Aberjona watershed in eastern Massachusetts. Figure 2-12a shows arsenic concentrations in three overlapping sediment cores, which all reflect the same high concentrations associated with particular events in the watershed. Figure 2-12b shows lead concentrations in two overlapping cores. These lead concentrations reflect both industrial activities in the watershed and regional lead deposition from the use of leaded gasoline [data from Spliethoff and Hemond (1996)] *(Figure continues)*.

and its watershed. Furthermore, it is often possible to assign actual dates to the watershed events recorded in the sediments. Among the several techniques used to date sediments (Lerman *et al.*, 1972), *lead-210 dating* is one of the most valuable. Lead-210 (^{210}Pb) dating is based on the relatively constant atmospheric deposition of this radionuclide onto surface waters, the subsequent sorption of ^{210}Pb by particles in the water, and ^{210}Pb deposition into the sediment by settling of those particles. Although several complicating factors must often be considered, the simplest case of constant annual ^{210}Pb deposition and no sediment mixing may be analyzed in a straightforward way. The ^{210}Pb concentration (as measured by its radioactivity) at any depth in the

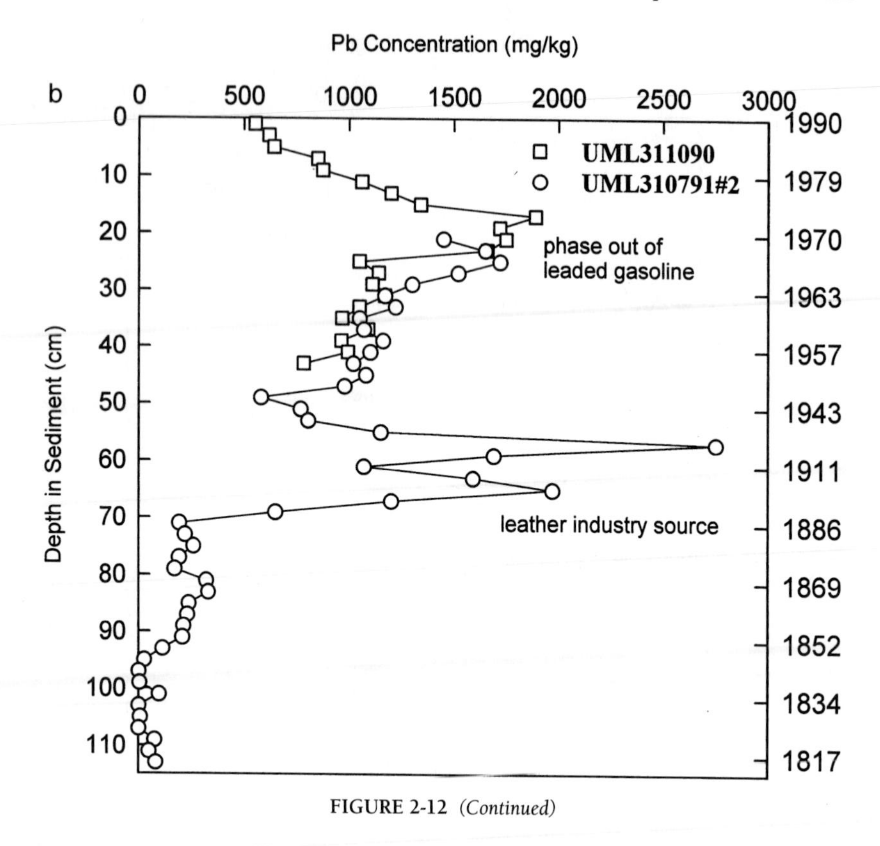

FIGURE 2-12 *(Continued)*

sediment is equal to its concentration in freshly deposited material multiplied by $\exp(-\lambda t)$, where λ is the radioactive decay constant for ^{210}Pb, which is approximately 0.03/year. (This first-order decay constant corresponds to a ^{210}Pb half-life of 22 years.) In this idealized case the age, t, of a layer of sediment at a depth d below the sediment–water interface is given by

$$t = \frac{-1}{\lambda} \cdot \ln\left(\frac{A_d}{A_0}\right), \qquad [2\text{-}26]$$

where A_0 is the ^{210}Pb activity (e.g., in becquerels per gram dry weight) at the sediment–water interface, and A_d is the activity at depth d. The amount of a nuclide that produces one nuclear disintegration event per second is 1 Bq (becquerel).

EXAMPLE 2-5

Sediment from a 10-cm depth in the bottom sediment of a lake has a ^{210}Pb activity of 2.5 disintegrations per minute (DPM). Sediment collected at the sediment–water interface has an activity of 4 DPM per gram. Assuming constant ^{210}Pb and sediment deposition rates, no sediment compression as the sediment ages, and no mixing or losses in the sediments, how rapidly does sediment accumulate in this lake?

Equation [2-26], the basic equation for radioactive decay, can be used:

$$t = \frac{-1}{0.03/\text{year}} \ln\left(\frac{2.5\ \text{DPM}}{4\ \text{DPM}}\right)$$

$$= 16\ \text{year}.$$

The sediment accumulation rate can then be estimated as

$$\frac{10\ \text{cm}}{16\ \text{year}} = 0.6\ \text{cm/year}.$$

2.3 AIR–WATER EXCHANGE

Thus far, the discussion of chemical removal from the water column has focused on incorporation of chemicals sorbed to particles into bottom sediment. However, chemicals dissolved in surface waters may also leave the water column and enter the atmosphere as gases or vapors. Conversely, chemicals present in the atmosphere may dissolve into a lake, river, or estuary. For volatile chemicals, which include most common industrial solvents and liquid fuels, the process of water-to-air exchange can be the most important mechanism of chemical removal from a surface water.

The concentration of a dissolved gas or vapor in a surface water *at equilibrium* with the atmosphere (C_{equil}) is determined by C_a, the concentration in air, and the Henry's law constant (H) of the chemical:

$$C_{\text{equil}} = C_a/H. \qquad [2\text{-}27]$$

If the concentration in water (C_w) is higher than C_{equil}, the chemical will volatilize from the water body into the atmosphere. The flux density is proportional to the product of the difference between the actual (C_w) and the equilibrium (C_{equil}) concentrations in the water:

$$J = -k_w(C_w - C_a/H), \qquad [2\text{-}28]$$

where J is the flux density [M/L^{2}T], k_w is the *gas exchange coefficient* [L/T], C_w is the chemical concentration in the water [M/L^3], C_a is the chemical concentration in the air [M/L^3], and H is the Henry's law constant (dimensionless). This gas exchange coefficient is sometimes called a *piston velocity* because in the special case where C_a is equal to zero in Eq. [2-28], the flux density J is equal to the flux that would result if a hypothetical piston were to move vertically through the water at speed k_w pushing the dissolved gas across the air–water interface. The magnitude of the piston velocity depends on the nature of the water flow and the air movement above the water. The flux density J is positive as defined in Eq. [2-28] if the flux is into the water. Note that C_a is an important term if the gas under consideration is a component of the atmosphere [e.g., nitrogen (N_2), oxygen (O_2), or carbon dioxide (CO_2)], but C_a is often essentially zero in the case of anthropogenic chemicals.

The most accurate determination of the gas exchange coefficient requires that careful field experiments be conducted. Although techniques for estimating k_w from measurable hydraulic attributes of a water body also exist, they are less accurate due to the state of incomplete understanding of the air–water gas exchange process and the fact that multiple factors may control the exchange rate for any particular chemical (Fig. 2-13). Two different models

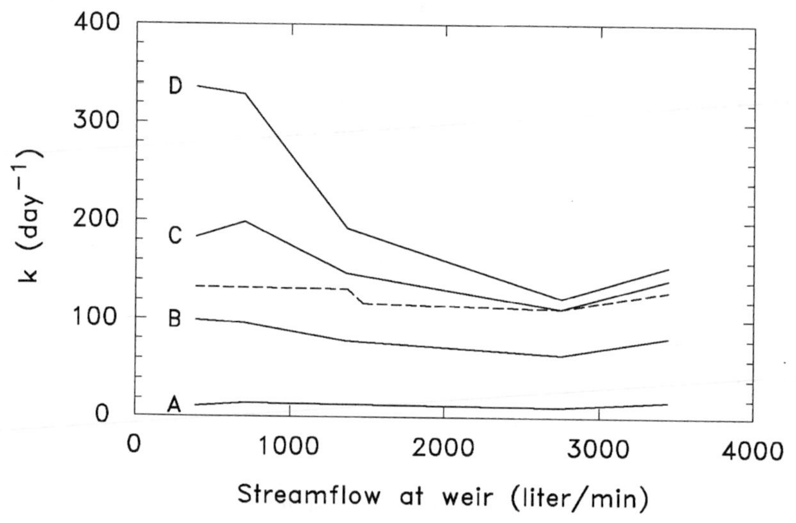

FIGURE 2-13 Air–water reaeration coefficients for oxygen in a reach of Walker Branch, a stream in Oak Ridge, Tennessee. Measured coefficients (dashed line) and calculated coefficients from several published predictive equations (solid lines) are shown. Until better predictive relationships are developed, highly accurate estimates of gas exchange appear to require experimental determination (data from Genereux and Hemond, 1992).

of the exchange process are currently in use: the *thin film model* and the *surface renewal model*. Each attempts to explain and predict the gas exchange coefficient on the basis of a different physical conceptualization of the micro-scale processes occurring at the air–water interface.

2.3.1 Thin Film Model

The thin film (or *stagnant layer*) model is based on the assumption that a dissolved chemical has a uniform concentration throughout a surface water body, due to turbulent diffusion, except in a very thin layer at the water's surface. A similar assumption is made concerning the chemical concentration in overlying air. Within a few micrometers or millimeters of the water–air interface, it is assumed that the eddies responsible for turbulent diffusion are suppressed; therefore, chemical transport in this thin layer (or film) can only occur by molecular diffusion, which is considered to be the rate-limiting step of air–water exchange (Fig. 2-14) (Liss and Slater, 1974).

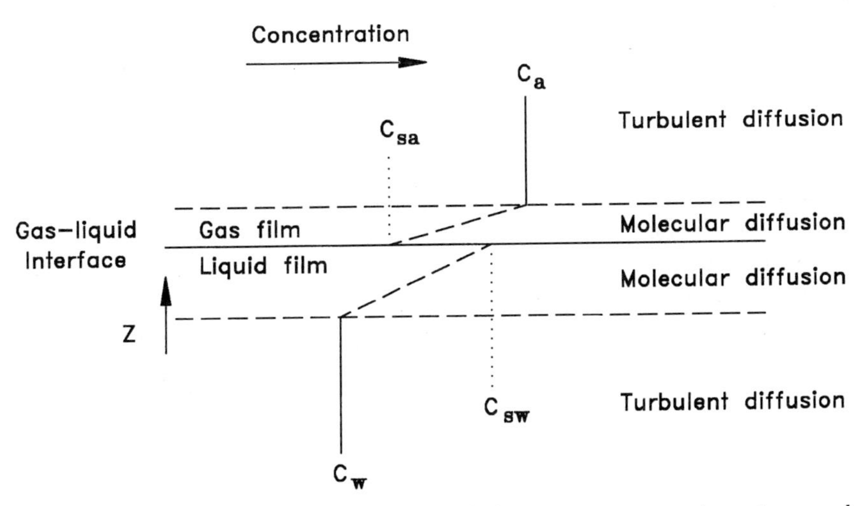

FIGURE 2-14 Schematic of the thin film model. All the resistance to gas exchange is assumed to lie in a thin stagnant (nonturbulent) layer of water and a similar stagnant layer of air. Within these films, transport occurs from higher to lower concentrations by molecular diffusion, governed by Fick's first law. Outside the films, in "bulk" air or water, Fickian transport coefficients are assumed to be much larger, due to turbulent diffusion; therefore, concentration gradients are negligible.

Water-Side Control

If the dimensionless Henry's law constant for a chemical, H, is much greater than 0.01—as is the case for a large number of solvents, fuels, and gases—resistance to gas exchange in the stagnant air layer immediately above the water can be neglected. The thin film model then describes the flux of a chemical into or out of the water by

$$J = -D_w(C_w - C_a/H)/\delta_w, \qquad [2\text{-}29]$$

where J is the flux density of the chemical [M/L^{2}T], D_w is the molecular diffusion coefficient for that chemical in water [L^2/T], C_w is the chemical concentration in the water [M/L^3], C_a is the chemical concentration in the air [M/L^3], H is the Henry's law constant (dimensionless), and δ_w is the thickness of the hypothetical thin boundary layer of water [L].

Under this theory, for water-side control, the gas exchange coefficient, k_w, is equal to the quotient D_w/δ_w. If the atmospheric concentration of a chemical is essentially zero, Eq. [2-29] simplifies to

$$J = \frac{-D_w}{\delta_w} C_w = -k_w C_w. \qquad [2\text{-}30]$$

It should be noted that film thickness cannot be measured directly; k_w, however, can be estimated, and δ_w can then be estimated from k_w and an independent knowledge of D_w. Typical values of δ_w are in the range of 20 to 200 μm.

Air-Side Control

If the dimensionless Henry's law constant for a chemical is much less than 0.01 [typically the case for polycyclic aromatic hydrocarbons (PAHs) and many pesticides], molecular diffusion through the stagnant boundary layer of air above the water surface becomes the most significant barrier to air–water gas exchange. In this situation the resistance due to the stagnant water film is considered to be negligible; thus, the concentration of chemical in the air at the water–air interface is given by the product $(C_w \cdot H)$. The flux density through the stagnant air film is then,

$$J = -(D_a/\delta_a) \cdot (C_w \cdot H - C_a)$$

or [2-31]

$$J = -(D_a \cdot H/\delta_a) \cdot (C_w - C_a/H),$$

where D_a is the molecular diffusion coefficient for the chemical in air [L^2/T] and δ_a is the thickness of the hypothetical stagnant air layer [L]. The quotient

D_a/δ_a may be considered to be an air-side gas exchange coefficient, k_a. Typical values of δ_a are on the order of 1 cm.

The General Situation

In the most general case, which must be invoked when the value of the dimensionless Henry's law constant is on the order of 0.01, both resistances contribute to limiting the gas exchange rate. In this case, the complete expression for flux density must be used:

$$J = -\left[\frac{1}{\delta_w/D_w + \delta_a/(D_a \cdot H)}\right]\left[C_w - \frac{C_a}{H}\right]. \qquad [2\text{-}32]$$

This equation is derived by setting the flux through the thin film in water, Eq. [2-29], equal to the flux through the stagnant air layer, Eq. [2-31]; the fluxes must be equal at steady state due to mass conservation.

Equation [2-32] reduces to Eq. [2-29] when δ_w/D_w is much greater than δ_a/D_aH, and to Eq. [2-31] when it is much less. Note that the molecular diffusion coefficient for molecules in air (D_a) is approximately 10^4 times greater than the corresponding molecular diffusion coefficient in water (D_w), whereas the air film thickness, δ_a, is generally much less than 10^4 times the water film thickness, δ_w.

Estimation of Gas Exchange Coefficients

A reasonably accurate gas exchange coefficient for a given chemical can be determined by introducing a tracer gas into the surface water body of interest and by observing the rate of loss of the tracer. In practice, such a tracer experiment is more feasible in a stream or river than in a lake or estuary. According to thin film theory, the ratio of the gas exchange coefficients of two volatile chemicals is equal to the ratio of their molecular diffusion coefficients in water; the ratio of molecular diffusion coefficients of two chemicals in turn is approximately equal to the inverse of the ratio of the square roots of their molecular weights. Thus, the measured gas exchange coefficient of a tracer gas (A) can be used to predict the gas exchange coefficient of another chemical (B):

$$\frac{k_A}{k_B} = \frac{D_A}{D_B} \approx \frac{\sqrt{\mathrm{MW}_B}}{\sqrt{\mathrm{MW}_A}}. \qquad [2\text{-}33]$$

In practice, a nontoxic, inexpensive, and easily measured substance such as propane (C_3H_8) is a convenient tracer; its measured gas exchange coeffi-

TABLE 2-5 Empirical Equations for Gas Exchange in Surface Waters

Equation	Reference	Line in Fig. 2-13
$K_{O_2} = 1.92 \cdot \left(\frac{V}{d}\right)^{0.85}$	Negelescu and Rojanski, 1969	A
$K_{O_2} = \frac{24.94 \cdot (1 + \sqrt{N}) \cdot u^*}{d}$	Thackston and Krenkel, 1969	B
$K_{O_2} = \frac{23.2\ V^{0.73}}{d^{1.75}}$	Owens *et al.*, 1964	C
$K_{O_2} = \frac{106\ V^{0.413} w^{0.273}}{d^{1.408}}$	Bennett and Rathbun, 1972	D

cient can be used to estimate gas exchange coefficients for many other volatile chemicals dissolved in a given surface water.

In the absence of tracer data, estimates of gas exchange coefficients in streams can be made from a number of empirical equations, which typically depend on a combination of the stream mean velocity and depth (V and d, respectively). Some equations contain other parameters, such as shear velocity, width, and *Froude number* (u^*, w, and N, respectively) of the stream. The Froude number is equal to ($V/\sqrt{gd}$), and is the ratio of stream velocity to the travel speed of a shallow-water surface wave. By convention, the empirical equations given for streams are usually for a *reaeration coefficient*, which is the gas exchange coefficient for oxygen divided by the average stream depth. Examples of empirical equations for reaeration coefficients are shown in Table 2-5.

Note that none of the equations in Table 2-5 include wind speed, because it is assumed that the turbulence generated within the stream due to velocity shear and turbulent diffusion primarily controls gas exchange. Unfortunately, the predictions of these equations are often not in good agreement with one another, and it is difficult to know which one is best in a given situation (see Fig. 2-13).

Gas exchange coefficients can also be estimated for lakes and estuaries using a variety of empirical equations, although the calculated values may differ from measured values by a factor of two or three in any given water body. Consistent with the assumption that turbulence is primarily driven by wind in such slowly flowing waters, the expressions typically have gas exchange coefficients as a function of wind speed. In the case of water-side control for slowly flowing waters, Schwarzenbach *et al.* (1993) suggest approximating k_w by

$$k_w(\text{cm/sec}) \approx 4 \times 10^{-4} + 4 \times 10^{-5} \cdot u_{10}^2, \quad [2\text{-}34]$$

where u_{10} is the wind speed (in m/sec) measured 10 m above the water surface. Typical values of k_w range from 1 to 10 cm/hr. Note that Eq. [2-34] estimates k_w in units of cm/sec.

In the case of air-side control, Schwarzenbach *et al.* (1993) suggest approximating k_a by

$$k_a(\text{cm/sec}) \approx 0.3 + 0.2 \cdot u_{10}(\text{m/sec}). \quad [2\text{-}35]$$

An even simpler formula for estimating k_a, which estimates k_a in units of centimeters per hour, is

$$k_a(\text{cm/hr}) \approx 1100 \cdot u(\text{m/sec}). \quad [2\text{-}36]$$

There are other empirical equations relating gas exchange to wind speed, such as those given by O'Connor (1983) and Yu and Hamrick (1984).

EXAMPLE 2-6

The dissolved concentration of trichloroethylene (TCE, C_2Cl_3H) in a lake is 1 ppb. Given a dimensionless Henry's law constant, H, of 0.4, and a measured gas exchange coefficient of 3×10^{-3} cm/sec in water, for propane (C_3H_8), what is the flux density of TCE from the lake?

First, estimate the molecular weights of TCE and propane:

MW TCE: 131 g/mol

MW propane: 44 g/mol.

Assume C_a is essentially zero. Diffusion through air is not a bottleneck due to the fairly high H. D_w for TCE is approximately equal to D_w for propane multiplied by the inverse of the ratio of the square roots of the molecular weights, Eq. [2-33]:

$$\frac{D_{\text{TCE}}}{\delta_w} = 3 \times 10^{-3} \text{ cm/sec} \cdot \frac{\sqrt{44}}{\sqrt{131}} = 1.7 \times 10^{-3} \text{ cm/sec.}$$

Then use Eq. [2-30]:

$$J_{\text{TCE}} = (1.7 \times 10^{-3} \text{ cm/sec}) \left(\frac{1\ \mu\text{g}}{\text{liter}}\right)\left(\frac{1 \text{ liter}}{1000 \text{ cm}^3}\right) = 1.7 \times 10^{-6} \frac{\mu\text{g}}{\text{cm}^2 \cdot \text{sec}}.$$

2.3.2 Surface Renewal Model

An alternative model of air–water gas exchange assumes that turbulent eddies in the water periodically bring small parcels of water to the surface, where they begin to equilibrate with the atmosphere. In this model, the average amount of time each water parcel spends at the surface determines the overall gas exchange rate. If parcels on average spend a long time at the surface, they may equilibrate with the atmosphere; after equilibrium is reached, no further chemical flux to the air occurs until the depleted parcel is replaced. If fresh parcels are more frequently brought to the surface, the average chemical flux across the interface is larger.

The surface renewal model, like the thin film model, yields a piston, or gas exchange, velocity that can be used to calculate chemical fluxes as previously described. In contrast with the thin film model, however, the surface renewal model predicts that the ratio of piston velocities for two different volatile chemicals depends on the *square root* of the ratio of their molecular diffusion coefficients (and thus approximately the *fourth root* of the inverse ratio of their molecular weights). Schwarzenbach *et al.* (1993) discuss molecular diffusion coefficients in more detail.

An explicit choice of an air–water exchange model must be made if gas exchange coefficients determined for one chemical (usually from a tracer experiment) are to be used to estimate gas exchange coefficients for another chemical. The existing literature is not adequate to make the choice of a model clear-cut. Commonly, the thin film model is considered to be more appropriate for relatively quiescent water bodies, such as lakes, while a surface renewal model is considered more appropriate for more highly turbulent surface waters, such as rivers. Usually, unavoidable experimental error, even in carefully conducted tracer experiments, prevents unequivocal endorsement of one model over the other. Often, field data on the gas exchange coefficients of two different chemicals are most consistent with a ratio of the chemical diffusion coefficients raised to some power between 0.5 and 1.0, suggesting that the actual mechanism of gas exchange contains some elements of each process and is more complex than either idealized model suggests (Genereux and Hemond, 1992).

EXAMPLE 2-7

Trichloroethylene has been spilled in a *river* so that the dissolved concentration is 1 ppb. Given a dimensionless Henry's law constant, H, of 0.4, and a piston velocity of 3×10^{-3} cm/sec in water for propane, what will be the flux density of TCE from the river?

MW TCE	131 g/mol
MW propane	44 g/mol.

Assuming C_a is essentially zero, diffusion through air is not a bottleneck due to the fairly high H, and D_w for TCE is approximately equal to D_w for propane multiplied by the square root of the inverse of the ratio of the square roots of the molecular weights:

$$\frac{D_{\mathrm{TCE}}}{\delta_w} = 3 \times 10^{-3}\ \mathrm{cm/sec} \cdot \sqrt{\frac{\sqrt{44}}{\sqrt{131}}} = 2.3 \times 10^{-3}\ \mathrm{cm/sec}.$$

Then use Eq. [2-30]:

$$J = (2.3 \times 10^{-3}\ \mathrm{cm/sec}) \left(\frac{1\ \mu\mathrm{g}}{\mathrm{liter}}\right)\left(\frac{1\ \mathrm{liter}}{1000\ \mathrm{cm}^3}\right) = 2.3 \times 10^{-6} \frac{\mu\mathrm{g}}{\mathrm{cm}^2 \cdot \mathrm{sec}}.$$

Note that this flux is higher than that obtained in Example 2-6, but in practice the difference might be masked by experimental variability.

2.3.3 The Reaeration Coefficient

In the expressions for the gas exchange coefficient employed previously, it is evident that the air–water gas exchange flux density is proportional to the difference between a chemical concentration in the water (C_w) and the corresponding equilibrium concentration ($C_w \cdot H$) in air. Consequently, the difference between actual and equilibrium concentration in the water tends to decay exponentially, as expected for any first-order process. In many situations, exponential decay may provide a useful model of a volatile chemical concentration in a surface water. A classic example is *degassing* of a dissolved gas from a stream; if the gas is present at concentration C_0 upstream, atmospheric concentration of the gas is negligible, and flow is steady and uniform along the stream, then the gas concentration in the stream is given by

$$C = C_0 \cdot e^{-k_r \tau}, \qquad [2\text{-}37]$$

where C is the downstream concentration [M/L^3], C_0 is the upstream concentration [M/L^3], k_r is a coefficient characterizing the gas transfer process [T^{-1}], and τ is the travel time from upstream location to downstream location [T]. Note that the units of k_r are [T^{-1}]; k_r is equal to the gas exchange coefficient k[L/T] divided by average stream depth [L]. As previously mentioned, if oxygen (O_2) is the gas being transferred between air and water, then k_r is called the reaeration coefficient, an essential parameter in classical dissolved oxygen modeling in streams (see Section 2.5).

2.3.4 Volatilization from Pure Phase Liquids

A special situation in air–water gas exchange occurs when a liquid such as gasoline or a solvent forms a floating slick or a layer of *nonaqueous phase liquid* (NAPL) on a water surface or on the ground. Comparison of this situation with the thin film model suggests that chemical molecules from the slick can enter the atmosphere much more rapidly than if they were dissolved in water, because chemicals do not need to pass through a thin stagnant water layer to enter the atmosphere. If the slick is composed of a pure chemical compound, the only significant barrier to transport into the atmosphere is diffusion through a thin stagnant film of air. As is the case with water, this resistance to volatilization can be expressed as a gas exchange coefficient that depends on site-specific conditions, particularly wind speed. The situation is the same as previously described for air-side control, except that the concentration of vapor at the interface must be derived from the NAPL's vapor pressure, instead of being given by ($C_w \cdot H$). The temperature gradient in air above the liquid also has an effect; see the discussion of atmospheric stability in Section 4.2.

The concentration of the chemical at the base of the stagnant air layer, just above the surface of the NAPL, is determined from the vapor pressure of the chemical as

$$C_a = \frac{P}{RT}\,(\mathrm{MW}), \qquad [2\text{-}38]$$

where C_a is the chemical concentration in the air [M/L^3], P is the vapor pressure of the chemical [M/LT^2], R is the universal gas constant, T is absolute temperature, and MW is the molecular weight of the chemical (g/mol).

The rate of volatilization from a NAPL surface is then given by the following expression, as compared with Eq. [2-32]:

$$J = \frac{-D_a}{\delta_a} \cdot C_a. \qquad [2\text{-}39]$$

The quantity D_a/δ_a is the same as previously discussed, and can be estimated from empirical equations. This velocity is larger for chemicals having larger diffusion coefficients in air, and is smaller for larger slicks or pools, because vapor advected over a downwind point has the effect of decreasing the concentration gradient over that point. This effect is reflected in the following expression for the gas exchange coefficient (also called the *gas phase transfer velocity*) given by Thibodeaux (1979),

$$v = 0.029 v_w L^{-0.11} S_c^{-0.67}, \qquad [2\text{-}40]$$

where v (the gas phase transfer velocity) and v_w (wind speed at 10 m height) are in meters per hour, and L, the spill diameter, is in meters. The molecular diffusion coefficient is contained in the *Schmidt number* (S_c), which is the ratio of the kinematic viscosity [L^2/T] to the molecular diffusion coefficient [L^2/T] of the vapor and lies in the range of one to two for many common solvent vapors. However, because the gas phase transfer velocity is not strongly dependent on the slick diameter or the Schmidt number, the empirical equation for k_a previously presented in Eq. [2-36] may be used for approximate purposes with common solvents, modest pool sizes, and moderate wind speeds.

A layer of NAPL floating on water can also lose mass by dissolution into the water body. If the NAPL is denser than water (in which case it is abbreviated DNAPL, for dense NAPL), it will sink through the water body to the bottom sediments. Most halogenated solvents are denser than water (see Table 1-3) and therefore have greatly diminished volatilization rates from water bodies relative to loss rates for floating NAPLs.

EXAMPLE 2-8

Benzene is spilled onto a lake from an overturned tanker truck. Given a 3 m/sec wind speed at a 10-m height, what will be the flux density from the slick?

From Table 1-3, benzene vapor pressure is 0.12 atm at 20°C. Use Eq. [2-38] to calculate the concentration of benzene at the air–NAPL interface:

$$\frac{(0.12 \text{ atm})}{(293 \text{ K})\left(0.082 \dfrac{\text{atm} \cdot \text{liter}}{\text{mol} \cdot \text{K}}\right)} \cdot \frac{78 \text{ g}}{\text{mol}} = 0.4 \text{ g/liter}.$$

Then use Eqs. [2-36] and [2-39]:

$$J = \left(\frac{3300 \text{ cm}}{\text{hr}}\right)\left(\frac{0.4 \text{ g}}{\text{liter}}\right)\left(\frac{1 \text{ liter}}{1000 \text{ cm}^3}\right) = \frac{1.3 \text{ g}}{\text{cm}^2 \cdot \text{hr}} \quad \text{or} \quad \frac{360 \ \mu\text{g}}{\text{cm}^2 \cdot \text{sec}}.$$

Compare this result with TCE volatilization shown in Examples 2-6 and 2-7. Although benzene and TCE have similar vapor pressures (Table 1-3), benzene evaporates much faster from a slick (approximately 100 million times faster) than dissolved TCE volatilizes from a lake or river.

2.4 CHEMICAL AND BIOLOGICAL CHARACTERISTICS OF SURFACE WATERS

2.4.1 Acid–Base Chemistry

All natural waters contain dissolved chemicals, many present as inorganic *ions*. Common inorganic ions in natural waters are sodium (Na^+), potassium (K^+), magnesium (Mg^{2+}), calcium (Ca^{2+}), ammonium (NH_4^+), sulfate (SO_4^{2-}), chloride (Cl^-), and nitrate (NO_3^-). Even pure water, containing no dissolved substances, ionizes to a certain extent to form hydrogen ions (H^+) and hydroxide ions (OH^-), as discussed in Section 1.6.3. The pH of most natural waters ranges from about 4 to 9; extreme environments, such as streams receiving acid mine drainage, may have a pH below 2, while some alkaline lakes may have a pH above 10.

pH often determines the suitability of a water body as a biological habitat or as a water supply, and also influences the chemical speciation of many dissolved compounds and the rates at which many pollutants degrade. As such, pH is often called a *master variable* in natural water chemistry, and the equilibrium of reactions that produce or consume H^+ is of special interest. *Acids* (such as the acetic acid of Example 1-6) ionize in water, producing H^+ and a negatively charged *anion; bases* produce OH^- and a positively charged *cation*. By definition, *strong acids* and *strong bases,* such as hydrochloric acid (HCl), nitric acid (HNO_3), sulfuric acid (H_2SO_4), and sodium hydroxide (NaOH), ionize completely in water under environmental conditions. Thus, if 1 mol of HNO_3 is put into water, 1 mol of H^+ and 1 mol of nitrate ion (NO_3^-) are formed. The contributions of H^+ from a mole of strong acid and OH^- from a mole of strong base cancel each other out; the H^+ and OH^- react to form H_2O. If strong acid is in excess of strong base, H^+ is formed in an amount equal to the difference. Thus, it is not necessary to use mass action expressions to determine how much H^+ or OH^- is produced in solution by a mixture of strong acids and bases; knowledge of the *difference* in their concentrations is sufficient. *Alkalinity* (Alk) is defined as the concentration of strong bases (C_B) minus the concentration of strong acids (C_A). Note that concentrations must be in equivalents per liter (Section 1.2):

$$\text{Alk} = C_B - C_A. \tag{2-41}$$

If the concentration of strong acids exceeds the concentration of strong bases, Alk is negative and in the absence of other bases is approximately equal to the negative of the H^+ concentration. For example, in a water containing only Alk of -10^{-4} equivalents per liter, $[H^+] = 10^{-4}$ *M*. If Alk is positive, the concentration of OH^-, in the absence of other acids or bases, is approximately equal to Alk.

Inevitably, surface waters also contain dissolved carbon dioxide (CO_2), an acid that reacts in water to form three *carbonate system* species: $H_2CO_3^*$, HCO_3^-, and CO_3^{2-}. The predominant species in any given water depends on the pH of the water. In this book, the notation $H_2CO_3^*$ represents the sum of dissolved CO_2 and its reaction product with water, carbonic acid (H_2CO_3). $H_2CO_3^*$ is the dominant carbonate system species when the pH of a surface water is below approximately 6.3. $H_2CO_3^*$ ionizes to form bicarbonate ion (HCO_3^-), which is the most abundant of the three species when the pH is between approximately 6.3 and 10.3. Because the ionization of $H_2CO_3^*$ is incomplete over a large pH range, $H_2CO_3^*$ is called a *weak acid*. Bicarbonate is also a weak acid, which ionizes to form carbonate ion (CO_3^{2-}), the most abundant carbonate species above a pH of approximately 10.3. The relevant reaction equations and equilibrium constants are:

$$CO_2 + H_2O \rightleftarrows H_2CO_3^* \qquad (K_H = 10^{-1.5}\ \text{mol/(atm} \cdot \text{liter)}). \qquad [2\text{-}42]$$

$$H_2CO_3^* \rightleftarrows HCO_3^- + H^+ \qquad (K_{a1} \approx 10^{-6.3}\ \text{mol/liter}). \qquad [2\text{-}43]$$

$$HCO_3^- \leftrightarrows CO_3^{2-} + H^+ \qquad (K_{a2} \approx 10^{-10.3}\ \text{mol/liter}). \qquad [2\text{-}44]$$

EXAMPLE 2-9

A river has a pH of approximately 6. If the river water is in equilibrium with atmospheric carbon dioxide (which has a pressure, P_{CO_2}, of approximately $10^{-3.5}$ atm), what are the concentrations of carbonate system species in the water?

Because the river water is in equilibrium with atmospheric CO_2 and the pH is fixed by other acids and bases present in the water, Eq. [2-42] can be used to determine $[H_2CO_3^*]$:

$$K_H = \frac{[H_2CO_3^*]}{P_{CO_2}} = \frac{10^{-1.5}\ \text{mol}}{\text{atm} \cdot \text{liter}}.$$

Therefore,

$$[H_2CO_3^*] = 10^{-3.5}\text{ atm}\cdot\frac{10^{-1.5}\ M}{\text{atm}}$$

$$= 10^{-5}\ M.$$

$[HCO_3^-]$ is then determined by using the mass action expression for the ionization of $[H_2CO_3^*]$ into $[H^+]$ and $[HCO_3^-]$, Eq. [2-43]:

$$K_{a1} = \frac{[HCO_3^-][H^+]}{[H_2CO_3^*]} = 10^{-6.3}.$$

Therefore,

$$[HCO_3^-] = \frac{(10^{-6.3})(10^{-5})}{10^{-6}} = 10^{-5.3}\ M.$$

By using Eq. [2-44],

$$K_{a2} = \frac{[CO_3^{2-}][H^+]}{[HCO_3^-]} = 10^{-10.3},$$

the carbonate ion concentration can be calculated as

$$[CO_3^{2-}] = \frac{(10^{-10.3})(10^{-5.3})}{(10^{-6})} = 10^{-9.6}\ M.$$

Recall that equilibrium constants are temperature dependent (see Section 1.6.3) and the values given previously are approximate. Although much of the CO_2 dissolved in surface waters originates from the atmosphere or from biological activity, some of it may also come from dissolution of underlying geologic formations, such as calcite, $CaCO_3$, and dolomite (Ca, Mg) · (CO_3).

Natural waters contain both carbonic acid and a mixture of strong acids and strong bases (i.e., alkalinity). Because weak acids such as $H_2CO_3^*$ and HCO_3^- ionize to a variable extent, calculation of the pH of such a mixture requires that mass conservation equations, the electroneutrality condition, and mass action equations for all weak acids be solved simultaneously.

As an alternative to algebraic manipulation of these equations, graphic solutions, such as shown in Fig. 2-15, have been developed for determining the pH of water as a function of Alk and C_T, where C_T is the sum of the concentration of carbonic acid and the concentrations of the two anions produced when it ionizes (i.e., HCO_3^- and CO_3^{2-}).

Such a graph, also known as a *Deffeyes diagram*, may be derived from the electroneutrality equation, written with Alk in place of the difference between

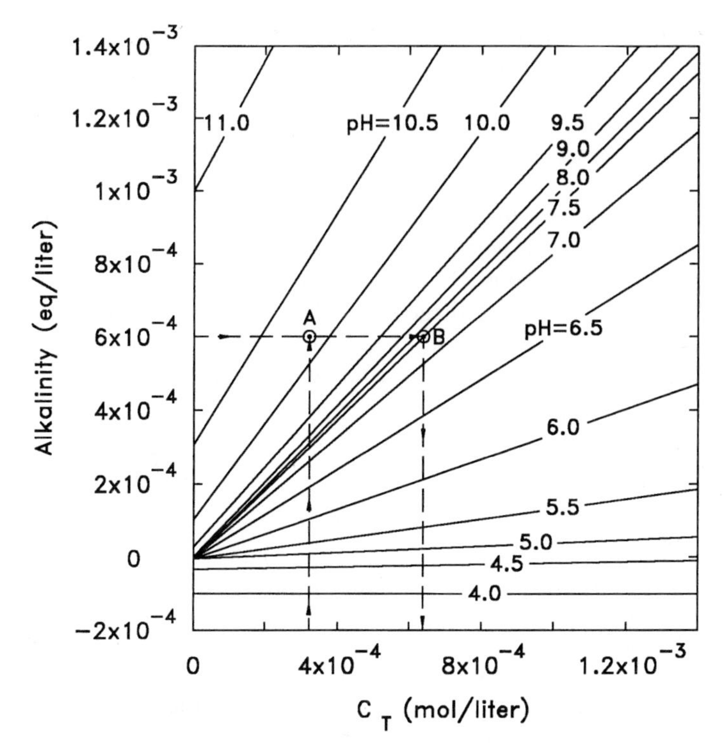

FIGURE 2-15 An alkalinity (Alk)–carbonate system plot, also called a *Deffeyes diagram.* In this figure, the relationship among Alk, C_T (the sum of all carbonate system species concentrations: $[H_2CO_3^*] + [HCO_3^-] + [CO_3^{2-}]$), and pH is shown. If any two of these quantities are known, the third may be immediately determined from the plot. For example, for a water in which Alk is 6×10^{-4} eq/liter, and C_T is 3.2×10^{-4} *M*, the pH is 10.2 (point A). If C_T is 6.4×10^{-4} *M*, the pH is 7.5 (point B) (see Example 2-10).

the strong base cation concentrations and the strong acid anion concentrations:

$$\text{Alk} + [H^+] - [OH^-] - [HCO_3^-] - 2[CO_3^{2-}] = 0 \qquad [2\text{-}45]$$

Eq. [2-45] may be rewritten as

$$\text{Alk} = \{-[H^+] + [OH^-]\} + \{[HCO_3^-] + 2[CO_3^{2-}]\} \qquad [2\text{-}46]$$

At any given pH, the first term on the right-hand side of Eq. [2-46] is a known constant (c_1), while the second term is equal to C_T multiplied by a second constant (c_2). Thus, the relationship between C_T and Alk at a given pH may be written as

$$\text{Alk} = c_1 + c_2 \cdot C_T \qquad [2\text{-}47]$$

The Deffeyes diagram of Fig. 2-15 is a family of such straight-line relationships, plotted for each of several pH values. Note that very similar graphs can be drawn in which P_{CO_2} is used in place of C_T; in this case, the carbonate species concentrations in the electroneutrality equation are equal to P_{CO_2} multiplied by a constant which is a function of pH.

The effects of additions of Alk or C_T to a water are easy to determine graphically using a Deffeyes diagram, as shown below in Example 2-10. For further detail on the acid–base chemistry of natural waters, the reader is referred to Stumm and Morgan (1996) or Morel and Hering (1993).

EXAMPLE 2-10

A lake water has an alkalinity of 6×10^{-4} eq/liter. In the early morning, a monitoring team measures the lake pH as part of an acid rain study and finds the pH to be 7.5.

(a) What is C_T in the lake water at this time?
(b) The survey team returns after lunch to recheck their data. By this time, algae and green plants have depleted the C_T of the lake to half of its morning value. What pH does the team find now?

The solution is based on the Alk–C_T graph (Deffeyes diagram) of Fig. 2-15.

In the early morning, C_T is found to be 6.4×10^{-4} *M* from the graph. [Follow the horizontal line from 6×10^{-4} on the Alk axis until it intersects pH 7.5 (point B); then follow the vertical line to the C_T axis.]

After lunch, C_T is approximately 3.2×10^{-4} *M*. Follow the vertical line up from the C_T axis until it intersects the horizontal line corresponding to the essentially unchanged Alk of 6×10^{-4} eq/liter (point A); the new pH is 10.2.

In addition to inorganic ions, surface waters contain *dissolved organic carbon* (DOC). A large fraction of DOC is made up of *humic* and *fulvic* acids, complex yellow-brown mixtures of organic chemicals originating primarily from decaying plant material. DOC also includes organic compounds leached or exuded from living organisms or released during cell *lysis* (rupture). DOC is measured as the total concentration of organically bound carbon, and ranges from less than 1 mg/liter to several tens of milligrams per liter in

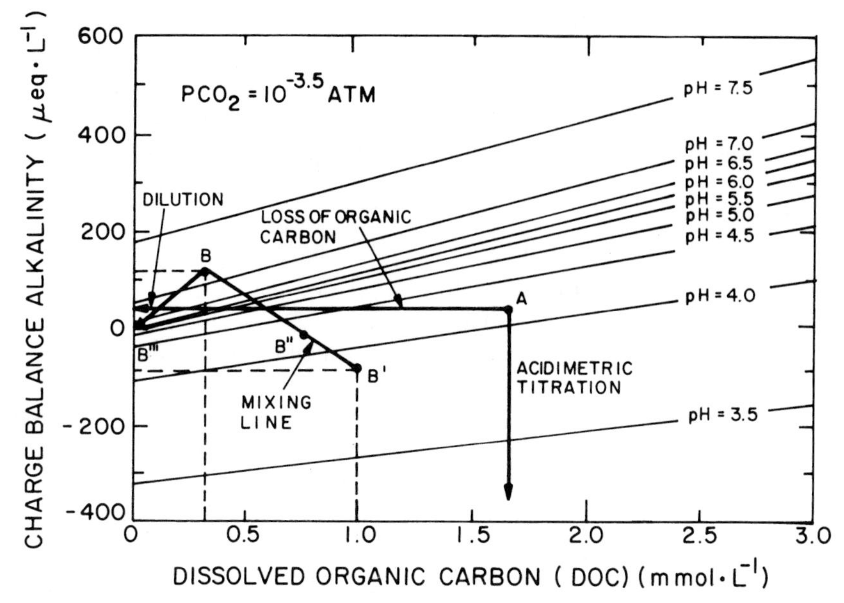

FIGURE 2-16 Plot of alkalinity (Alk) versus organic acid concentration over a range of pH values. This graph is analogous to the Deffeyes diagram of Fig. 2-15. Point A indicates that water having a DOC concentration of 1.7 mmol/liter (20 mg/liter) and an Alk of approximately 40 μeq/liter will have a pH of 4.2. Changes in pH resulting from decreasing Alk (as when adding strong acid) are shown along the vertical line pointing downward from point A, while the horizontal line extending to the left from the point shows the pH increase resulting from a decrease in DOC. The pH of a mix of two waters having compositions given by points B and B′ can be found along line BB′. If water of composition B is mixed with pure water (i.e., water with zero Alk and zero DOC), the resultant composition lies along the dilution line from B to the graph's origin. Note that this plot applies to a P_{CO_2} of $10^{-3.5}$ atm and an organic acid that ionizes according to one particular empirical model. Further details are given in Hemond (1990).

natural waters. *Dissolved organic material* (DOM) is sometimes measured and reported; DOC is about half of DOM, because organic carbon constitutes about half of most organic material in soils and sediments [58%; Lyman *et al.* (1990)]. DOC is significant as a source of organic carbon to bacteria, as an absorber of light and participant in photochemical reactions, as a natural acid, and as a *complexing agent* that binds with metal ions to form metal–organic *complexes*.

As their name implies, humic and fulvic acids have a significant impact on the pH of natural waters. The graphic approach described earlier for determining the pH of water as a function of Alk and C_T has been extended to include DOC effects, as shown in Fig. 2-16. The defining equation is again electroneutrality, with the inclusion of a term for organic acid anions. The full graphic relationship among pH, C_T (or P_{CO_2}), Alk, and DOC concentration must be represented as surfaces of constant pH in a three-dimensional graph whose axes are Alk, C_T (or P_{CO_2}), and DOC [Hemond, (1990)]. Fig. 2-16 is a slice of such a graph for water in equilibrium with the P_{CO_2} of the atmosphere.

2.4.2 Aquatic Ecosystems

Surface waters are more than just physical entities; most surface waters teem with an incredible variety of living organisms. To better understand the behavior of pollutants in these waters, it is essential to understand the nature of the biota and their relationships to the physical and chemical processes occurring in the water bodies. To do this, surface waters must be considered as *ecosystems*. An ecosystem comprises a physical environment and the populations of organisms such as plants, animals, and bacteria (the biological community) that inhabit it. Biological communities and their physical environments interact in ways that are exceedingly complex; however, the fundamental principles driving ecosystems can be described in fairly general terms. Understanding the nature and functioning of aquatic ecosystems is crucial for two reasons: to understand chemical fate and transport, because biota transport and transform a great number of chemicals in the environment; and to recognize potential detrimental effects from pollutant chemicals on aquatic organisms themselves.

This section describes only the most general attributes of aquatic ecosystems. All ecosystems can be functionally characterized in terms of their *processing of energy* and their *cycling of mass* (e.g., carbon and nutrients). These functions, as well as the "life history" aspects of ecosystems, are discussed in the following subsections. [For a more complete discussion of ecosystems, the reader is referred to several texts, such as Odum (1971), Ricklefs (1990), or Curtis (1983)].

Energy Flow

All biota require energy to maintain themselves, to grow and reproduce, and to perform other activities. Therefore, there must be a constant flow of energy through an ecosystem. Sunlight is the source of this energy in nearly all ecosystems. Organisms use the light energy of the sun in the process of *photosynthesis* to form organic material from carbon dioxide and water; oxygen is released as a product. In a surface water ecosystem, organic material such as fallen leaves and insects can also be imported from other ecosystems; such *allochthonous* organic matter augments *autochthonous* organic matter, which is formed within the surface water ecosystem. In the presence of oxygen, organic material is energy-rich and its energy can be harvested by *heterotrophs,* biota that are unable to use sunlight directly as an energy source. Photosynthetic organisms, often called *autotrophs* or, more specifically, *photoautotrophs,* thereby enable the heterotrophic populations of an ecosystem to exist. The organic material fixed by photoautotrophs is a complex mixture of compounds, but is often simply represented by an empirical chemical formula for starches and sugars, CH_2O. A very general chemical representation of the photosynthetic process is

$$CO_2 + H_2O \xrightarrow{h\nu} CH_2O + O_2, \quad [2\text{-}48]$$

where $h\nu$ represents the energy of photons of sunlight (see also Section 2.7.1).

The reverse of this process is *respiration,* in which biota release the energy stored in CH_2O by reacting it with oxygen or another oxidant. The energy released is then used in activities such as maintenance, growth, mobility, or reproduction of the biota.

Energy is transferred from photosynthetic organisms to nonphotosynthetic organisms via a *food chain,* as depicted in Fig. 2-17. Most ecosystems contain a population of animals that, because they obtain organic material by eating plants, are called *primary consumers* or herbivores. Animals called *secondary consumers* obtain organic carbon by eating primary consumers; these secondary consumers may in turn be eaten by other organisms further up the food chain. Each level in the food chain is referred to as a *trophic level.*

Because energy cannot be transferred with 100% efficiency from one trophic level to the next, and because at each level organisms also use energy for their own maintenance, there is a diminishing flow of energy from each trophic level to the next. A pyramid is a helpful representation of the ever-decreasing amount of available energy with increasing trophic level. Often, but not always, the mass of biota present at each trophic level also decreases at successively higher levels (Fig. 2-18). The ultimate fate of energy harvested by the primary and secondary consumers is to be "lost" as heat—thus energy must be constantly fed into an ecosystem for it to keep working. In most ecosystems, the simple concept of the food chain is more realistically represented as a *food web,* as shown in Fig. 2-19; nonetheless, the basic principles of energy flow are unchanged.

Mass Cycling

Because energy is transferred through an ecosystem via its association with organic material, it is evident that there must be a movement of mass from organism to organism through the ecosystem. Much of this mass is organically bound carbon, both the major structural building block and the primary energy source for biota. Many other elements and compounds move in asso-

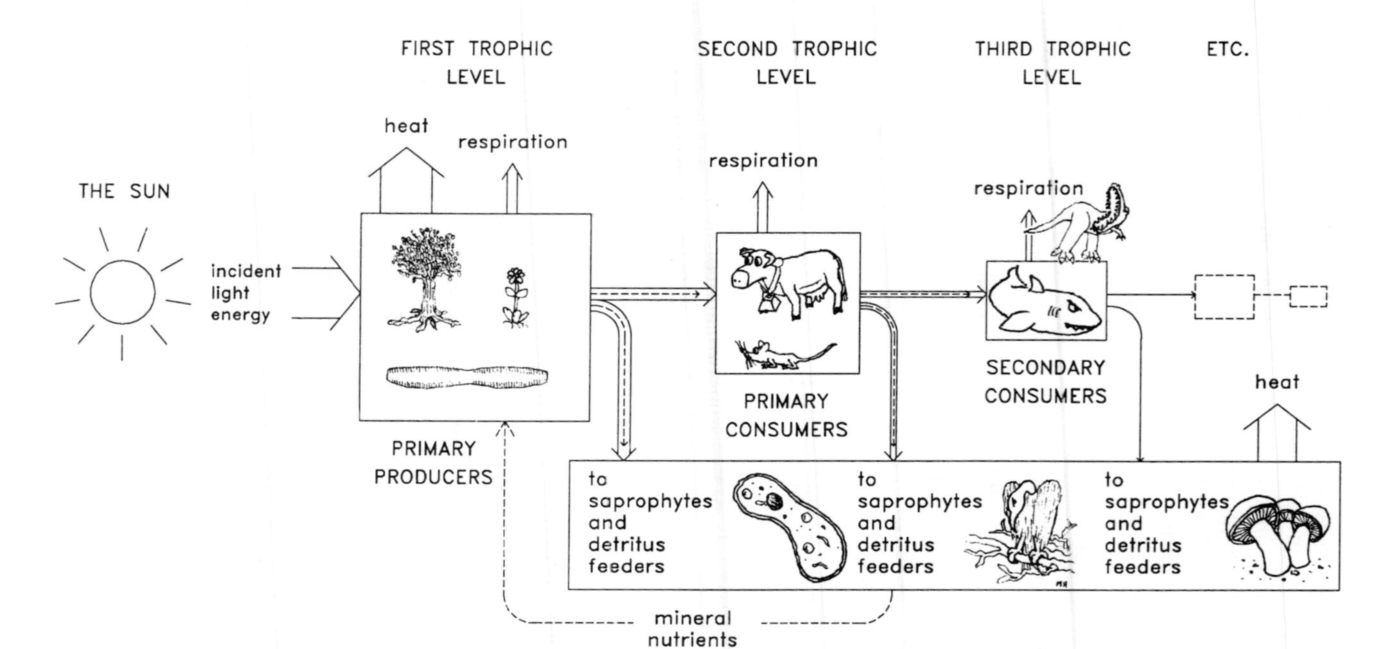

FIGURE 2-17 A simple energy flow diagram, or food chain, for an ecosystem. Energy input to the system comes from sunlight, of which only a fairly small fraction is captured as chemical energy in the biomass of the primary producers. Organisms at the second trophic level (herbivores, or primary consumers) typically utilize only a small portion of this chemical energy; a large portion goes directly to saprophytic microorganisms and detritus-feeding animals as dead organic matter (detritus). The amount of chemical energy available per unit time to the third trophic level (carnivores, or secondary consumers) is lower still, due to energy loss via the respiration of the herbivores and due to the large fraction of herbivore biomass that goes directly to saprophytes and detritus feeders.

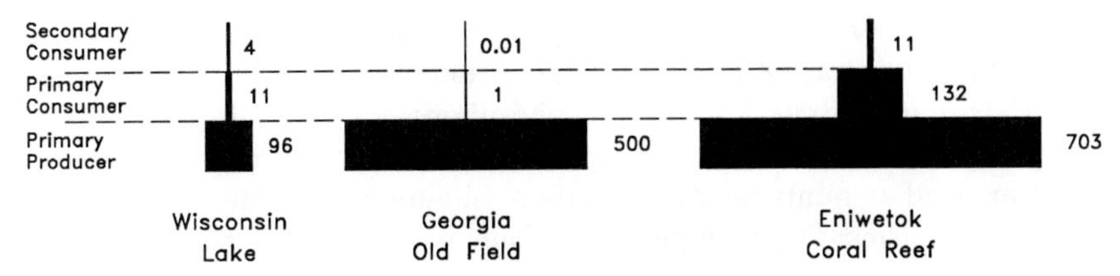

FIGURE 2-18 Pyramids of biomass (g/m^2) in three ecosystems. Typically, the biomass of a given trophic level is less than the biomass of lower trophic levels, giving rise to a "pyramid" when box sizes are drawn proportional to biomass. This is the consequence of decreasing energy availability at successively higher trophic levels [data are from Odum (1971)].

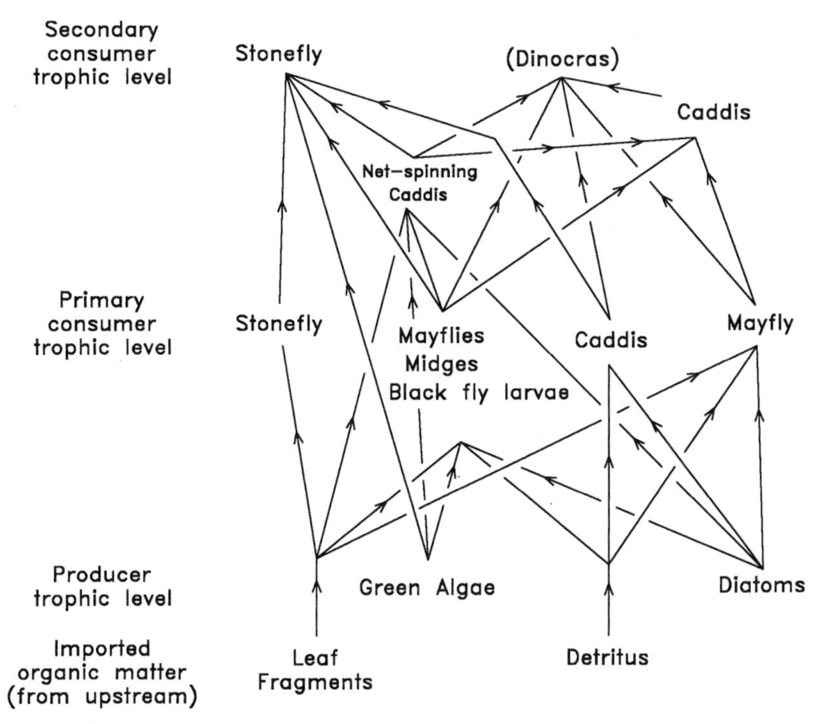

FIGURE 2-19 Portion of an aquatic food web in a small stream in South Wales. Note that some organisms (such as the net-spinning caddis shown here) actually feed at more than one trophic level [adapted from Jones (1949). Reproduced with permission of Blackwell Science Ltd.].

ciation with carbon. In an unpolluted environment, these include chemicals recognized as mineral nutrients, such as various forms of nitrogen [ammonium (NH_4^+) and nitrate (NO_3^-)] and phosphorus (typically in the form of orthophosphate species, such as $H_2PO_4^-$ and HPO_4^{2-}). Such nutrients are required in small amounts relative to carbon, but are essential to the structure of living organisms. For example, *phytoplankton*—free-living photosynthetic algae in surface waters—on average incorporate carbon, hydrogen, oxygen, nitrogen, and phosphorus according to the Redfield ratio: $C_{106}H_{263}O_{110}N_{16}P_1$. The simplified photosynthesis reaction shown in Eq. [2-48] can be expanded to reflect the Redfield ratio. If nitrate is the predominant nitrogen source, the photosynthesis reaction can be written as

$$106CO_2 + 16NO_3^- + H_2PO_4^- + 122H_2O + 17H^+ \rightarrow C_{106}H_{263}O_{110}N_{16}P + 138O_2. \quad [2\text{-}49]$$

If ammonium is the predominant nitrogen source, the photosynthesis reaction can be written as

$$106CO_2 + 16NH_4^+ + H_2PO_4^- + 106H_2O \rightarrow C_{106}H_{263}O_{110}N_{16}P + 106O_2 + 15H^+. \quad [2\text{-}50]$$

Many pollutant chemicals, when introduced into surface waters, move through the food chain in association with naturally occurring substances. Thus, an appreciation of nutrient cycling is very important in understanding the fate of anthropogenic pollutants in surface waters.

Mass is typically *recycled* in ecosystems; a given atom of phosphorus or nitrogen may pass through dozens and dozens of organisms during its residence in a particular ecosystem. Chemical substances are passed from one organism to another via the food chain. Eventually, either through excretion or the death of these organisms, the carbon and nutrients stop moving up the food chain and are released from dead biomass and converted back to their inorganic forms through the process of *decomposition*. Decomposition is often described, from a chemical cycling standpoint, as being equivalent to the process of *mineralization* (or, sometimes, *remineralization*), in which organic compounds are broken down into their inorganic components. The agents of decomposition are bacteria and fungi, often in association with worms, insects, and other organisms that aid the process by breaking up organic material both chemically and mechanically. The resulting inorganic chemical species can then be reused by the primary producers, as summarized in Fig. 2-20.

Life Histories of Species Composing the Biota

The foregoing discussion intentionally makes few references to specific kinds of organisms and no reference to individual species of plants or animals in aquatic systems. The principles of energy flow and mass cycling are universal; virtually every surface water body is an ecosystem with primary producers, primary consumers, and a food chain or food web. The specific organisms that fulfill the roles of producers, consumers, and decomposers, however, vary

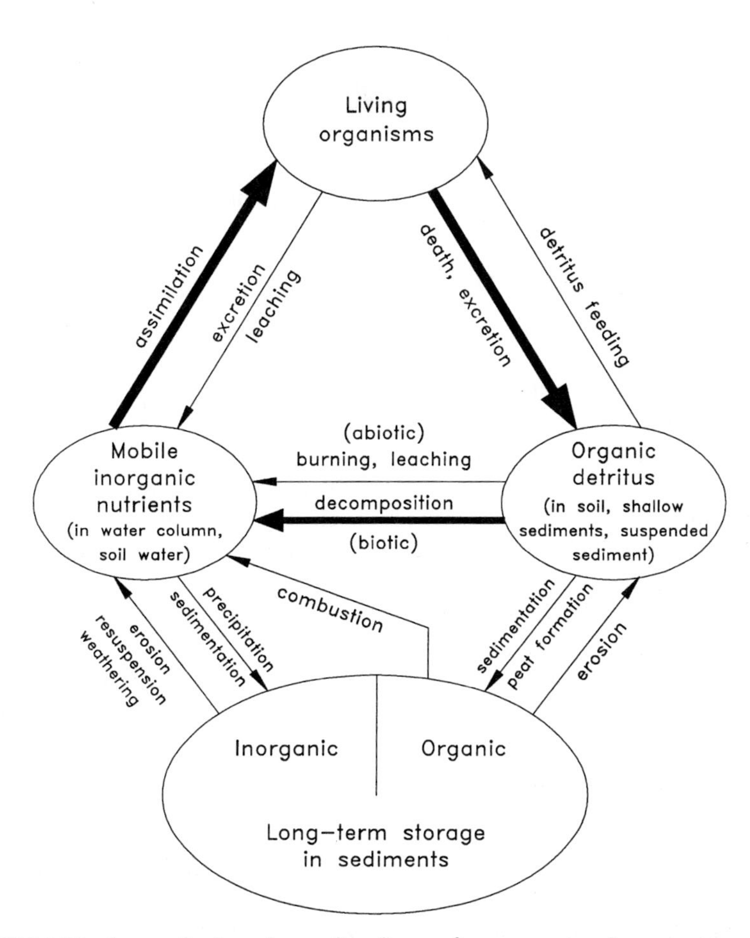

FIGURE 2-20 A generalized nutrient cycling diagram for an ecosystem. In contrast to energy, which flows through an ecosystem, nutrients are mostly recycled within the system. Inorganic forms are taken up by primary producers, converted to organic forms, passed up the food chain, and finally reconverted to inorganic forms by decomposers.

enormously from place to place. In any given surface water, numerous variables, in addition to the presence or absence of organic material and oxygen, determine what plant and animal species are present. Temperature, precipitation, cover and nesting sites, availability of mineral nutrients, absence of toxins, and presence or absence of other species of prey or predators are but a few factors influencing the composition of the biological communities that form an ecosystem.

The biota of an ecosystem interact in numerous fascinating and complex ways. Each species follows its own life history and has specific requirements necessary to complete its life cycle. A population ecologist may approach an ecosystem almost entirely from a life history standpoint, with the goal of understanding why specific plants and animals occur where they do. Invariably, however, no matter how intricate, interwoven, and seemingly improbable the life histories of plants and animals in a surface water ecosystem are, one can be sure that the system is organized in a manner consistent with the previously described principles of energy flow and mass cycling.

2.4.3 Reduction–Oxidation Chemistry: Power for Ecosystems

The overview of ecosystem functions presented in Section 2.4.2 emphasized the key ecological processes of energy flow and element cycling, which are enacted via the food chain; by eating, organisms obtain energy and nutrients. The chemical process of photosynthesis is the mechanism by which energy from sunlight is stored by primary producers as food, while respiration is the mechanism by which food provides energy to consumer organisms. From a chemical standpoint, photosynthesis and respiration are examples of *reduction–oxidation (redox)* reactions, reactions that involve the loss or gain of electrons by their reactants, and often involve relatively large free energy changes. Redox reactions mediate not only most energy flow in ecosystems but also the fate of many environmental chemicals.

Half-Reactions and Oxidation States

In reduction–oxidation reactions, electrons are transferred from one atom to another. Because the free energy change associated with these electron transfers is often considerably larger than the change associated with most other environmental chemical reactions, such as acid–base reactions, it is not surprising that most energy processing by organisms depends on redox reactions. People also exploit redox chemistry for nonphysiological purposes when they burn fuels for cooking, heating, and transportation.

TABLE 2-6 Common Environmental Redox Half-Reactions[a]

Half-reaction	log K, (ΔG°, kcal/mol)
$\frac{1}{4}O_2(g) + H^+ + e^- \rightarrow \frac{1}{2}H_2O$	20.75 (−28.22)
$\frac{1}{5}NO_3^- + \frac{6}{5}H^+ + e^- \rightarrow \frac{1}{10}N_2(g) + \frac{3}{5}H_2O$	21.05 (−28.63)
$\frac{1}{2}MnO_2(s) + 2H^+ + e^- \rightarrow \frac{1}{2}Mn^{2+} + H_2O$	21.00 (−28.56)
$Fe(OH)_3(s) + 3H^+ + e^- \rightarrow Fe^{2+} + 3H_2O$	16.5 (−22.44)
$\frac{1}{8}SO_4^{2-} + \frac{5}{4}H^+ + e^- \rightarrow \frac{1}{8}H_2S(g) + \frac{1}{2}H_2O$	5.25 (−7.14)
$\frac{1}{8}SO_4^{2-} + \frac{9}{8}H^+ + e^- \rightarrow \frac{1}{8}HS^- + \frac{1}{2}H_2O$	4.25 (−5.78)
$\frac{1}{8}CO_2(g) + H^+ + e^- \rightarrow \frac{1}{8}CH_4(g) + \frac{1}{4}H_2O$	2.87 (−3.90)
$\frac{1}{4}CH_2O + \frac{1}{4}H_2O \rightarrow e^- + H^+ + \frac{1}{4}CO_2(g)$	1.2 (−1.63)

[a]Stumm and Morgan (1981).

Chemical *reduction* is defined as the gain of electrons, and *oxidation* as the loss of electrons. If, in a reaction involving atoms A and B, atom A gains electrons, and is thus reduced, then atom B, which lost or "donated" the electrons, is called the *reductant*. Because atom B loses electrons, B itself is oxidized, while atom A is called the *oxidant*. Each reaction involving loss or gain of an electron by a chemical species is termed a *half-reaction;* Table 2-6 shows several common environmental half-reactions. For these reactions, Table 2-6 indicates an equilibrium constant (K) and the standard free energy change ($\Delta G°$) associated with the transfer of 1 mol of electrons [1 equivalent (eq)].

Any complete redox reaction in an aqueous medium is a combination of two half-reactions, because electrons can be neither stored in nor removed from water; free electrons are too unstable to persist for long as isolated species in water. When two half-reactions are combined into one reaction, the $\Delta G°$s are added. The $\Delta G°$ value for the overall redox reaction may be used to calculate the equilibrium constant, as discussed in Section 1.6.3.

To determine the extent of electron transfer during a chemical reaction, it is necessary to determine the oxidation state of each atom before and after the reaction. The pure element, which has no charge, is the reference state or the zero (0) oxidation state. Nitrogen gas (N_2), metallic iron (Fe), metallic lead (Pb), and oxygen (O_2) are examples of elements in the zero oxidation state. Free, ionized atoms (ions) have an oxidation state equal to their actual charge. Thus, hydrogen, in the form of the hydrogen ion, H^+, is in the (+I) oxidation state, while chlorine, in the form of the chloride ion, Cl^-, is in the (−I) oxidation state. Note that by convention, the oxidation state of an atom is represented by Roman numerals.

When several atoms are combined, the oxidation state of each atom can usually be determined by using the following simple rules: the oxidation state of hydrogen, when combined in common environmental compounds, is (+I); the oxidation state of oxygen in most common environmental compounds is (−II). (Due to its high electron affinity, oxygen is formally considered to have completely captured two electrons from the other atoms, although oxygen typically forms bonds that involve some sharing of electrons. Likewise, hydrogen is assumed to have given up its electron completely, although it may in fact share an electron with another atom in an unequal partnership.) The halogens, chlorine, fluorine, bromine, and iodine, are, like oxygen, very *electronegative;* halogens may be assumed to be in the (−I) oxidation state in most compounds. For further information on oxidation states of specific chemicals, the reader is referred to Bodek *et al.* (1988).

EXAMPLE 2-11

Determine the oxidation states of (a) nitrogen in nitrate (NO_3^-); (b) sulfur in H_2S gas; and (c) carbon in chloroform ($CHCl_3$).

(a) In a nitrate ion, the oxidation state of the nitrogen can be determined by simple algebra:

$$1 \cdot (\text{nitrogen oxidation state}) + 3 \cdot (-\text{II}) = -1.$$

The oxidation state of nitrogen is therefore (+V).

(b) In H_2S, assume each H atom has an oxidation state of (+I). Then,

$$2 \cdot (+\text{I}) + 1 \cdot (\text{sulfur oxidation state}) = 0.$$

Sulfur is in the (−II) oxidation state. [Note that if this H_2S gas, a weak acid, were dissolved in water and reacted to form HS^-, the calculation would still yield a (−II) oxidation state for sulfur. Acid–base reactions can proceed entirely independently of electron transfer.]

(c) The oxidation state of carbon in chloroform can be calculated as follows:

$$3 \cdot (-\text{I}) + 1 \cdot (+\text{I}) + 1 \cdot (\text{carbon oxidation state}) = 0.$$

The oxidation state of the carbon is therefore (+ II).

Now consider the photosynthesis reaction presented in Eq. [2-48] as the sum of two half-reactions from Table 2-6:

(a) $e^- + H^+ + \frac{1}{4}CO_2(g) \rightarrow \frac{1}{4}CH_2O + \frac{1}{4}H_2O$ ($\Delta G° = 1.63$ kcal/mol).

(b) $\frac{1}{2}H_2O \rightarrow \frac{1}{4}O_2(g) + H^+ + e^-$ ($\Delta G° = 28.22$ kcal/mol).

Adding (a) and (b) results in the complete redox equation:

$$\tfrac{1}{4}CO_2(g) + \tfrac{1}{4}H_2O \longrightarrow \tfrac{1}{4}CH_2O + \tfrac{1}{4}O_2(g) \quad (\Delta G° = 29.9 \text{ kcal/mol}).$$

(Note that for uniformity with Table 2-6, the $\Delta G°$ value of 29.9 kcal/mol refers to the transfer of 1 mol of electrons, which in this case corresponds to the reaction of $\frac{1}{4}$ mol of CO_2. Likewise, the $\Delta G°$ values for subsequent complete redox equations in this chapter are for the transfer of 1 mol of electrons.)

In photosynthesis, carbon is reduced from the (+ IV) oxidation state to (0) while oxygen is oxidized from the (− II) state to (0). When these two reactions are added together, the electrons shown in each half-reaction cancel each other out. Note that the $\Delta G°$ for this reaction (29.9 kcal/mol) is positive, meaning that energy must be put into the system for the reaction to occur.

A variant of photosynthesis occurs in certain environments where sulfide (S^{2-}) is present and anaerobic conditions exist (i.e., there is no molecular oxygen, O_2, present). In this case, sulfide can be used as a reductant in place of water, and the photosynthetic process is performed by colored (green or purple) sulfur bacteria:

$$\tfrac{1}{4}CO_2(g) + \tfrac{1}{2}H_2S(g) \xrightarrow{h\nu} \tfrac{1}{4}CH_2O + \tfrac{1}{4}H_2O + \tfrac{1}{2}S(s) \quad (\Delta G° = 5.6 \text{ kcal/mol}) \qquad [2\text{-}51a]$$

or

$$\tfrac{1}{4}CO_2(g) + \tfrac{1}{4}H_2O + \tfrac{1}{8}HS^- \xrightarrow{h\nu} \tfrac{1}{4}CH_2O + \tfrac{1}{8}SO_4^{2-} + \tfrac{1}{8}H^+ \quad (\Delta G° = 7.2 \text{ kcal/mol}). \qquad [2\text{-}51b]$$

In the environment, some of the "CH_2O" produced by photosynthesis gets buried (e.g., as coal) and does not get reoxidized. As a result, oxygen is left over; this has resulted, over the millennia, in the accumulation of oxygen in the atmosphere, with profound implications for both environmental chemistry and biology.

Photoautotrophs are not the only organisms that can convert carbon dioxide to organic carbon. Several kinds of nonphotosynthetic bacteria, called *chemoautotrophs,* can also reduce carbon dioxide to organic material by using energy from the oxidation, typically with oxygen, of certain inorganic compounds. In such cases, the reaction is the same as shown in Eq. [2-48], except light energy ($h\nu$) is replaced by chemical energy from the oxidation of hydrogen sulfide [or ammonia, nitrite (NO_2^-), Fe^{2+}, or even arsenite, AsO_3^{3-}]. Deep sea hydrothermal vent ecosystems are powered almost entirely by the oxidation of hydrogen sulfide (although the oxygen used as the oxidant is a product of photosynthesis).

Oxidation of Organic Matter and the Ecological Redox Sequence

In all ecosystems, the oxidation of organic material by organisms during respiration releases the energy stored by photosynthesis. Writing the respiration reaction so that the stoichiometry reflects the transfer of 1 mol of electrons gives

$$\tfrac{1}{4}CH_2O + \tfrac{1}{4}O_2(g) \longrightarrow \tfrac{1}{4}CO_2(g) + \tfrac{1}{4}H_2O \qquad (\Delta G^\circ = -29.9 \text{ kcal/mol}). \qquad [2\text{-}52]$$

Note that the release of energy is represented by a negative ΔG°. Recall from Section 1.6.3 that the overall change in Gibbs free energy (ΔG) depends not only on ΔG°, but also on the reaction quotient Q. Calculation of a typical change in free energy, ΔG_t, based on typical environmental conditions, yields an estimate of the amount of free energy available from a particular reaction ($\Delta G_t = \Delta G^\circ + RT \ln Q$).

For comparative purposes, consider a concentration of 10^{-3} M for an idealized organic carbon source, CH_2O. Then, using approximate atmospheric pressures of oxygen (0.2 atm) and carbon dioxide ($10^{-3.5}$ atm) in the respiration reaction, the reaction quotient Q can be expressed as

$$Q = \frac{(P_{CO_2})^{1/4}}{(P_{O_2})^{1/4}[CH_2O]^{1/4}} = \frac{(10^{-3.5})^{1/4}}{(0.2)^{1/4}(10^{-3})^{1/4}} = 1.12.$$

Therefore, at a temperature of 25°C,

$$\Delta G_t = -29.9\,\frac{\text{kcal}}{\text{mol}} + \left(0.00199\,\frac{\text{kcal}}{\text{mol}\cdot\text{K}}\right)(298\text{ K})\ln(1.12)$$

$$\Delta G_t = -29.8 \text{ kcal/mol}.$$

Although in this instance $\Delta G°$ by itself is a good approximation of the overall change in free energy under typical environmental conditions (ΔG_t), this is not always the case. See, for example, the calculations pertaining to denitrification, in Eq. [2-53].

The consumption of oxygen by respiration leads to anoxia if oxygen cannot be replenished rapidly enough from the air or by photosynthesis. This may occur in bottom sediments, in the hypolimnia of stratified lakes, and in waterlogged wetland sediments. Oxygen can only diffuse slowly into such locations, due to its relatively low molecular diffusion coefficient in water (Section 1.4.2); degradation of organic material by bacteria may consume oxygen at a faster rate than it can be replaced. Without oxygen, energy to support life cannot come from the oxygen-consuming respiration reaction depicted in Eq. [2-52]; many aerobic organisms die under low-oxygen conditions. Those that can survive anoxia, mostly bacteria, must shift to oxidants other than oxygen to oxidize organic material and release energy. These alternative reactions occur in a commonly observed sequence, often called the *ecological redox sequence.*

In the first anoxic step of this sequence, anaerobic microorganisms oxidize organic material by using nitrate ions (NO_3^-) as the oxidant. Nitrate, which occurs naturally as part of the environmental nitrogen cycle [Nielsen and MacDonald (1978)], is thereby reduced to nitrogen gas according to the following reaction:

$$\tfrac{1}{4}CH_2O + \tfrac{1}{5}NO_3^- + \tfrac{1}{5}H^+ \longrightarrow \tfrac{1}{10}N_2(g) + \tfrac{1}{4}CO_2(g) + \tfrac{7}{20}H_2O \quad (\Delta G° = -30.3 \text{ kcal/mol}). \qquad [2\text{-}53]$$

This process is called *denitrification.* Initially it appears that denitrification is more energetically favorable than aerobic respiration. Calculation of ΔG_t, however, shows that this is not usually the case. Assuming 10^{-3} *M* CH_2O as before, 10^{-4} *M* NO_3^- (a generous level for a surface water), and a pH of 7, with approximate atmospheric pressures for N_2 (0.8 atm) and CO_2 ($10^{-3.5}$ atm) at 25°C,

$$\Delta G_t = -30.3\,\frac{\text{kcal}}{\text{mol}} + \left(0.00199\,\frac{\text{kcal}}{\text{mol}\cdot\text{K}}\right)(298\text{ K})\cdot\ln\left(\frac{(10^{-3.5})^{1/4}(0.8)^{1/10}}{(10^{-3})^{1/4}(10^{-4})^{1/5}(10^{-7})^{1/5}}\right)$$

$$\Delta G_t = -27.5 \text{ kcal/mol}.$$

In typical environmental settings, then, denitrification is not as energetically favorable as aerobic respiration.

When all nitrate is depleted, iron oxyhydroxides, $Fe(OH)_3$, and manganese oxyhydroxides, $Mn(OH)_4$, often represented as MnO_2, serve as oxidants. These oxides are very common in soils and sediments. Iron reduction proceeds as follows:

$$\tfrac{1}{4}CH_2O + Fe(OH)_3(s) + 2H^+ \longrightarrow Fe^{2+} + \tfrac{1}{4}CO_2(g) + \tfrac{11}{4}H_2O \quad (\Delta G^\circ = -24 \text{ kcal/mol}). \qquad [2\text{-}54]$$

As is the case with denitrification, typical $-\Delta G_t$ values under ordinary conditions are not quite as large as $-\Delta G^\circ$; for the same concentration of CH_2O as previously, a pH of 7, and 10^{-5} *M* Fe^{2+}, $\Delta G_t = -12$ kcal/mol at 25°C.

A similar reaction using manganese oxyhydroxides can occur; the ΔG° is -30.3 kcal/mol. For the same concentrations of CH_2O and H^+ as previously, and with 10^{-5} *M* Mn^{4+}, ΔG_t is -24.3 kcal/mol at 25°C. Although manganese reduction is energetically more favorable than iron reduction, manganese is much less abundant than iron in most environments, and therefore iron (III) is likely to be the more important oxidant.

Reduction plays a major role in the behavior of iron and manganese as well as in the behavior of pollutant chemicals in the environment. The solids $Fe(OH)_3$ and MnO_2 are strong adsorbents of many chemicals, especially metals. When particles of these oxides form in, or are transported into, surface waters, they can sorb metals; the suspended particles may then be removed from the water column by settling, as previously discussed. Such a process may lead to either temporary or long-term deposition of the metals into the sediment. Retention in the sediment is only short term if iron reduction and manganese reduction subsequently lead to dissolution of the oxides and release of the metals back into the water column.

After iron oxides and manganese oxides have been consumed, bacterial respiration can continue through the use of an even less energetically favorable oxidant, sulfate (SO_4^{2-}). The process of sulfate reduction proceeds as follows:

$$\tfrac{1}{4}CH_2O + \tfrac{1}{8}SO_4^{2-} + \tfrac{1}{8}H^+ \longrightarrow \tfrac{1}{8}HS^- + \tfrac{1}{4}CO_2(g) + \tfrac{1}{4}H_2O \quad (\Delta G^\circ = -7.4 \text{ kcal/mol}). \qquad [2\text{-}55]$$

This reaction leads to the production of sulfide species, such as H_2S (hydrogen sulfide), HS^- (bisulfide), and S^{2-} (sulfide ion), which are the most reduced forms of sulfur. Sulfides are toxic; most popularly known is H_2S, which forms in rotten eggs and gives them their characteristic pungent odor. Sulfides are also important in causing the chemical precipitation of many metals, such as iron, copper, lead, and zinc, which form solids (such as FeS, CuS, PbS, ZnS) on reaction with S^{2-}.

Finally, after all of the sulfate is consumed, organic material can be *fermented* into methane and carbon dioxide in the process of *methanogenesis*. *Fermentative reactions*, of which there are many examples in the environment, differ from respiration in that they do not change the average oxidation state of carbon. The reaction for methanogenesis is

$$\tfrac{1}{4}CH_2O \longrightarrow \tfrac{1}{8}CO_2(g) + \tfrac{1}{8}CH_4(g) \qquad (\Delta G^\circ = -5.5 \text{ kcal/mol}). \qquad [2\text{-}56]$$

It has been repeatedly observed in different environments that after oxygen is consumed, the preceding reactions occur in the order described. A common prerequisite for the depletion of oxygen is for the environment to be waterlogged, thus hindering resupply of oxygen from the atmosphere except by the slow process of molecular diffusion through water. An adequate concentration of degradable organic material, the reducing agent, is also needed. Thus, one expects—and finds—denitrification, iron reduction, sulfate reduction, and methanogenesis in the deeper water layers and bottom sediments of surface waters, where settling organic matter accumulates.

EXAMPLE 2-12

A 0.1-g roach is accidentally bottled in a 1-liter bottle of soda, which initially contains 10 mg/liter oxygen and 1.2 mg/liter NO_3^-. Assume 10% of the roach's fresh weight is CH_2O. If microbes capable of completely mineralizing the roach are present, does the bottle become anaerobic? Is any of the nitrate consumed?

First convert the masses of CH_2O, O_2, and NO_3^- to moles:

$$0.01 \text{ g } CH_2O \cdot \frac{1 \text{ mol}}{30 \text{ g}} = 3.3 \times 10^{-4} \text{ mol.}$$

$$10 \text{ mg } O_2 \cdot \frac{1 \text{ g}}{1000 \text{ mg}} \cdot \frac{1 \text{ mol}}{32 \text{ g}} = 3.1 \times 10^{-4} \text{ mol.}$$

$$1.2 \text{ mg } NO_3^- \cdot \frac{1 \text{ g}}{1000 \text{ mg}} \cdot \frac{1 \text{ mol}}{62 \text{ g}} = 1.9 \times 10^{-5} \text{ mol.}$$

Then estimate whether the CH_2O portion of the roach can be completely mineralized with the available O_2. From Eq. [2-52], in aerobic respiration, 1 mol of O_2 oxidizes 1 mol of CH_2O to CO_2. Therefore, 2×10^{-5} mol of CH_2O will be left after all the oxygen is consumed and the bottle becomes anaerobic.

Further oxidation then can occur using nitrate. From Eq. [2-53], 1 mol of NO_3^- can oxidize 5/4 mol of CH_2O to CO_2. Calculate how much NO_3^- is needed to oxidize the remaining 2×10^{-5} mol of CH_2O:

$$2 \times 10^{-5} \text{ mol } CH_2O \cdot \frac{1 \text{ mol } NO_3^-}{5/4 \text{ mol } CH_2O} = 1.7 \times 10^{-5} \text{ mol } NO_3^-.$$

This concentration of NO_3^- is less than the available amount in the bottle; therefore the CH_2O portion of the roach can be fully oxidized, with approximately 3×10^{-6} mol (3 μmol) of NO_3^- remaining.

The Redox Scale

Because the preceding sequence of redox reactions is consistently observed as an environment becomes increasingly reduced, it is convenient and appropriate to describe the intensity of reducing conditions in terms of the chemical species present. For example, if large amounts of soluble iron are found in the pore waters of a river sediment, but no sulfide is present, that environment could be characterized as an *iron-reducing* environment. If the sediment becomes more strongly reducing, such that sulfide or methane are formed, it could be characterized as a *sulfidic* or as a *methanogenic* environment. It is also possible to define quantitative indices of the intensity of reduction in these environments; one such index is the activity of electrons. In a very *oxidizing* environment (e.g., where molecular oxygen, O_2, is present) the activity of electrons is low. In a methanogenic environment, the activity of electrons is high—sufficiently high, in fact, to react with carbon dioxide to produce methane. Electron activity can be characterized by using the lowercase p notation (just as pH is characterized):

$$p\epsilon = -\log\{e^-\}. \qquad [2\text{-}57]$$

The $p\epsilon$ of a surface water at pH 7, in equilibrium with atmospheric oxygen, is calculated to be 13.6; it decreases to approximately 4 in an environment where both oxidized and reduced iron are present and drops to approximately −4 where sulfide or methane are being produced. $p\epsilon$ can be calculated from the measured concentrations of products and reactants in a redox half-reaction. A scale equivalent to $p\epsilon$ is the E_h scale, which is expressed in volts and is based on the determination of electron activity using electrochemical methods. E_h is related to $p\epsilon$ by the following equation,

$$E_h = \frac{2.3\,RT}{F} \cdot p\epsilon, \qquad [2\text{-}58]$$

where R is the gas constant, T is the absolute temperature, and F is the *Faraday constant*. The Faraday constant is the electric charge carried by

1 mol of electrons and has a value of 96,485 coulombs/mol. At 25°C, Eq. [2-58] can be written as

$$E_h = 0.059 \cdot p\epsilon, \quad [2\text{-}59]$$

where E_h is in volts.

EXAMPLE 2-13

HS^- is measured at a concentration of 0.1 ppm (as sulfur) in a sediment pore water whose sulfate concentration is 10^{-6} *M*. What are the theoretical pϵ and E_h if these species are in equilibrium? The pH is 7.3.

First, convert the HS^- concentration to a molar concentration:

$$\frac{0.1 \text{ mg}}{\text{liter}} \cdot 10^{-3}\frac{\text{g}}{\text{mg}} \cdot \frac{1 \text{ mol HS}^-}{33 \text{ g}} = 3.0 \times 10^{-6}\frac{\text{mol HS}^-}{\text{liter}}.$$

From Table 2-5,

$$\tfrac{1}{8}SO_4^{2-} + e^- + \tfrac{9}{8}H^+ \longrightarrow \tfrac{1}{8}HS^- + \tfrac{1}{2}H_2O: \qquad K = 10^{4.25}.$$

Therefore,

$$\frac{[HS^-]^{1/8}}{[SO_4^{2-}]^{1/8}\{e^-\}\{H^+\}^{9/8}} = 10^{4.25}$$

$$\{e^-\} = \frac{(3.0 \times 10^{-6})^{1/8}}{(10^{-6})^{1/8}(10^{-7.3})^{9/8}(10^{4.25})} = 1.1 \times 10^4.$$

By using Eq. [2-57],

$$p\epsilon = -\log\{e^-\} = -4.0.$$

By using Eq. [2-59],

$$E_h = 0.059 \cdot p\epsilon = -0.24 \text{ V}.$$

Figure 2-21 presents a graph of environmental pϵ and E_h ranges as a function of the dominant redox species in the environment. This graph must be regarded as approximate; strictly, the pϵ (or E_h) of an environment is an equilibrium notion and is therefore not rigorously applicable in natural waters where chemical transformations may be actively occurring and have not

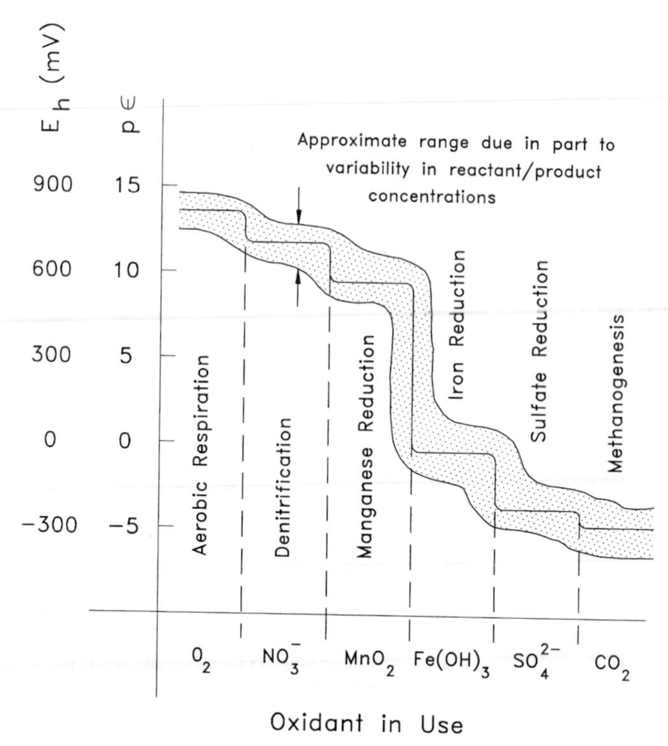

FIGURE 2-21 The ecological redox sequence. In an organic-rich environment that becomes isolated from the atmosphere, bacteria, after first consuming any available oxygen, utilize alternative oxidants in the sequence shown from left to right. As each oxidant is being utilized, the pε and E_h of the system lie in the approximate ranges shown on the vertical axis. The broad and indefinite ranges of pε and E_h associated with each oxidant are intended to reflect both variation in the oxidant and reductant concentrations and the fact that while pε and E_h are calculated on the basis of equilibrium, natural redox systems are usually not at equilibrium.

reached equilibrium. Moreover, the definitions of pε and E_h are ambiguous in locations where multiple reductants and oxidants are present.

E_h is often confused with the closely related *redox potential* or *oxidation–reduction potential* (ORP) measurement. ORP is measured by placing a redox electrode into water or sediment; the redox electrode is a piece of metallic platinum, which acquires a more negative potential with respect to its reference electrode under reducing conditions where electron activities are higher. ORP is the voltage measured between this redox electrode and a reference electrode placed in the same environment. It provides a useful, approximate characterization of redox conditions in the aquatic environment, although it lacks precise theoretical definition. Although ORP and E_h are both measured in volts and do show some rough correspondence, they are determined differently and should not be treated as synonymous.

EXAMPLE 2-14

An ORP electrode is inserted to a point 0.3 cm below the sediment–water interface of a stratified lake in late August. A voltage of approximately 0.25 V is recorded. Assuming for the moment that ORP is a crude estimate of E_h, what reactions are most likely to be occurring in the sediment at this depth?

Use Eq. [2-59]:

$$\text{ORP} \approx p\epsilon = \frac{0.25 \text{ V}}{0.059} = 4.2.$$

As shown in Fig. 2-21, iron oxides and manganese oxides are most likely the chemical species being used as oxidants in the degradation of organic matter at this $p\epsilon$.

2.5 DISSOLVED OXYGEN MODELING IN SURFACE WATERS

Given that many organisms require oxygen for respiration, it is not surprising that the concentration of oxygen in a surface water is a critical attribute of the ecosystem. The suitability of natural waters for many types of organisms, including fish, is often characterized by the concentration of *dissolved oxygen* (DO). In a water body at equilibrium with the atmosphere, the concentration of DO is approximately 10 ppm; see Table 2-4 for oxygen solubility values as a function of temperature and salinity.

One of the first adverse effects of organic pollutants in surface waters to be recognized was the decrease in DO that resulted when sewage or other organic wastes were discharged to surface waters. In surface water, bacteria degrade and ultimately mineralize organic waste, consuming oxygen and releasing carbon dioxide in the process; see Eq. [2-52]. If oxygen is consumed at a rate that exceeds the replenishment rate of DO from the atmosphere or from photosynthesis, the DO concentration can decrease to less than a few milligrams per liter, and organisms such as fish suffocate. Therefore, reducing the severity of oxygen depletion in the vicinity of pollutant discharge points, such as sewage outfalls, was an early concern of environmental engineers, and models were devised to predict the amount of oxygen depletion that would occur under any given set of stream conditions and pollutant loading rates.

EXAMPLE 2-15

Upstream of a sewage outfall, a river contains 7 mg/liter DO. Some distance downstream of the outfall, however, DO has been diminished to 4 mg/liter due to organic waste decomposition by microbes. What is the approximate amount of organic matter ("CH_2O") that must have been degraded to account for this consumption of DO?

Neglecting O_2 diffusion into the water from the atmosphere and O_2 production by photosynthesis, and assuming no O_2 consumption by organisms other than the microbes, 3 mg/liter O_2 are consumed. This corresponds to

$$\frac{3\text{ mg } O_2}{1\text{ liter}} \cdot \frac{1\text{ mol } O_2}{32{,}000\text{ mg}} \cdot \frac{1\text{ mol } CH_2O}{1\text{ mol } O_2} \cdot \frac{30{,}000\text{ mg}}{1\text{ mol } CH_2O} = \frac{2.8\text{ mg } CH_2O}{1\text{ liter}}.$$

The degradation of 2.8 mg/liter of organic matter thus consumes 3 mg/liter of dissolved oxygen. Actually, some O_2 from the atmosphere will have dissolved into the stream, so 3 mg/liter is a minimum value for O_2 consumption.

Biochemical or *biological oxygen demand* (BOD) is a measure of the amount of oxygen required by bacteria to degrade the dissolved and suspended organic matter in a volume of water. Therefore, BOD is an indirect measure of the organic content of a water. [In ammonia-rich waters, some oxygen is also consumed by the oxidation of ammonia to nitrate in the process of *nitrification* (see Section 4.7.2).] Commonly, a 5-day test (BOD_5) is conducted, in which a water sample is fully aerated and then incubated over a 5-day period at 20°C. At the end of the test, the total amount of DO that has been consumed is measured. The duration and temperature of the test are historically based on the maximum travel time to the sea of some British rivers and the mean summer temperature of those rivers, respectively. BOD is expressed as milligrams of DO needed to oxidize the organic waste contained in 1 liter of water or wastewater (mg/L).

In the past, the primary goals of wastewater engineers were the lowering

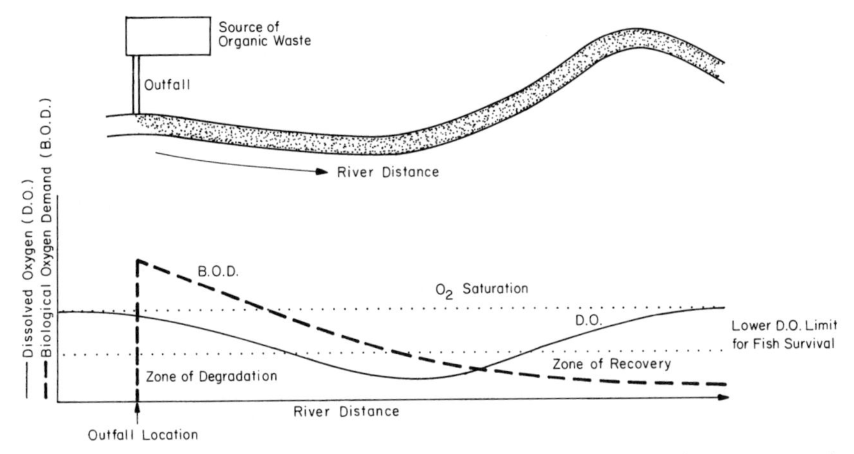

FIGURE 2-22 "DO sag" induced in a river by inputs of organic waste such as sewage. In the zone of degradation, oxygen is consumed more rapidly by biodegrading microorganisms than it can be replenished from the atmosphere. Sufficiently heavy loadings can cause oxygen concentrations to fall below the minimum required for many desirable forms of aquatic life (e.g., fish) or can even cause waters to become completely anoxic. Recovery begins downstream after much of the organic waste is degraded, and the river becomes reaerated.

of BOD to acceptable levels and the inactivation of disease-causing organisms *(pathogens)*. Later, more advanced techniques for the removal of nutrients as well as heavy metals and other pollutants in wastewater were developed, as more subtle effects of wastewater discharges on surface waters were discovered. In a river with a sewage outfall, a plot of DO versus distance downstream of the outfall (Fig. 2-22) is influenced by the DO consumption rate (which is maximum where BOD is maximum), the dissolution of oxygen from the atmosphere into the river, photosynthesis, and O_2 consumption in the bottom sediment *(benthic O_2 demand)*. The classic analysis of this situation is summarized in the *Streeter–Phelps* model, which is based on the mass balances of O_2 and BOD in a river or stream (Metcalf and Eddy, 1991).

The Streeter–Phelps model allows the estimation of the DO "sag" in the stream as a function of distance. This equation may be developed by considering a stream, initially saturated with oxygen, which receives a wastewater discharge containing BOD. As the water moves downstream, the BOD is assumed to decay at a first-order rate, K_{BOD}, typically on the order of 0.2/day:

$$\frac{d\,\mathrm{BOD}}{d\tau} = (-K_{\mathrm{BOD}})(\mathrm{BOD}), \qquad [2\text{-}60]$$

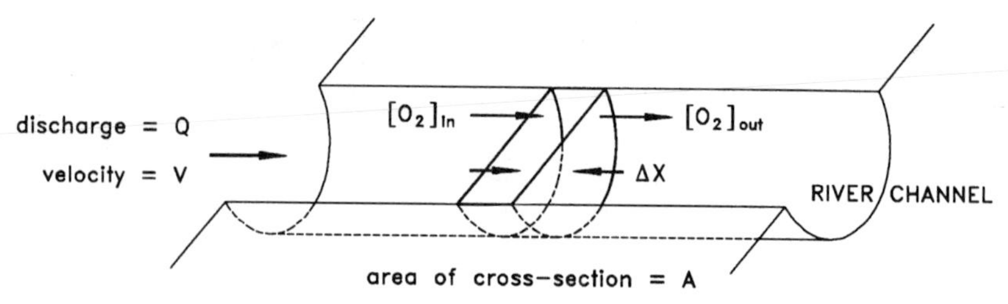

FIGURE 2-23 Definition of the control volume for which the DO mass balance is expressed in Eq. [2-62]. The control volume is a slice of thickness Δx and cross-sectional area A and is stationary in the river.

where τ is travel time. The solution to Eq. [2-60], comparable to the solution to Eq. [1-18] shown in Eq. [1-19], is

$$\text{BOD} = \text{BOD}_0 e^{-K_{\text{BOD}}\tau}, \qquad [2\text{-}61]$$

where BOD_0 is the initial BOD at the location of the wastewater discharge.

To analyze the mass balance of DO, a useful control volume is a stationary segment of river Δx units long, as shown in Fig. 2-23. The steady-state mass conservation expression for oxygen in this slice is

$$\begin{aligned}&\text{Rate of } O_2 \text{ inflow} - \text{rate of } O_2 \text{ outflow} + \text{rate of } O_2 \text{ reaeration}\\ &\qquad - \text{rate of } O_2 \text{ consumption by BOD} = 0.\end{aligned} \qquad [2\text{-}62]$$

The first two terms of Eq. [2-62], advective inflow and outflow, can be written as

$$V \cdot A \cdot [O_2]_{in} - V \cdot A \cdot [O_2]_{out} \approx -V \cdot A \cdot \frac{d[O_2]}{dx} \cdot \Delta x, \qquad [2\text{-}63]$$

where V is river velocity [L/T], A is the cross-sectional area of the river [L^2], and Δx is the thickness of the river slice [L]. The flux density of oxygen from the atmosphere into the control volume, the third term in Eq. [2-62], is proportional to the O_2 deficit, i.e., the difference between the saturated O_2 concentration and the actual O_2 concentration, and the piston velocity, or gas exchange coefficient. Recall that the reaeration coefficient for oxygen is the piston velocity for oxygen divided by stream depth; thus, piston velocity equals the product of the reaeration coefficient and depth. Total flux of atmospheric O_2 into the control volume is

$$\{\text{Surface area}\} \cdot \{\text{piston velocity}\} \cdot \{O_2 \text{ deficit}\},$$

or

$$\{(\Delta x)(\text{width})\} \cdot \{(K_{O_2})(\text{depth})\} \cdot \{[O_2]_{sat} - [O_2]\},$$

which equals

$$K_{O_2} \cdot ([O_2]_{sat} - [O_2])(A)(\Delta x), \tag{2-64}$$

where K_{O_2} is the reaeration coefficient $[T^{-1}]$, $[O_2]_{sat}$ is the saturated oxygen concentration $[M/L^3]$, and $[O_2]$ is the actual oxygen concentration $[M/L^3]$.

The fourth term in Eq. [2-62], the rate of O_2 consumption by BOD, is equal to the BOD decay rate, $K_{BOD} \cdot \text{BOD}$, multiplied by the volume of water in the slice, $A\,\Delta x$. Thus, Eq. [2-62] can be rewritten as

$$-V \cdot A \cdot \frac{d[O_2]}{dx}\Delta x + K_{O_2} \cdot ([O_2]_{sat} - [O_2]) \cdot A\,\Delta x - K_{BOD} \cdot \text{BOD} \cdot A\,\Delta x = 0. \tag{2-65}$$

The corresponding differential equation is

$$-V\frac{d[O_2]}{dx} + K_{O_2} \cdot ([O_2]_{sat} - [O_2]) = K_{BOD} \cdot \text{BOD}. \tag{2-66}$$

For the upstream conditions of BOD_0 and $[O_2]_{sat}$, Eq. [2-66] has the solution:

$$[O_2]_{sat} - [O_2] = \frac{K_{BOD} \cdot \text{BOD}_0}{K_{O_2} - K_{BOD}}\left(e^{-K_{BOD}\tau} - e^{-K_{O_2}\tau}\right). \tag{2-67}$$

The minimum DO occurs at τ_{max}, the travel time at which the derivative of the preceding equation is zero:

$$\tau_{max} = \frac{1}{K_{O_2} - K_{BOD}} \ln \frac{K_{O_2}}{K_{BOD}}. \tag{2-68}$$

The minimum $[O_2]$ value, $[O_2]_{min}$, is given by

$$[O_2]_{sat} - [O_2]_{min} = \frac{K_{BOD}}{K_{O_2}} \text{BOD}_0 e^{-K_{BOD}\tau_{max}}. \tag{2-69}$$

This special case of the Streeter–Phelps model neglects photosynthesis, as well as oxygen consumption by the bottom sediments and by nitrification (see Section 4.7.2).

EXAMPLE 2-16

A stream is in equilibrium with atmospheric oxygen upstream of a waste outfall, which creates a BOD_0 of 20 mg/liter immediately downstream. K_{BOD} is 0.4/day, and K_{O_2} is 1.4/day. The stream temperature is 15°C. How far downstream, in terms of travel time, is the maximum DO sag, and what is the minimum DO in the river?

The travel time to the maximum DO sag can be estimated by Eq. [2-68]:

$$\tau_{\max} = \left(\frac{1}{1.4/\text{day} - 0.4/\text{day}}\right) \ln \left(\frac{1.4/\text{day}}{0.4/\text{day}}\right)$$

$$\tau_{\max} = (1 \text{ day})(1.25) = 1.25 \text{ days}.$$

The maximum O_2 sag, or deficit, which occurs at the location corresponding to the preceding travel time, is given by Eq. [2-69]:

$$[O_2]_{\text{sat}} - [O_2]_{\min} = \frac{0.4/\text{day}}{1.4/\text{day}} \cdot 20 \text{ mg/liter} \cdot e^{-(0.4/\text{day})(1.25 \text{ day})} \approx 3.5 \text{ mg/liter}.$$

From Table 2-4, $[O_2]_{\text{sat}}$ is approximately 10 mg/liter at 15°C. Therefore, $[O_2]_{\min}$ is approximately 6.5 mg/liter, which is probably enough oxygen to keep some hardy fish alive.

An alternative to BOD_5 is a test known as *chemical oxygen demand* (COD), which also measures the milligrams of oxygen needed to oxidize organic waste in 1 liter of water. This test takes only a few hours to perform on a water sample, but does not reflect natural conditions because a strong chemical oxidant is added to the water sample to cause the oxidation of organic matter. The relationship between BOD and COD varies from water body to water body.

2.6 BIOTRANSFORMATION AND BIODEGRADATION

The primary purpose of the Streeter–Phelps model is to predict dissolved oxygen concentrations in a stream or river based on BOD loadings. In more general terms, the Streeter–Phelps model predicts the impact on the stream of the *biotransformation* of a chemical substance (BOD). Numerous biological processes in surface waters can transform chemicals into other chemicals. The term *biodegradation* is often used to describe the biotransformation of an organic pollutant into other compounds. Although initial transformation products can occasionally be more toxic to humans or aquatic organisms than the original parent compound, eventually successive biological transformations in oxic waters tend to convert organic pollutants into carbon dioxide, water, and mineral salts. The overall process by which organic compounds are converted into simple inorganic compounds is called *mineralization*. Mineralization of organic pollutants often occurs by the same processes that are involved in the degradation of natural organic matter in ecosystems. The many intermediate biotic transformations of organic pollutants that do not produce purely inorganic compounds represent *partial biodegradation.* Note that inorganic pollutants that contain intrinsically toxic elements, such as mercury and arsenic, may also be transformed from one chemical species to another [e.g., metallic mercury (Hg^0) to the highly toxic monomethyl mercury (CH_3Hg^+)]. Unlike the case with organic contaminants, however, the toxic elements themselves cannot be destroyed.

Biotransformation processes are mediated by microorganisms (microbes), especially bacteria and fungi. These two groups are remarkably diverse, although the metabolic capabilities of bacteria as a group tend to be greater; while fungi are aerobic, bacteria are active in both aerobic and anaerobic environments, obtaining energy from a tremendous number of chemicals.

Although the range of chemical transformations of which bacteria and fungi are capable is extensive, a few general principles hold. First, microbes often mediate biotransformations that are energetically favorable. Such reactions result in a net decrease in the Gibbs free energy of the chemical system, and the microbes harvest some of the released energy for their own use. There are important exceptions, such as photosynthesis (in which a cell will use energy obtained elsewhere to force an energetically unfavorable chemical reaction to occur) and *cometabolism* (in which a cell has enzymes that transform a chemical even though the transformation yields no energy to the cell). Often, however, the biodegradation potential of a given compound is related to the free energy changes that can be accomplished by its reaction with other chemicals simultaneously present in the same environment. For example, the energy available from a molecule of sugar is large if oxygen is present, but is much smaller in a system devoid of oxygen.

Second, microbial chemical transformations are accomplished by means of *enzymes,* proteins that act as *catalysts.* Catalysts bind with reactants and hold them in such an orientation that they more readily react. The products of the reaction are then released, leaving the catalyst ready to facilitate another transformation. (It is possible for an enzyme to be destroyed if a chemical mimics the proper substrate sufficiently to bind, but fails to react and subsequently release from the enzyme.) Because each enzyme is produced in response to a section of the genetic code (DNA) in the organism and many enzymes are extremely specific, it is possible that some strains of a species of bacteria may accomplish a certain chemical transformation while other individuals cannot. By using modern techniques of molecular biology, scientists can insert specific biotransformation capabilities into bacteria by means of genetic transfer. This procedure is easiest if the genetic material is associated with *plasmids,* which are small circular molecules of DNA that can exist independently within a bacterial cell.

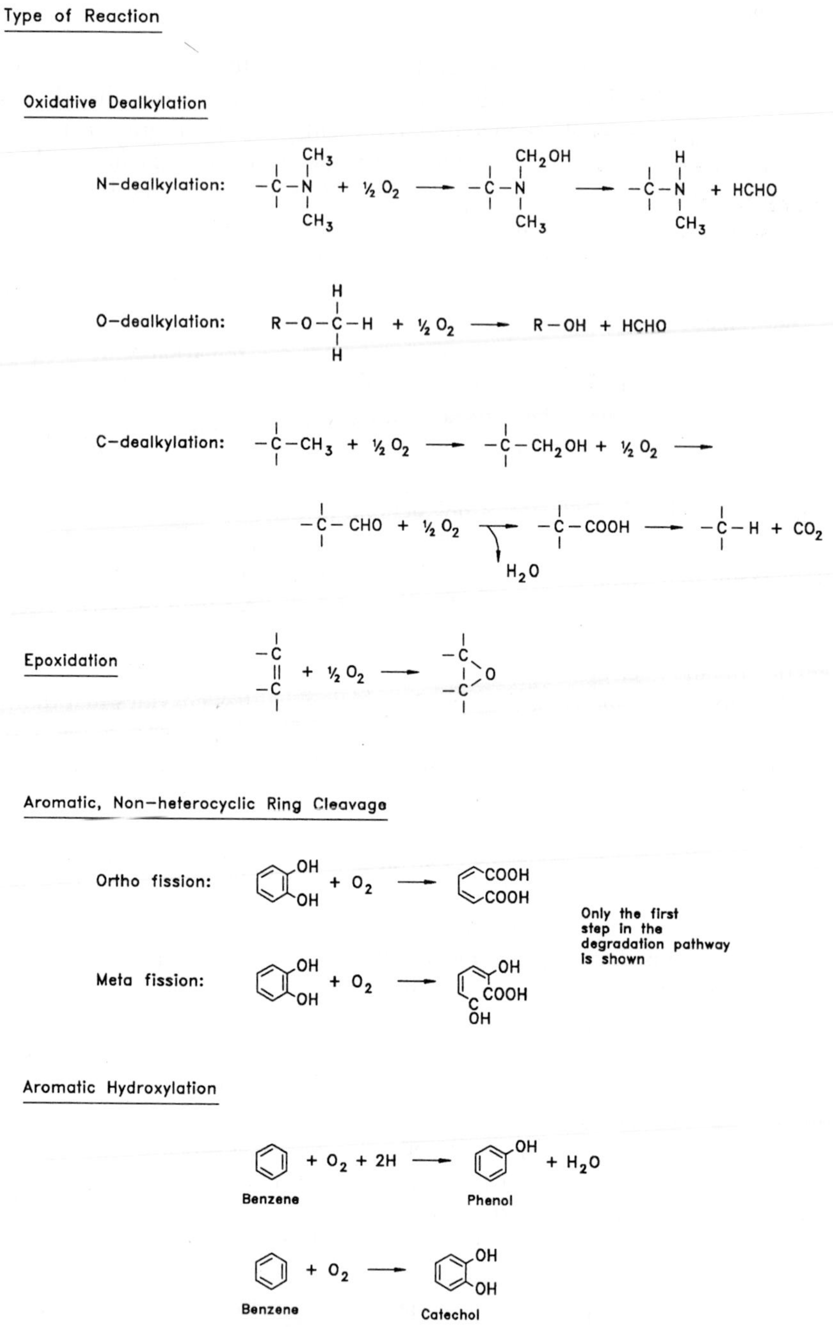

Type of Reaction

Oxidative Dealkylation

N–dealkylation: $-\overset{|}{\underset{|}{C}}-N(CH_3)_2 + \frac{1}{2}O_2 \longrightarrow -\overset{|}{\underset{|}{C}}-N(CH_2OH)(CH_3) \longrightarrow -\overset{|}{\underset{|}{C}}-NH(CH_3) + HCHO$

O–dealkylation: $R-O-CH_3 + \frac{1}{2}O_2 \longrightarrow R-OH + HCHO$

C–dealkylation: $-\overset{|}{\underset{|}{C}}-CH_3 + \frac{1}{2}O_2 \longrightarrow -\overset{|}{\underset{|}{C}}-CH_2OH + \frac{1}{2}O_2 \longrightarrow$

$-\overset{|}{\underset{|}{C}}-CHO + \frac{1}{2}O_2 \xrightarrow{-H_2O} -\overset{|}{\underset{|}{C}}-COOH \longrightarrow -\overset{|}{\underset{|}{C}}-H + CO_2$

Epoxidation

$>C=C< + \frac{1}{2}O_2 \longrightarrow >C(-O-)C<$

Aromatic, Non–heterocyclic Ring Cleavage

Ortho fission: catechol $+ O_2 \longrightarrow$ (ring-opened) COOH, COOH

Meta fission: catechol $+ O_2 \longrightarrow$ (ring-opened) OH, COOH, C(=O)H

Only the first step in the degradation pathway is shown

Aromatic Hydroxylation

$\text{Benzene} + O_2 + 2H \longrightarrow \text{Phenol} + H_2O$

Benzene — Phenol

$\text{Benzene} + O_2 \longrightarrow \text{Catechol}$

Benzene — Catechol

FIGURE 2-24 Several examples of organic compounds biodegraded by *aerobic* microorganisms, and their associated biodegradation reactions. Each of the reactions shown involves the *oxidation* of the organic compound.

2.6.1 Aerobic Biodegradation of Organic Compounds

Although BOD is an aggregate measure of the concentration of biologically degradable material in water, BOD conveys no information about the identities of specific organic compounds or their individual degradation rates. One factor that partially determines the degree to which a particular chemical is readily biodegraded is the extent to which it is a readily usable source of energy for microbes. Most organic pollutants contain carbon in a more reduced state than the (+IV) oxidation state found in carbon dioxide, and oxidation of the organic pollutants to carbon dioxide is often a viable means of aerobic biodegradation. In an aerobic environment, the most energy is usually available from oxidation of the most highly reduced carbon atoms.

Given their enormous variety, biochemical diversity, and rapid growth rates, microbes can oxidize many anthropogenic chemicals, such as hydrocarbon fuels and solvents, as well as the detrital organic material produced by ecosystems. Low-molecular-weight and soluble organic compounds such as alcohols and organic acids are utilized particularly rapidly, perhaps because these classes of compounds also occur naturally in the environment and microorganisms have evolved to degrade them efficiently. The rate of microbial oxidation generally is lower for compounds of high molecular weight, compounds having low water solubilities, and compounds that possess aromatic rings, a large amount of branching, and/or halogen atoms (chlorine, fluorine, bromine, and iodine) in their chemical structure. Figure 2-24 shows typical compound types commonly oxidized by aerobic microorganisms.

Oxidation also can be promoted by nonbiological processes in the environment. For example, fire brings about the rapid oxidation of chemicals, at elevated temperature, using oxygen from the atmosphere as the electron acceptor. The high temperature obviates the need for the catalytic action provided by organisms in aerobic biodegradation. Although obviously not an issue in most surface waters, fire can oxidize large quantities of organic material in wetlands during times of severe drought. Oxidation also can be promoted by light, as discussed in Section 2.7.1. A few chemicals can be oxidized spontaneously and abiotically: one example is the oxidation of soluble reduced iron (Fe^{2+}) to Fe^{3+} by dissolved oxygen in water at room temperature, a process that is common in iron-rich well waters and is responsible for the brown staining of porcelain and clothes. As another example, chlorophenols ($C_6OHCl_xH_{5-x}$) have been found to be abiotically oxidized at the surface of manganese oxide (MnO_2) particles in water (Ulrich and Stone, 1989).

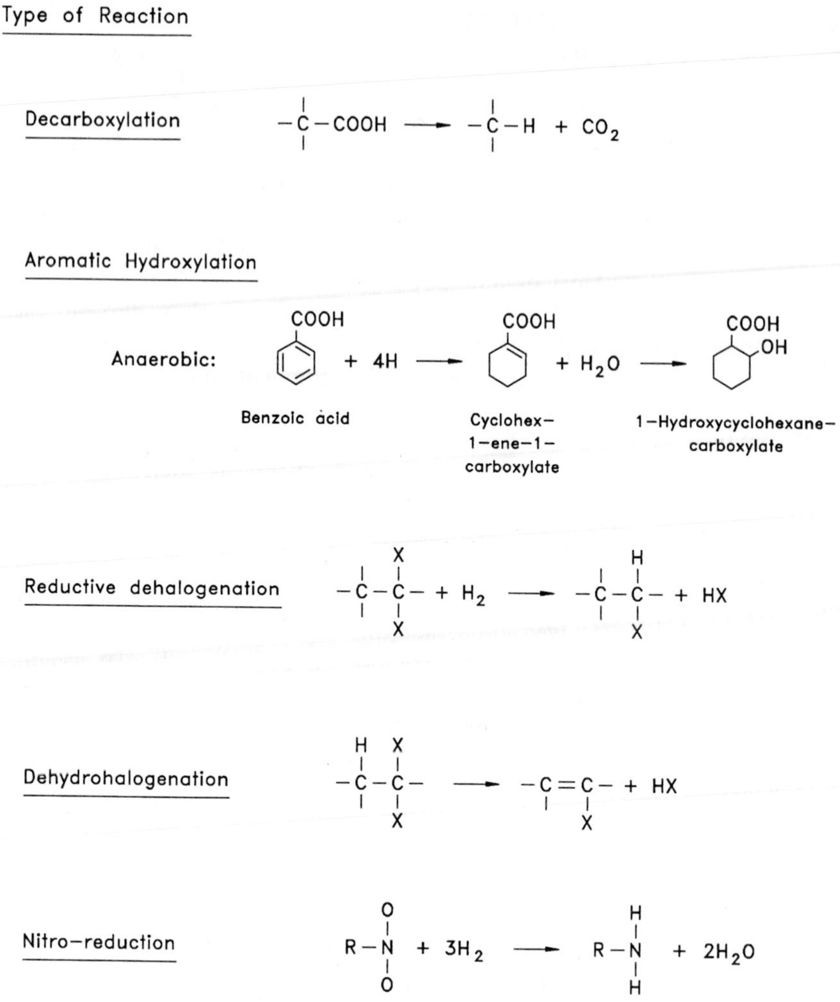

Type of Reaction

Decarboxylation

$$-\overset{|}{\underset{|}{C}}-COOH \longrightarrow -\overset{|}{\underset{|}{C}}-H + CO_2$$

Aromatic Hydroxylation

Anaerobic:

$$\text{Benzoic acid} + 4H \longrightarrow \text{Cyclohex-1-ene-1-carboxylate} + H_2O \longrightarrow \text{1-Hydroxycyclohexane-carboxylate}$$

Benzoic acid; Cyclohex-1-ene-1-carboxylate; 1-Hydroxycyclohexane-carboxylate

Reductive dehalogenation

$$-\overset{|}{\underset{|}{C}}-\overset{X}{\underset{X}{C}}- + H_2 \longrightarrow -\overset{|}{\underset{|}{C}}-\overset{H}{\underset{X}{C}}- + HX$$

Dehydrohalogenation

$$-\overset{H}{\underset{|}{C}}-\overset{X}{\underset{X}{C}}- \longrightarrow -\underset{|}{C}=\underset{X}{C}- + HX$$

Nitro-reduction

$$R-\overset{O}{\underset{O}{N}} + 3H_2 \longrightarrow R-\overset{H}{\underset{H}{N}} + 2H_2O$$

FIGURE 2-25 Several examples of organic compounds biodegraded by *anaerobic* microorganisms and their associated biodegradation reactions occurring in reducing, anaerobic environments. Note that most of these reactions involve the *reduction* of the organic compound.

2.6.2 Anaerobic Biodegradation of Organic Compounds

Because the free energy released in respiration decreases as oxygen is depleted and the microbial community shifts to the use of less favorable oxidants such as $Fe(OH)_3$ and SO_4^{2-}, the tendency for oxidative biodegradation tends to decrease as the ecological redox sequence proceeds and conditions become increasingly reducing. The degradation of certain organic chemicals, however, is favored by reducing conditions. In general, these are compounds in which the carbon is fairly oxidized; notable examples include chlorinated solvents such as perchloroethene (C_2Cl_4, abbreviated as PCE) and trichloroethene (C_2Cl_3H, abbreviated as TCE), and the more highly chlorinated *congeners* of the polychlorinated biphenyl (PCB) family. [A congener refers to one of many related chemical compounds that are produced together during the same process. In the case of PCBs, each congener is a biphenyl molecule containing a certain number and arrangement of added chlorine atoms. There are many commerically marketed products (e.g., Aroclor) containing varying mixtures of PCB congeners.] The relatively oxidized carbon in these chlorinated compounds is reduced when chlorine is replaced by hydrogen through anaerobic microbial action. For example, when TCE is partially dechlorinated to *trans*-1,2-dichloroethene, *cis*-1,2-dichloroethene, or 1,1-dichloroethene (all having the formula $C_2Cl_2H_2$, abbreviated DCE), the carbon is reduced from the (+I) oxidation state to the (0) oxidation state:

$$\tfrac{1}{2}CH_2O + \tfrac{1}{2}H_2O + C_2Cl_3H \longrightarrow C_2Cl_2H_2 + \tfrac{1}{2}CO_2 + Cl^- + H^+. \qquad [2\text{-}70a]$$

The dichloroethene *isomers* (compounds with the same formula but different structures) can be further degraded under reducing conditions into chloroethene (vinyl chloride), and the oxidation state of the carbon is thereby reduced to (−I):

$$\tfrac{1}{2}CH_2O + \tfrac{1}{2}H_2O + C_2H_2Cl_2 \longrightarrow C_2H_3Cl + \tfrac{1}{2}CO_2 + Cl^- + H^+. \qquad [2\text{-}70b]$$

Reductions such as these usually do not completely mineralize a pollutant. Their greatest significance lies in the removal of chlorine or other halogen atoms, rendering the transformed chemical more subject to oxidation if it is ultimately transported back into an aerobic environment. Figure 2-25 shows some types of anaerobically degraded compounds.

2.6.3 Modeling Biodegradation

Kinetics of Microbial Transformations of Chemicals

A simple biodegradation model is one in which microorganisms are in contact with water containing a dissolved organic chemical that serves as the energy substrate. Because chemical uptake into a cell is followed by enzymatic transformation, biodegradation and uptake rates are equivalent in this model. It is

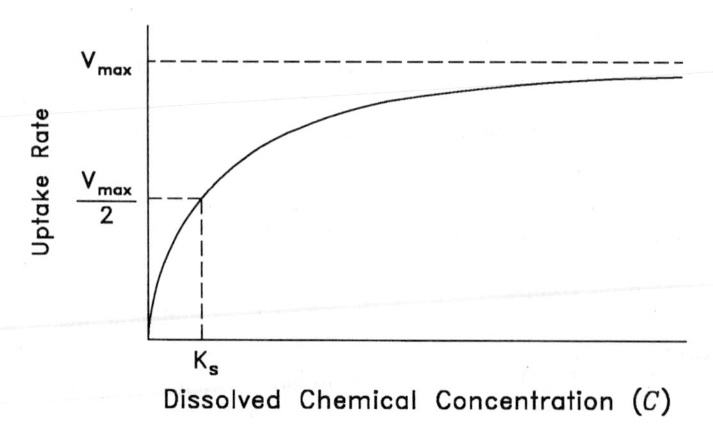

FIGURE 2-26 Microbial uptake rate of a chemical, according to Michaelis–Menten kinetics [see Eq. [2-71a]]. The rate cannot exceed V_{max} no matter how high the chemical concentration becomes. K_s, the half-saturation constant, is the chemical concentration at which uptake equals half of V_{max}. At low concentrations, uptake is nearly proportional to concentration and may be approximated as a first-order process.

customary to model chemical uptake by a cell according to *Michaelis–Menten enzyme kinetics,*

$$V = V_{max} \cdot \frac{C}{C + K_s}, \qquad [2\text{-}71a]$$

where V is the rate of chemical uptake per cell [M/(cell · T)], V_{max} is the maximum possible chemical uptake rate per cell [M/(cell · T)], C is the concentration of dissolved chemical [M/L^3], and K_s is the *half-saturation constant* [M/L^3]. Equation [2-71a] is often written with an S (for substrate concentration) in place of C, the symbol used in this text to represent chemical concentration. The mathematical form of Eq. [2-71a] also appears in other applications; for example, it shows up as the Langmuir isotherm, which is used similarly to the Freundlich isotherm to model certain nonlinear sorption equilibria, as discussed in Section 1.8.3. When K_s equals C, the uptake (and hence transformation) occurs at one-half of its maximum possible rate, V_{max}. When the rate of chemical uptake is plotted against the dissolved chemical concentration, the curve shown in Fig. 2-26 is obtained. Note that V approaches zero when there is no chemical present and reaches a plateau at V_{max} for high concentrations. When K_s is much greater than C (i.e., at low concentrations), the rate of uptake becomes nearly proportional to the chemical concentration, thereby approximating first-order kinetics,

$$V \approx \left[\frac{V_{max}}{K_s}\right] \cdot C, \qquad [2\text{-}71b]$$

where V is the rate of chemical uptake per cell [M/(cell · T)], V_{max} is the maximum possible chemical uptake rate per cell [M/(cell · T)], C is the concentration of dissolved chemical [M/L^3], and K_s is the half-saturation constant [M/L^3].

When C is much greater than K_s, V approaches independence of C and the rate approximates zero-order kinetics (i.e., there is no dependence on the chemical concentration),

$$V \approx V_{max}, \qquad [2\text{-}71c]$$

where V is the rate of chemical uptake per cell [M/(cell · T)] and V_{max} is the maximum possible chemical uptake rate per cell [M/(cell · T)].

The rate of uptake of the chemical per unit of water is proportional to both

TABLE 2-7 Aerobic Biodegradation Rates Observed in Incubations of River Water Samples[a]

Compound	Rate constant (per day)
Anthracene	0.007–0.055[b]
Atrazine (*N*-phosphorylated)	0.22
Benz[*a*]anthracene	None observed
Benzene	0.11
Benzo[*a*]pyrene	None observed
Chlorobenzene	0.0045
Glucose	0.24
Mirex	None observed
Nitrilotriacetate (NTA)	0.05–0.23[c]
Parathion	<0.00016
Phenol	0.079
2,4,5-T	0.001
1,4,5-Trichlorophenoxyacetic acid	0.0005

[a] Adapted from Lyman *et al.* (1990).
[b] First value is mean for days 0–15; second is for days 20–65.
[c] Dissolved concentrations ranging from 0.2 mg/liter to saturation.

TABLE 2-8 Anaerobic Biodegradation Rates Observed in Soil[a]

Compound	Rate constant (per day)
Carbofuran	0.026
DDT	0.0035
Endrin	0.03
Lindane	0.0046[b]
Pentachlorophenol	0.07[b]
Trifluralin	0.025
Trichloroethene	0.009[c]
1,1-Dichloroethene	0.0063[d]

[a] Rates from die-away studies in soil presented in Lyman *et al.* (1990), unless otherwise noted.
[b] Rates from $^{14}CO_2$ evolution studies in soil presented in Lyman *et al.* (1990).
[c] Rate from die-away studies in soil in Bouwer and McCarty (1983).
[d] Rate from die-away studies in soil in Barrio-Lage *et al.* (1986).

the chemical uptake rate and the cell density,

$$\frac{dC}{dt} = V \cdot X, \tag{2-72}$$

where dC/dt is the change in chemical concentration with time $[M/(L^3 \cdot T)]$, V is the rate of chemical uptake per cell $[M/(\text{cell} \cdot T)]$, and X is the cell density $[\text{cells}/L^3]$.

Thus, the simplest quantitative model for biodegradation in a surface water is one in which the dissolved organic chemical concentration is significantly less than K_s, such that Eq. [2-71b] applies, and cell density X is assumed constant. The change in chemical concentration with time, Eq. [2-72], is then proportional to the chemical concentration, and first-order kinetics may be applied. Many rate constants have been published for surface waters (Table 2-7). Note that V_{max}, K_s, and X are not individually measured; their effects are lumped into a single empirical rate constant. Because degradation rates are highly dependent on the nature and abundance of the microbial population present in the surface water at the time the experiment was conducted, these values are in no way absolute, but instead provide an approximate indication of some typical rates of aerobic biodegradation in surface waters. Table 2-8 provides some very approximate rates of anaerobic degradation observed for certain specific compounds in soil, and may be applicable to degradation in the sediments of some surface waters.

EXAMPLE 2-17

Spilled benzene (C_6H_6) dissolves into a river flowing at an average velocity of 0.3 m/sec. Will biodegradation significantly decrease the concentration of benzene in the river over a 20-mi reach?

The travel time of the river is

$$\tau = 20 \text{ mi} \cdot \frac{1609 \text{ m}}{1 \text{ mi}} \cdot \frac{\text{sec}}{0.3 \text{ m}} \cdot \frac{1 \text{ hr}}{3600 \text{ sec}} \approx 30 \text{ hr}.$$

From Table 2-7, an approximate aerobic degradation rate for benzene is 0.11/day. By assuming first-order decay,

$$C/C_0 = e^{-kt}.$$

In 30 hr, which is approximately 1.2 days,

$$C/C_0 = e^{-(0.11/\text{day})(1.2\text{ day})}$$

$$C/C_0 = 0.87.$$

Therefore, over 10% of the benzene may degrade in a 20-mi. reach. Due to the large uncertainty in the aerobic degradation rate estimate, this calculation provides only a crude approximation of the amount of benzene remaining. It is sufficient, however, to indicate that biodegradation may be significant over this reach of river.

EXAMPLE 2-18

A 1 *M* solution of DCE is accidentally spilled into a stratified lake whose bottom waters are anaerobic. Because the mixture is denser than water, it will tend to sink. Assume the DCE becomes dissolved in the pore waters of the bottom sediments. After 2 months, what compounds would you expect to find in the bottom sediments, and in roughly what ratio?

Equation [2-70b] shows that the anaerobic degradation pathway for DCE is conversion to vinyl chloride. From Table 2-8, a rough estimate of the anaerobic biodegradation rate for DCE in soil is 0.0063/day. Vinyl chloride is very resistant to anaerobic degradation; therefore, its net accumulation is essentially equal to the amount of DCE reduced. By assuming first-order kinetics and 30 days in a month,

$$C = C_0 e^{-kt}$$

$$C/C_0 = e^{-(0.0063/\text{day})(60\text{ days})}$$

$$C/C_0 = 0.7.$$

A very approximate estimate is that 70% of DCE will remain after 2 months; 30% will have been converted to vinyl chloride.

In the case in which X is constant but C is not significantly less than K_s, Eq. [2-71a] must be used, as shown in the following example.

EXAMPLE 2-19

A *metalimnion* (the transitional layer between the epilimnion and the hypolimnion) contains *methanotrophic* bacteria (bacteria that aerobically oxidize methane) at a cell concentration of 10^5 cells per milliliter. There is adequate oxygen available and the cells have a V_{max} for CH_4 of 10^{-19} mol/(cell · sec) and a half saturation constant, K_s, of 10^{-4} mol/liter. At what rate, R_{CH_4}, is methane degraded if it is present at a concentration of $1.5 \times 10^{-5}\ M$?

By using Eq. [2-71a]:

$$V = 10^{-19}\,\frac{\text{mol}}{\text{cell}\cdot\text{sec}}\cdot\frac{1.5\times 10^{-5}\,M}{1.5\times 10^{-5}\,M + 10^{-4}\,M} = 0.13\times 10^{-19}\,\frac{\text{mol}}{\text{cell}\cdot\text{sec}}.$$

The rate at which CH_4 is degraded is a function of both the CH_4 uptake rate and the population density:

$$R_{CH_4} = \left(0.13\times 10^{-19}\,\frac{\text{mol}}{\text{cell}\cdot\text{sec}}\right)\cdot\left(\frac{10^5\text{ cells}}{\text{ml}}\right)$$

$$= 1.3\times 10^{-15}\,\frac{\text{mol}}{\text{ml}\cdot\text{sec}} \quad\text{or}\quad 1.3\times 10^{-12}\,\frac{\text{mol}}{\text{liter}\cdot\text{sec}}.$$

The model illustrated in Example 2-19 may also be applied to a flowing stream if the microorganisms are attached to the surfaces of the channel, have a relatively steady cell density, and are exposed to the full chemical concentration in the stream (Kim *et al.*, 1995; Cohen *et al.*, 1995). Microorganisms attached to solid surfaces form *biofilms,* as populations of attached microbes accumulate on top of one another, building up a layer of microbes and extracellular "glue" made of polysaccharide material. Within biofilms, X corresponds to the number of attached microorganisms divided by the volume of the biofilm. A biofilm may also be called a bacterial *slime*, for reasons that are evident to anyone who has tried to cross a stream by walking on the smooth submerged rocks; the slipperiness is due to the biological layers that have accumulated on the rocks. It appears to be advantageous for microbes to remain attached in one place in biofilms and harvest nutrients that are transported to them primarily by advection, instead of being free floating in the water and forced to rely solely on Fickian transport to supply nutrients. Numerous larger scale examples of this attachment strategy also occur; for example, some stream-dwelling insect larvae, such as caddis fly larvae, anchor themselves to rocks and trap small particles of organic detritus for food from the flowing streamwater. Biofilms will be considered in more detail in Chapter 3.

Monod Growth Kinetics

In many situations, the use of a constant cell density in a biodegradation model is not justified. For example, a spill of an easily biodegraded chemical into a surface water may greatly stimulate bacterial growth. *Monod kinetics* allow a model to account for a changing population of microbial cells when cell growth is stimulated by energy released from the biotransformation process. Monod growth kinetics can be inferred from Michaelis–Menten kinetics, together with the assumption that a certain number of new cells grow per unit mass of chemical transformed. The ratio of the number of new cells to the mass of chemical taken up (transformed) is called the *cell yield,* y [cells/M]. The *specific growth rate,* μ [T^{-1}], is the product of the chemical uptake rate, V [M/(cell · T)], and the cell yield. The maximum specific growth rate, μ_{max}, is equal to the product ($V_{max} \cdot y$). The specific growth rate corresponding to a concentration of chemical, C, is therefore,

$$\mu = \mu_{max} \cdot \frac{C}{C + K_s}, \qquad [2\text{-}73]$$

where μ is the specific growth rate [T^{-1}], μ_{max} is the maximum specific growth rate [T^{-1}], C is the concentration of a chemical [M/L^3], and K_s is the half-saturation constant [M/L^3]. Note that Eq. [2-73], which describes Monod growth kinetics, has the same mathematical form as Eq. [2-71a], which describes Michaelis–Menten enzyme kinetics applied to chemical uptake.

EXAMPLE 2-20

A test tube is filled with water containing bacteria and a chemical that supports cell growth. The chemical concentration is 3×10^{-5} mol/liter and the bacteria have a cell yield of 10^{12} cells/mol. What will be the specific growth rate μ if V_{max} is 10^{-15} mol/(cell · min) and K_s is 5×10^{-5} mol/liter?

The specific growth rate is equal to the product of the chemical uptake rate and the cell yield:

$$\mu = V \cdot y.$$

First, estimate V from Eq. [2-71a]:

$$V = 10^{-15} \frac{\text{mol}}{\text{cell} \cdot \text{min}} \cdot \frac{3 \times 10^{-5}\, M}{(3 \times 10^{-5} + 5 \times 10^{-5})M} = 3.8 \times 10^{-16} \frac{\text{mol}}{\text{cell} \cdot \text{min}}.$$

Then estimate μ as the product of the chemical uptake rate and the cell yield:

$$\mu = 3.8 \times 10^{-16} \frac{\text{mol}}{\text{cell} \cdot \text{min}} \cdot 10^{12} \frac{\text{cell}}{\text{mol}} = 3.8 \times 10^{-4}/\text{min}.$$

Monod kinetics can be used to formulate models that account for both chemical transformations and changes in the microbial population. One such model applies to the *batch culture,* a volume into which a biodegradable, growth-supporting chemical and an initial population of suitable bacteria are introduced. The batch culture could correspond to a chemical spill into a pond, to an industrial process, or even to a vat of brewing beer. Biotransformation is strongly influenced by the increase in cell density, which typically follows three stages (Fig. 2-27). The first stage is the *lag phase,* an interval during which cell growth is much less than predicted by Monod kinetics because the cells are becoming acclimated to the new environment. In the *exponential phase* (assuming C is much greater than K_s), cell density increases according to the equation,

$$X = X_0 e^{\mu t}, \qquad [2\text{-}74]$$

where X is the cell density [cells/L^3] at some time t, X_0 is the initial cell density [cells/L^3], and μ is the specific growth rate [T^{-1}]. If Eq. [2-74] is

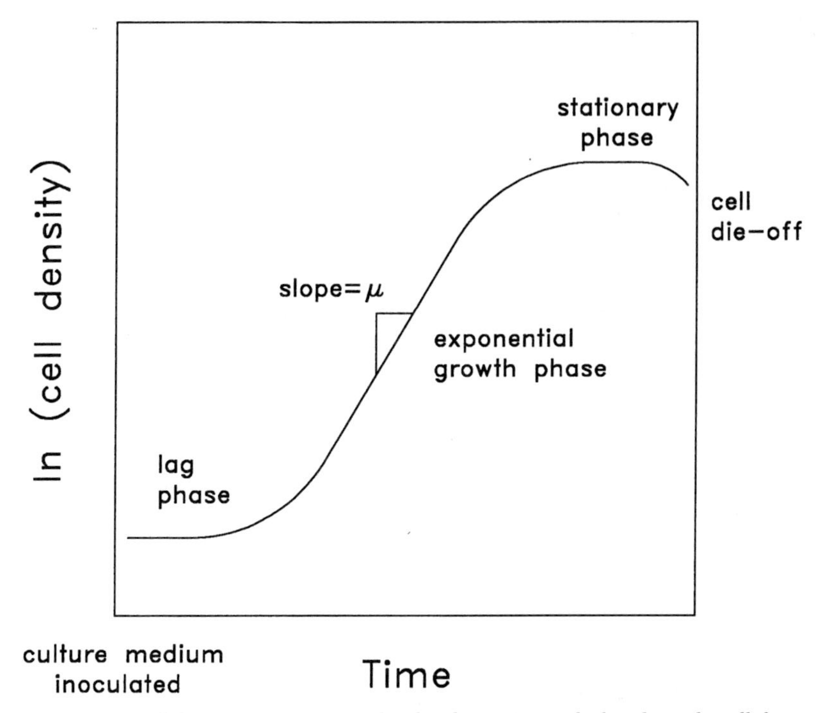

FIGURE 2-27 Cell density versus time in a batch culture. During the lag phase, the cells become acclimated. During the exponential phase, the number of cells increases exponentially as described by Monod growth kinetics, see Eq. [2-73]. During the stationary phase, cell density is constant for some time before the culture declines due to substrate depletion, waste accumulation, and/or excessive cell density.

substituted into Eq. [2-72], it is evident that the chemical degradation rate can increase very rapidly during this phase, with a very large fraction of biodegradation occurring near the end of this phase. As nutrients are depleted and/or excessive cell densities or waste accumulation suppress growth, the cells enter the third stage, a *stationary phase* in which cell density is constant for a period of time prior to decline of the culture.

A second type of culture described by Monod kinetics is the *continuous culture,* in which a chemical is constantly fed into a vessel and both microbial cells and the chemical are constantly lost from the vessel at a given rate. This culture is often called a *chemostat* when operated under steady-state conditions. Like the batch culture, a continuous culture may be a useful model of certain environmental systems, such as lakes receiving continuous discharges of pollutants. Continuous cultures are common in industrial processes as well, including many wastewater treatment processes, in which the large-scale, well-mixed culture vessel is often called a *continuously stirred tank reactor* (CSTR). In a chemostat, the growth rate μ of the cells must equal or exceed the reciprocal of the hydraulic residence time, or X will decline and the system will suffer *washout,* after which no further biodegradation will occur. For further discussion of batch and continuous cultures the reader is referred to Brock and Madigan (1991).

The models discussed here are all simplifications of the real world, in which numerous factors influence microbial populations, which in turn comprise a mixture of numerous microbial strains having a great range of growth and uptake parameters. However, even the most complex computer-based models that attempt to predict biodegradation rates are usually based on the uptake and growth expressions described here, and may be recognized as variations of batch, continuous, or biofilm models. (Biofilm models, although they are also applicable to rivers and streams, will be discussed further in Chapter 3.)

2.6.4 Bioconcentration and Bioaccumulation in Aquatic Organisms

Aquatic biota not only degrade pollutant chemicals but also may accumulate them. If aquatic organisms accumulate chemicals only from the water, the process is called *bioconcentration;* if they accumulate chemicals from both water and food, the process is called *bioaccumulation.* In surface waters, bioconcentration and bioaccumulation are of particular concern in relatively large aquatic organisms that may be ingested by humans or by other nonaquatic organisms, such as birds or bears. A current example of the detrimental effects of bioaccumulation is the occurrence of mercury poisoning among natives of the Hudson Bay area of Canada due to hydroelectric development. This poisoning is a result of the release of mercury from flooded soils, its transformation into methylmercury (CH_3Hg^+), and its subsequent bioaccumulation in fish that are consumed by people. The near extinction of many birds of prey, which resulted from reproductive failure caused by the accumulation of DDT pesticide residues from fish, is another, historical example of the effects of bioaccumulation.

In surface waters, the *bioconcentration factor* (BCF) is the ratio of a chemical's concentration in an organism to the chemical's aqueous concentration. BCF is often expressed in units of liter per kilogram (i.e., the ratio of mg of chemical per kg of organism to mg of chemical per liter of water). The BCF may be merely an observed ratio, or it may be the prediction of a *partitioning model.*

Partitioning models are founded on the assumptions that pollutant chemicals partition, in a more or less passive fashion, between water and aquatic organisms and that chemical equilibrium exists between the organisms and the aquatic environment. Such assumptions are most justifiable in the case of hydrophobic chemicals that are more rapidly exchanged between an organism and the water than they are excreted or metabolized by the organism. An organism such as a fish is in effect modeled as a bag of oil and tissue water; the chemical partitions between the bag's contents and the surrounding water according to its hydrophobicity (as reflected by its K_{ow} or the reciprocal of its water solubility) and the percentage of fish that is oil and fat (the lipid content). Exchange is facilitated by a large surface to volume ratio (as is found in smaller organisms) and by organs that facilitate exchange between the organism and water, such as the gills of fish. Several empirical formulae for establishing BCF on the basis of partitioning are shown in Table 2-9 (see also Hawker and Connell, 1989). Although partitioning models are very simple, some are reasonably successful in appropriate circumstances (Fig. 2-28).

TABLE 2-9 Regression Equations for Estimating BCF for Fish

Equation[a]	N[b]	r^2[c]	Species used
$\log BCF = 0.76 \log K_{ow} - 0.23$	84	0.823	Fathead minnow Bluegill sunfish Rainbow trout Mosquitofish
$\log BCF = \log K_{ow} - 1.32$[d]	44	0.95	Various
$\log BCF = 2.791 - 0.564 \log S$ (S in ppm)	36	0.49	Brook trout Rainbow trout Bluegill sunfish Fathead minnow Carp
$\log BCF = 3.41 - 0.508 \log S$[e] ($S$ in μM)	7	0.93	Rainbow trout
$\log BCF = 1.119 \log K_{oc} - 1.579$	13	0.757	Various

[a] Adapted from Lyman *et al.* (1990) unless otherwise noted. BCF, bioconcentration factor; K_{ow}, octanol–water partition coefficient; S, water solubility; K_{oc}, organic carbon–water partition coefficient.

[b] N = number of chemicals used to obtain regression equation.

[c] r^2 = correlation coefficient for regression equation.

[d] Mackay (1982).

[e] Chiou *et al.* (1977).

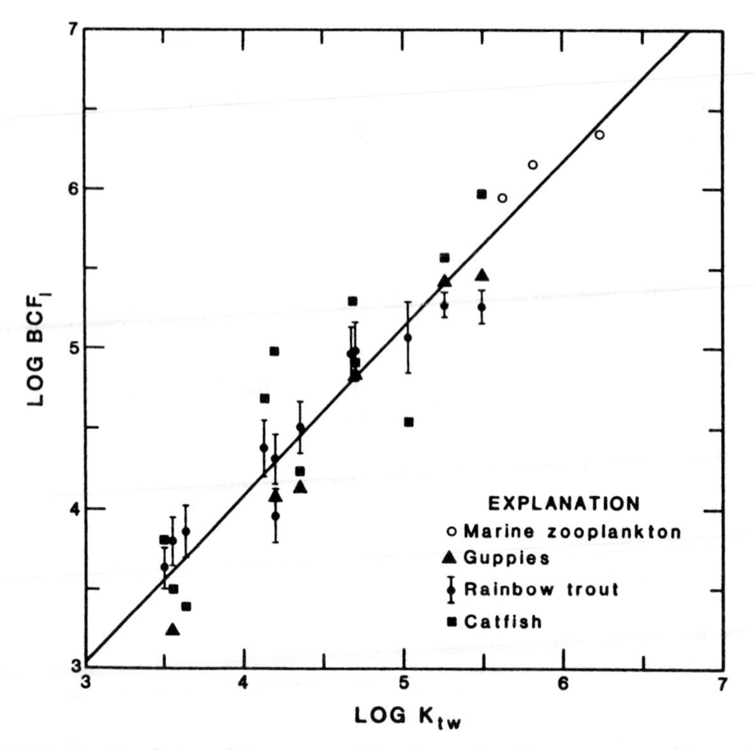

FIGURE 2-28 Correlation of the measured, lipid-normalized bioconcentration factor (BCF) with the triolene–water partition coefficient (K_{tw}) for a suite of 16 nonpolar compounds. (The partition coefficient K_{tw} is similar to K_{ow}.) Note that the overall correlation is very strong, but any single prediction of BCF from K_{tw} may be in error by half a log unit (a factor of three) [Smith *et al.* (1988). Reproduced with permission of Springer-Verlag New York, Inc.].

It is important to note that partitioning models *do not imply* that an increase in chemical concentration occurs as one moves up a food chain; in fact, partitioning models predict that the concentration of a chemical in an organism is not dependent on what the organism eats. Partitioning models are not appropriate for terrestrial ecosystems; for those ecosystems, models for chemical accumulation must be based on the food chain (i.e., bioaccumulation).

Bioaccumulation can be estimated by a *kinetic model.* In kinetic models (sometimes called *physiological models* or *physiologically based pharmacokinetic models*), consideration is given to the dynamics of ingestion, internal transport, storage, metabolic transformation, and excretion processes that occur in each type of organism for each type of chemical. In kinetic models,

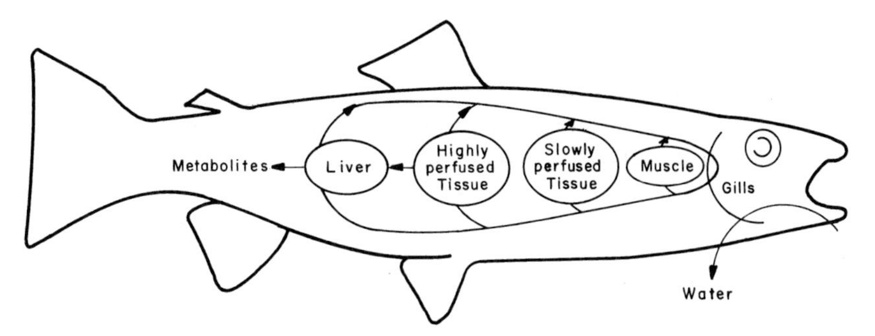

FIGURE 2-29 Schematic representation of a physiologically based kinetic model for bioaccumulation of a chemical that is absorbed through the gills, transported by blood flow, stored in various body tissues, and metabolized by the liver. Such a model requires much more detailed information on the fish than does a partitioning model; however, it may be necessary to use this more complex approach for chemicals that are metabolized or excreted by the fish more rapidly than they are exchanged with the water [adapted from Barron (1990). Reprinted with permission. © 1990 American Chemical Society].

organisms in a surface water ecosystem may ingest food containing a particular chemical and absorb the chemical from the water. The ingested and absorbed chemical is subject both to elimination through excretion and to metabolic transformation to other chemicals (often, in higher organisms, the liver carries out oxidative transformations catalyzed by cytochrome P450 enzymes; Fig. 2-29). Kinetic models can be reasonably accurate representations of what actually occurs in an organism, even a large and complex one. Kinetic models do, however, require considerable information on the uptake, transformation, storage, and excretion processes. The amount of data required to parameterize kinetic models becomes enormous when a whole ecosystem is considered, because an understanding of chemical accumulation by organisms high on the food chain requires understanding each lower step in the food chain, all the way down to the lowest trophic level at which chemicals become incorporated into organisms. For a detailed description of a bioaccumulation model for nonmetabolized organic chemicals, the reader is referred to Barber *et al.*, 1991.

EXAMPLE 2-21

A catfish metabolizes and/or excretes 2,4′,5-trichlorinated biphenyl (a PCB congener) with a hypothetical first-order rate constant of 0.021/day. How long will it take for fish from a contaminated stream, on being placed in clean water, to undergo *depuration* (cleansing of pollutants) if the levels of the biphenyl exceed safe levels by a factor of three (i.e., how long will it take for them to become "safe")?

First-order kinetics are suggested by the problem statement:

$$C = C_0 e^{-kt}$$

$$\tfrac{1}{3} = e^{-(0.021/\text{day})(t\ \text{days})}$$

$$-1.1 = -0.021\ t$$

$$t = 52\ \text{days}.$$

Note also that knowledge of the nature of the compound into which the biphenyl is being metabolized and the nature of other PCB congeners present is necessary before concluding that the fish are "clean" and safe for consumption!

2.7 ABIOTIC CHEMICAL TRANSFORMATIONS

2.7.1 Degradation of Chemicals by Light

Nature of Light

Light is a form of electromagnetic radiation, in which energy is transmitted through space by the interaction of electric and magnetic fields. Light can be described both in terms of particles and in terms of waves; different effects are better described by one or the other model. In surface waters, light penetrates to depths of at least several meters in all but the most darkly colored or turbid waters; numerous light-driven reactions can lead to the *photodegradation* of organic chemicals.

Light is characterized in part by its *wavelength* distribution. Visible light consists of radiation having wavelengths (λ) ranging from approximately 400 to 700 nm. The distribution of wavelengths in extraterrestrial solar radiation occupies a much wider range, from approximately 100 nm to greater than 3000 nm (Fig. 2-30). Some of these wavelengths are absorbed strongly as the radiation passes through Earth's atmosphere (as discussed more fully in Section 4.7); the amount of energy remaining in wavelengths less than 290 nm is small by the time the light reaches Earth's surface. All electromagnetic radiation travels at a constant speed, c, of 3.0×10^8 m/sec in free space. Thus,

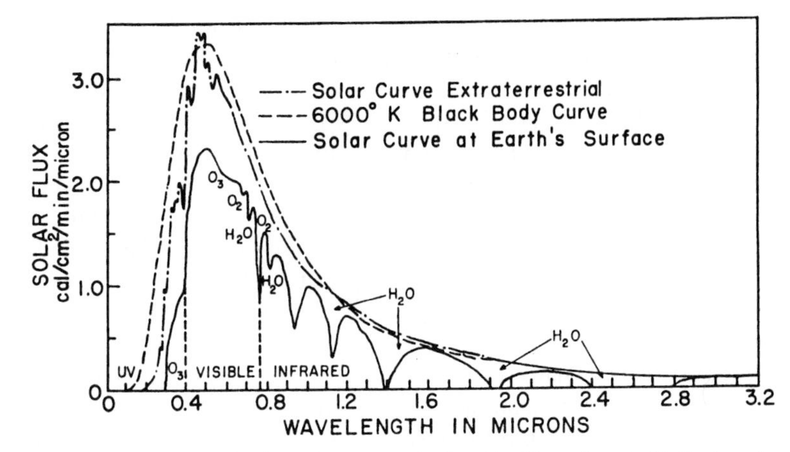

FIGURE 2-30 Spectrum of solar radiation received at Earth's surface (solid line) and at the outer edge of the atmosphere (dashed and dotted line). Radiation from the Sun approximates radiation from a surface having an emissivity of one (blackbody) at 6000 K (dashed line); see also Section 4.7.1. Note that significant amounts of radiation are absorbed by the atmosphere, and that this absorption is concentrated in discrete regions of the spectrum, a result of wavelength-specific absorption by gases and vapors. Absorption by water vapor, oxygen (O_2), and ozone (O_3) are shown here. Note also the large amount of energy that lies outside the visible range [Gates (1962)].

the *frequency*, ν, of light of a given wavelength is

$$\nu = c/\lambda = 3 \times 10^8/\lambda \text{ sec}^{-1}, \tag{2-75}$$

when λ is expressed in meters. When applied to electromagnetic radiation, the unit sec^{-1} is called the *hertz* (Hz).

Other aspects of the effects of light are best described by invoking its particle nature. Each quantum of light, called a *photon*, possesses an energy equal to E,

$$E = h \cdot \nu, \tag{2-76}$$

where h is Plank's constant, 6.6×10^{-34} J · sec. Note that the shorter the wavelength becomes, the higher the energy per photon becomes. Each individual photon of light at the blue end of the visible spectrum (λ = 400 nm) has an energy of 5.0×10^{-19} J, while each photon at the red end of the visible spectrum (λ = 700 nm) has an energy of 2.8×10^{-19} J.

Light Distribution in Surface Waters

The actual distribution of light in water is complex, affected not only by the light's wavelength composition and *intensity* (intensity can be thought of as the number of photons hitting a unit surface per unit time) but by the angle at which light enters the water (*angle of incidence*) and the optical properties of the water itself. Light intensity decreases with depth in water due to absorbance by the water and by substances in it. The extent of absorbance varies with the concentrations and types of substances dissolved in the water, such as dissolved humic substances, and with the concentration of suspended particles in the water [turbidity or *total suspended solids* (TSS)]. Humic substances include humic and fulvic acids, which are formed during the degradation of natural organic material (see also Section 2.4.1). Humic substances are of high molecular weight [on the order of 500 to 10,000 *daltons* (Da); a dalton is approximately the mass of a hydrogen atom], contain aromatic rings, and are at least somewhat resistant to further biodegradation. As an approximation, light intensity in a surface water body is typically modeled based on Beer's law. Beer's law, as applied to water, states that the absorbance (fraction of light absorbed) in traveling a path is equal to the product of chemical concentration, the chemical's molar absorptivity, and the path length. Absorptivity varies with wavelength; however, for practical purposes in surface waters where the mixture of absorbing chemicals often has a broad collective absorbance spectrum, an approximate overall *extinction coefficient* that is not wavelength dependent is often used. Light intensity is then modeled by

$$I = I_0 e^{-\eta z}, \qquad [2\text{-}77]$$

where I is the light intensity at a given depth z in the water expressed either as number of micromoles of photons per unit area per unit time [microeinsteins/($m^2 \cdot sec$)], or power per unit area (watts/m^2), I_0 is the light intensity at the water surface, and η is an experimentally measured extinction coefficient [L^{-1}].

In reality, the red and infrared wavelengths are preferentially absorbed, even in pure water, and the blue-violet–ultraviolet wavelengths are also selectively attenuated if the water contains appreciable concentrations of dissolved organic material. Thus, the midspectrum green and yellow wavelengths often penetrate most deeply into a surface water. In addition to neglecting the wavelength composition of light, Eq. [2-77] does not explicitly consider the angles at which the light travels through the water; this direction is not vertical for all of the photons. The greater attenuation that occurs with depth for light traveling nonvertical paths is accounted for in the empirical extinction coefficient.

The extinction coefficient η may be calculated from measurements of I made by a light meter. The Secchi disk is another time-honored device for estimating light penetration in a lake. In 1865, P. A. Secchi developed this method, which was subsequently used aboard the papal yacht, S. S. L'immacolata Concezione. The Secchi disk is about 20 cm in diameter, colored all white or white with two 90 degree sectors painted black. Suspended horizontally on a line, it is lowered into the water until it cannot be seen, and the depth of its disappearance is recorded. It is then raised and the depth of its reappearance recorded. The average of the two depths is an approximation of the depth above which 90 to 95% of the light initially entering the lake is absorbed.

EXAMPLE 2-22

What is the light intensity at 1-m depth in a lake, given an intensity of 3000 microeinsteins (μE)/($m^2 \cdot$ sec) just beneath the lake's surface and an extinction coefficient of 0.6/m? If an aquatic plant has a light *compensation point* (the light intensity at which respiration rate equals photosynthetic rate) of 150 μE/($m^2 \cdot$ sec), what is the maximum depth at which the plant may be expected to grow?

To estimate the light intensity at 1-m depth, use Eq. [2-77]:

$$I = \frac{3000\ \mu\mathrm{E}}{\mathrm{m}^2 \cdot \mathrm{sec}}\, e^{-(0.6/\mathrm{m} \cdot 1\ \mathrm{m})} = \frac{1650\ \mu\mathrm{E}}{\mathrm{m}^2 \cdot \mathrm{sec}}.$$

To calculate the maximum depth at which the plant can grow, use Eq. [2-77] and solve for z:

$$z = \frac{-1}{\eta} \cdot \ln\left(\frac{I}{I_0}\right) = \frac{-1}{0.6/\mathrm{m}} \cdot \ln\left(\frac{150\ \mu\mathrm{E/m}^2 \cdot \mathrm{sec}}{3000\ \mu\mathrm{E/m}^2 \cdot \mathrm{sec}}\right) = 5\ \mathrm{m}.$$

The plant should be able to barely survive at 5 m below the water surface.

Photodegradation

Everyday examples of photochemical degradation by sunlight are common; they include the fading of colors and dyes of objects exposed to the sun, and the embrittlement and cracking of plastic objects left outdoors. In surface waters, photodegradation of chemicals depends on both the intensity and wavelength spectrum of light. If the energy per photon is sufficient to break a specific chemical bond or otherwise induce a chemical reaction, then increased light intensity will cause the chemical reaction to proceed at a faster rate. If the energy required to initiate a reaction is greater than the energy per photon for light of a given wavelength, then that light will not break the chemical bond, regardless of its intensity. Due to the greater amount of energy possessed by photons at shorter wavelengths, ultraviolet light is particularly effective in degrading many materials. Often, photodegradation is called *photolysis,* in reference to the breaking of chemical bonds by light.

For light to cause chemical reactions, its energy must first be transferred to the chemical system (i.e., the light must be *absorbed*). Evidence for absorption of light by natural waters is often visible to the human eye. The color of a lake or river, as seen from above the surface, is due in part to selective absorption of certain wavelengths. To a diver, the phenomenon of light extinction in a surface water, discussed previously, is clearly evident. When light is absorbed by an atom or a molecule, it may cause movement of an electron from a *nonexcited* or *ground* state, to a higher energy level or *excited* state. According to quantum theory, only a finite number of energy levels are available. Prior to photoexcitation, most species are in the *singlet* state (i.e., for each electron of one spin, there is an electron with opposite spin). (O_2 is an exception; in the ground state it has two electrons with unpaired spins, making it a *triplet.*) Initial photoexcitation of most atoms or molecules is usually to a more energetic singlet state; photoexcitation of O_2 often causes transition from a triplet state to a singlet state. The excited atom or molecule may then lose its energy by one of several processes:

1. The energy may be lost as heat in the process of *internal conversion.*
2. The electron may lose energy by electromagnetic radiation as it returns to ground state, in the process of *fluorescence.*
3. The electron may undergo *intersystem crossing* into a *triplet* state; to go between a singlet and a triplet state, the spin of an electron is reversed. The triplet state is usually longer lived than the original excited state, and may decay radiatively to ground state in the process of *phosphorescence.*
4. The energy may initiate a chemical reaction within the molecule, in the process of *direct photodegradation.* Common examples are the *photodissociation* of H_2O_2 into $OH\cdot$ or Cl_2 into $Cl\cdot$.
5. The energy may be transferred to another molecule.

The probability that the initial excited state will decay via any given pathway is called the *quantum yield* for that pathway.

Because direct photodegradation (process 4) can only occur in chemicals that are capable of absorbing light energy, it often occurs in compounds that have light-absorbing double bonds between carbon atoms, as in alkenes or aromatic rings, although other structures also can absorb photons (Table 2-10). Note that absorption is strongly wavelength dependent. First-order photodegradation rate constants or half-lives for many compounds have been empirically determined (Table 2-11; see also Marcheterre *et al.*, 1988). Often such tabulated rates are measured under natural daylight; consequently, al-

TABLE 2-10 Several Chemical Structures That Absorb Light at Wavelengths Greater than 290 nm[a]

Group	λ_{max} (nm)	Molar absorptivity (liter/(mol·cm))
>C=O (aldehyde, ketone)	295	10
>C=S	460	Weak
—N=N—	347	15
$—NO_2$	278	10
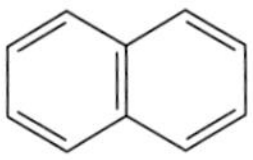 Naphthalene	311 270	250 5000
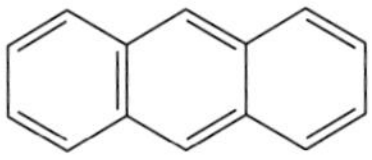 Anthracene	360	6000
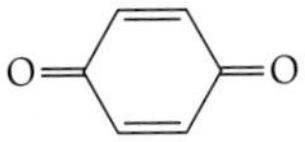 O=C₆H₄=O (quinone)	440 300	20 1000
>C=C—C=O	330	20

[a]Lyman *et al.* (1990).

though they are environmentally relevant, they do not reveal wavelength dependency. Photodegradation in daylight may be due primarily to the blue or ultraviolet fraction of the spectrum because of the higher energies associated with shorter wavelengths.

Even those chemicals that do not themselves absorb photons of light can be degraded in the environment through *indirect photodegradation.* One process of indirect photodegradation is *sensitized photodegradation,* which may occur when other, light-absorbing molecules in the water, such as plant pigments or humic substances, absorb photons and subsequently transfer energy, sometimes in association with electrons or hydrogen ions, to the chemical of interest (process 5 cited previously). The light-absorbing molecules serve as *chromophores.* Natural chromophores are often made of humic substances (Faust and Hoigné, 1987), although certain metal oxides, such as titanium dioxide (TiO_2) particles, are also known to serve as effective inter-

TABLE 2-11 Half-Lives for Disappearance via Direct Photolysis in Aqueous Media[a]

Compound	λ (nm)[b]	$t_{1/2}$
Pesticides		
Carbaryl	S	50 hr
2,4-D,butoxyethyl ester	S	12 days
2,4-D,methyl ester	S	62 days
DDE	S	22 hr (calc)
Malathion	S	15 hr
Methoxychlor	S	29 days
Methyl parathion	S	30 days
Mirex	S	1 year
N-Nitrosoatrazine	S	0.22 hr (calc)
Parathion	S	10 days (calc)
	S	9.2 days
Sevin	S	11 days
Polycyclic aromatic hydrocarbons (PAHs)		
Anthracene	366	0.75 hr
Benz[*a*]anthracene	S	3.3 hr
Benzo[*a*]pyrene	S	1 hr
Chrysene	313	4.4 hr
Fluoranthene	313	21 hr
Naphthalene	313	70 hr
Phenanthrene	313	8.4 hr
Pyrene	313, 366	0.68 hr
Miscellaneous		
Benzo[*f*]quinoline	S	1 hr
p-Cresol	S	35 days
Dibenzothiophene	S	4–8 hr
Quinoline	S	5–21 days

[a]Adapted from Lyman *et al.* (1990).
[b]Wavelength(s) at which photolysis rate was measured. S, sunlight.

mediates in photodegradation. Most natural chromophores are not specifically identified as to their origin and chemical structure and therefore are sometimes called *unknown photoreactive chromophores* (UPC).

Indirect photodegradation can also occur when highly reactive (usually oxygen-containing) species are formed photochemically, and subsequently attack and degrade chemical compounds. One of the most important species is the *hydroxyl radical*, $OH\cdot$, which can be formed by several processes. In one process, a chromophore absorbs light and reacts with water to form hydrogen peroxide (H_2O_2). H_2O_2 in turn breaks into two hydroxyl radicals on absorption of a photon of sufficient energy ($\lambda < 335$ nm). H_2O_2 can also react with Fe^{2+} to form Fe^{3+}, OH^-, and $OH\cdot$ in the *Fenton reaction*. $OH\cdot$ can also be formed by the photolysis of the nitrate ion (NO_3^-). OH· is present at levels of about 10^{-17} *M* in many illuminated surface waters. It is an exceedingly powerful oxidant in both air and water, capable of causing the degradation of many organic compounds. Another reactive species, *singlet oxygen* ($^1O_2$), is formed by the interaction of light with a chromophore and dissolved oxygen (Haag and Hoigné, 1986). A variety of other indirect photolysis possibilities exist, as shown in Table 2-12. For further information on photodegradation, the reader is referred to Lyman *et al.* (1990), Stumm and Morgan (1996), Mill (1989), Schwarzenbach *et al.* (1993), and Malkin (1992).

EXAMPLE 2-23

Benzo[*a*]pyrene, or B[*a*]p, a polycyclic aromatic hydrocarbon, is measured in a facility's wastewater lagoon 2.5 hr after a release at a concentration of 3 μg/liter. If direct photodegradation is the only degradation process occurring, what was the initial concentration of B[*a*]p in the lagoon?

From Table 2-10, it is evident that polycyclic aromatic hydrocarbons such as B[*a*]p are likely to directly photodegrade because double bonds in aromatic rings can absorb light. Indirect degradation of B[*a*]p can also occur via attack by $^1O_2$.

From Table 2-11, an approximate half-life for B[*a*]p due to direct photodegradation is 1 hr. From Eq. [1-20], the corresponding first-order decay constant is 0.69/hr. Then using Eq. [1-19]:

$$C_0 = C_t \cdot e^{kt}$$

$$C_0 = (3\ \mu g/\text{liter}) \cdot e^{(0.69/\text{hr}) \cdot 2.5\ \text{hr}}$$

$$C_0 = 17\ \mu g/\text{liter}.$$

Therefore, the initial concentration of B[*a*]p in the lagoon was approximately 17 μg/liter.

2.7.2 Degradation of Chemicals by Water

Hydrolysis

Even in the absence of microorganisms or light, some chemical pollutants undergo degradation in water due to a variety of abiotic reactions, including

TABLE 2-12 Several Examples of Light-Driven Chemical Processes Occurring in Surface Waters[a]

Environment	Substrates	Products	Probable mechanisms	Likely effects
Freshwaters	Natural organic chromophores and pigments, C·	C· + HO_2 or C· + AH·	H atom transfer to O_2 or A	
		$C^+ + \cdot O_2^-$	Electron transfer to O_2	Numerous
		C + O_2	Energy transfer to O_2	
		HOOH	$\cdot O_2^-$ disproportionation; other?	
	NO_2^-	·NO + ·OH	Direct photolysis	Changes N speciation
	R·	ROO·	O_2 addition	Oxidizes organic radicals
	Fe(III)–organic complex	Fe(II) + CO_2	Uncertain; consumes O_2	Oxidation of organics; reduction of O_2
	Fe(III)–organic–PO_4 complex	Fe(II) + PO_4	Unknown	Dissolution of colloidal Fe; bioavailability of P
Polluted waters				
Oil spills	RH, ArH, R_2S	R=O, RCO_2^-, ArOH, R_2SO	Free radicals; direct photolysis; singlet oxygen	Changes spreading emulsification and toxicity of oil
Herbicides	2,4-D	Oxidation, reduction, hydrolysis products	Direct photolysis	Complex
Pesticides	Disulfoton	Disulfoton sulfoxide	Singlet oxygen	Product more soluble and toxic in some tests
Preservatives	Pentachlorophenol (PCP)	Phenols, quinones, acids, CO_2, Cl^-	Initiated by direct photolysis of PCP	Complex
Domestic waste	Fe(III)–NTA	Fe(II) + amine + CO_2 + CH_2O	Charge transfer to metal	Degrades NTA; induces Fe(II) autooxidation

[a] Adapted from Zafiriou *et al.* (1984).

hydrolysis reactions. Hydrolysis literally means breaking of water. The net result of hydrolysis is that both a pollutant molecule and a water molecule are split, and the two water molecule fragments join to the two pollutant fragments to form new chemicals. Often hydrolysis is catalyzed by H^+ or OH^-. If either H^+ or OH^- is involved in the rate-limiting step of the overall process of hydrolysis, the hydrolysis reaction rate will be sensitive to the pH of the water. Other abiotic reactions include elimination reactions and nucleophilic substitutions; for further discussion of these reaction types, the reader is referred to Schwarzenbach *et al.* (1993).

Types of Compounds Undergoing Hydrolysis

The members of two classes of common pollutant chemicals are likely to undergo hydrolysis. One class includes the alkyl halides, which are straight-chain or branched hydrocarbons in which one or more hydrogen atoms have been replaced by a chlorine, fluorine, bromine, or iodine atom. Using "X" to represent a halogen atom and "R" to represent the hydrocarbon group, the overall hydrolysis reaction can be written

$$H_2O + R{-}X \longrightarrow R{-}OH + H^+ + X^-. \qquad [2\text{-}78]$$

In this reaction, an alkyl halide has been converted to an alcohol, the halogen has been released as a negatively charged halide ion, and a hydrogen ion has been released. The alcohol, in turn, is highly biodegradable. Although alcohols are frequently the products of alkyl halide hydrolysis, other products may be formed instead. For example, chloroform ($CHCl_3$) can hydrolyze to formic acid (HCOOH) and hydrochloric acid (HCl).

A second class of compounds that may undergo hydrolysis includes *esters* and ester analogs. An ester is a compound containing a modified carboxylic acid group (—COOH), in which the acidic hydrogen atom has been replaced by some other organic functional group. For example, if the acidic hydrogen atom of acetic acid (CH_3COOH) is replaced by an ethyl group (C_2H_5), the result is ethyl acetate ($CH_3COOC_2H_5$), an ester.

Hydrolysis converts esters into the "parent" organic acid and an alcohol. Therefore, continuing the example, ethyl acetate hydrolyzes to acetic acid and ethanol:

$$H_2O + CH_3COOC_2H_5 \longrightarrow CH_3COOH + C_2H_5OH. \qquad [2\text{-}79]$$

The hydrolysis reaction as written in the preceding equations is the overall reaction. Many steps are involved in a hydrolysis reaction, and a number of intermediate chemical species are formed. The reader is referred to Schwarzenbach *et al.* (1993) and Tinsley (1979) for a discussion of the actual reaction pathways.

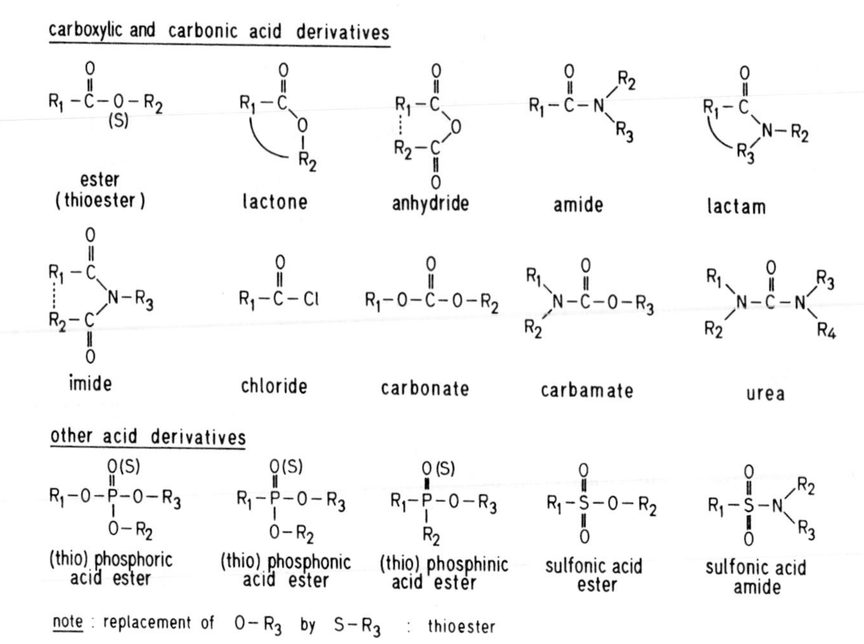

FIGURE 2-31 A variety of esters and ester analogs that tend to degrade by hydrolysis. [From *Environmental Organic Chemistry,* by R. P. Schwarzenbach, P. M. Gschwend, and D. M. Imboden. Copyright © 1993, John Wiley & Sons, Inc. Reprinted by permission of John Wiley & Sons, Inc.].

Hydrolysis can also occur in *ester analogs,* chemicals having ester-like structures. These chemicals include compounds in which the oxygen of the "parent" carboxylic acid can be thought of as being replaced by another electronegative element, such as sulfur or nitrogen. If the oxygen atom of the ester linkage is replaced by a nitrogen atom, the compound is called an amide or a substituted amide; if it is replaced by a sulfur atom, it is called a thioester. Some of these compounds are shown in Fig. 2-31, along with other compounds such as carbamates, phosphoric acid esters, and thiophosphoric acid esters, which also may undergo hydrolysis. Many current pesticides have ester-like structures to shorten their environmental lifetime and thereby to help limit their impact on ecosystems and their residues in the human food supply.

Rates of Hydrolysis Reactions

As discussed in Section 1.6.6, at any given temperature the rate of a reaction between two different chemicals depends not only on the chemicals involved, but also on their concentrations. Concentration affects the probability that, over any given time interval, a molecule of one reactant will collide with a molecule of the other reactant, thereby allowing the reaction to proceed. For a hydrolysis reaction that involves the direct action of the neutral species H_2O on an alkyl halide, ester, or ester analog molecule, the two concentrations of interest are the concentration of H_2O and the concentration of the ester. Because the concentration of H_2O is essentially constant in any dilute solution (approximately 55.4 *M*), the rate of this hydrolysis reaction varies only with changes in the concentration of the ester,

$$dC/dt = -k_{H_2O} \cdot C \cdot [H_2O] \qquad [2\text{-}80a]$$

or

$$dC/dt = -k'_n \cdot C, \qquad [2\text{-}80b]$$

where C is the concentration of the hydrolyzable compound $[M/L^3]$, k'_n is the pseudo-first-order neutral rate constant $[T^{-1}]$, and t is time [T]. Eq. [2-80b] is an example of pseudo-first-order kinetics because the constant k'_n is actually a product of a reactant concentration $[H_2O]$ that is nearly constant and an intrinsic rate constant k_{H_2O}. If the initial concentration of the hydrolyzable compound is C_0 and there are no other sources or sinks of the compound, the familiar exponential solution is obtained:

$$C_t = C_0 e^{-k'_n \cdot t}. \qquad [2\text{-}81]$$

In the case where the hydrolysis rate is controlled by a reaction step involving the hydroxide ion OH^- (*base-catalyzed hydrolysis*), the expression for the change in chemical concentration is

$$dC/dt = -k_b \cdot C \cdot [OH^-], \qquad [2\text{-}82]$$

where $[OH^-]$ is the concentration of hydroxide ions and k_b is the rate constant for the base-catalyzed hydrolysis reaction $[T^{-1} \cdot L^3 \cdot M^{-1}]$.

Recall that if pH is known, $[OH^-]$ can be directly calculated from the mass action expression for the ionization of water (see Section 1.6.3):

$$[OH^-] = 10^{-14}/[H^+]. \qquad [2\text{-}83]$$

It can be seen from Eqs. [2-82] and [2-83] that pH exerts a strong effect on base-catalyzed hydrolysis. For example, hydrolysis would be 1 million times faster at a pH of 10 than at a pH of 4. As long as pH is known and constant, however, the rate constant k_b can be combined with the concentration of hydroxide ions, resulting in a pseudo-first-order conditional rate constant k'_b

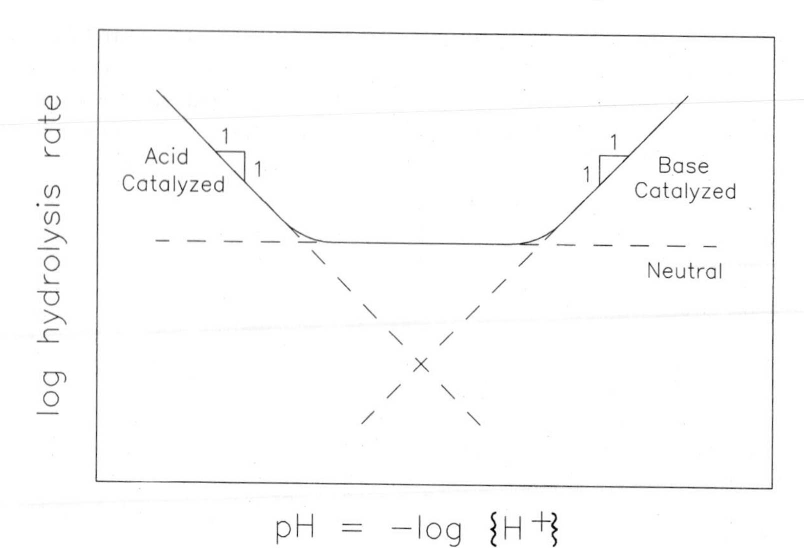

FIGURE 2-32 Effect of pH on hydrolysis rate. The graph shows, in dashed lines, the rate of each hydrolysis process (acid-catalyzed, neutral, and base-catalyzed) as it varies with pH. Because reaction rates are proportional to $[H^+]$ and $[OH^-]$ for the acid-catalyzed and base-catalyzed reactions, respectively, the slopes of the log (rate) versus pH lines for acid-catalyzed and base-catalyzed hydrolysis are -1 and $+1$, respectively.

that is valid only for that particular pH:

$$k_b' = k_b \cdot [OH^-]. \tag{2-84}$$

A parallel treatment applies for hydrolysis when reaction with the hydrogen ion is rate controlling *(acid-catalyzed hydrolysis)*. In this case, $k_a' = k_a \cdot [H^+]$. Pseudo-first-order rate constants, such as k_a' and k_b', can only be used as first-order rate constants as long as the pH is constant.

Hydrolysis by all three mechanisms can occur simultaneously and independently for some classes of compounds such as esters (Fig. 2-32); therefore, it is possible to define an overall conditional pseudo-first-order rate constant k_T',

$$k_T' = k_n' + k_b' + k_a' = k_n' + k_b \cdot [OH^-] + k_a \cdot [H^+]. \tag{2-85}$$

Hydrolysis rate constants are tabulated for many chemicals; a few are listed in Table 2-13. In certain environments, hydrolysis can be accelerated by microbial catalysis.

TABLE 2-13 Hydrolysis Rates for Some Esters and Six Ester Analogs[a]

Compound $R_1-C(=O)-O-R_2$: R_1	R_2	k_a ($M^{-1}sec^{-1}$)	k'_n (sec^{-1})	k_b ($M^{-1}sec^{-1}$)	$t_{1/2}$ (pH 7)
CH_3-	$-CH_2CH_3$	1.1×10^{-4}	1.5×10^{-10}	1.1×10^{-1}	2 years
CH_3-	$-C(CH_3)_3$	1.3×10^{-4}		1.5×10^{-3}	140 years
CH_3-	$-CH=CH_2$	1.4×10^{-4}	1.1×10^{-7}	1.0×10^{1}	7 days
CH_3-	phenyl	7.8×10^{-5}	6.6×10^{-8}	1.4×10^{0}	38 days
CH_3-	2,4-dinitrophenyl (O_2N, $-NO_2$)		1.1×10^{-5}	9.4×10^{1}	10 hr
CH_2Cl-	$-CH_3$	8.5×10^{-5}	2.1×10^{-7}	1.4×10^{2}	14 hr
$CHCl_2-$	$-CH_3$	2.3×10^{-4}	1.5×10^{-5}	2.8×10^{3}	40 min
$CHCl_2-$	phenyl		1.8×10^{-3}	1.3×10^{4}	4 min

(continues)

TABLE 2-13 *(continued)*

Compound $R_1-C(=O)-NR_2R_3$: R_1	R_2	R_3	k_a ($M^{-1}sec^{-1}$)	k'_n (sec^{-1})	k_b ($M^{-1}sec^{-1}$)	$t_{1/2}$ (pH 7)
CH_3-	$-H$	$-H$	8.4×10^{-6}		4.7×10^{-5}	4000 years
CH_2Cl-	$-H$	$-H$	1.1×10^{-5}		1.5×10^{-1}	1.5 years
CH_3-	$-CH_3$	$-H$	3.2×10^{-7}		5.5×10^{-6}	40,000 years
CH_3-	$-CH_3$	$-CH_3$	5.2×10^{-7}		1.1×10^{-5}	20,000 years
Parathion $(CH_3CH_2O)_2P(=S)-O-C_6H_4-NO_2$				8.3×10^{-8}	5.7×10^{-2}	89 days
Disulfoton[b] $(CH_3CH_2O)_2P(=S)-SCH_2CH_2SCH_2CH_3$				1.4×10^{-7}	2.0×10^{-3}	57 days

[a] Adapted from Schwarzenbach *et al.* (1993). All rates at 25°C except where otherwise noted.
[b] Rate measured at 20°C.

EXAMPLE 2-24

Ethyl acetate is spilled at an industrial site and runs into a pond, where it accumulates to a concentration of approximately 20 ppb. Assuming the water pH is 6 and hydrolysis is the only degradation process acting to diminish the concentration, what is the half-life of ethyl acetate?

From Table 2-13, $k_a = 1.1 \times 10^{-4}/(M \cdot \text{sec})$, $k'_n = 1.5 \times 10^{-10}/\text{sec}$, and $k_b = 1.1 \times 10^{-1}/(M \cdot \text{sec})$.

The overall rate constant from Eq. [2-85] is

$$k'_T = [1.5 \times 10^{-10} + (1.1 \times 10^{-1})(10^{-8}) + (1.1 \times 10^{-4})(10^{-6})]/\text{sec}$$

$$k'_T = 1.4 \times 10^{-9}/\text{sec} \quad \text{or} \quad 0.043/\text{year}$$

Use Eq. [1-20] to obtain:

$$t_{1/2} = 0.693/k_T = 16 \text{ years.}$$

2.8 CONCLUSION

Despite their enormous variability, surface waters share many common features, including an interface with the atmosphere, a sediment layer capable of both retaining and releasing chemicals, active communities of living organisms, and the presence of sunlight. Through knowledge of these and other aspects of surface waters, it is possible to make reasonable judgments about the expected behavior of individual chemicals in specific surface water ecosystems.

In the following chapter, the subsurface environment is discussed. In many respects the subsurface environment is radically different from rivers, lakes, and estuaries; yet, numerous analogies with surface waters can be drawn. Readers are urged to make their own comparisons between these two media, as an aid to fully appreciating the fundamental physical, chemical, and biological processes that govern the fate and transport of chemicals in each medium.

EXERCISES

1. On August 10, 1992, at 1535, a pulse injection of food-grade table salt (sodium chloride, NaCl) was made at a point on an experimental stream in the Bickford watershed (see Fig. 2-2) in central Massachusetts. The molarity of chloride as a function of time was measured at "site 1" 20 m downstream.

a. Estimate the average stream velocity in this reach.
b. Estimate a dispersion coefficient for this reach (assume the cloud is "frozen" as it passes site 1).

Is the time versus concentration curve *expected* to be exactly Gaussian in this experiment?

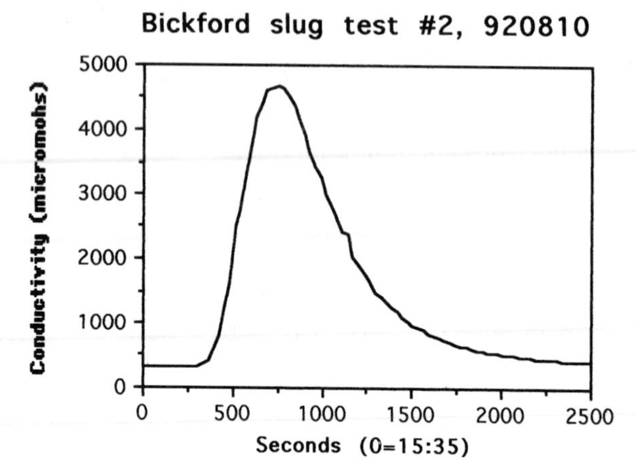

2. Figure 2-7 shows the dissolved oxyen and temperature profiles in Upper Mystic Lake on the Aberjona Watershed in eastern Massachusetts on September 30, 1991.
   a. Estimate the *minimum* possible flux density of downward Fickian oxygen transport through the thermocline at a depth of approximately 7.5 m.
   b. How could the actual flux density be greater than this estimate, assuming the gradient data are accurate?
   c. If oxygen is being transported downward, why is there essentially no oxygen in the hypolimnion?
3. A typical set of measurements of arsenite [in which arsenic is in the (+ III) oxidation state and is more toxic than in arsenate, where it is in the (+ V) oxidation state] for the Upper Mystic Lake in midsummer is as follows [Aurilio (1992)].

Depth (m)	Arsenite (n*M*)
1	10
3	10
5	8
8	5
12	2.5
18	1.5

a. Assume a thermocline at 5 m (a bit higher than in the fall data used in Exercise 2) and estimate the minimum possible flux density of arsenite from the epilimnion to the hypolimnion.
b. If the arsenic that enters this lake via stream flow is entirely in the (+V) oxidation state, what can you conclude about arsenic transformation in this lake?

4. The suspended solid content of a 1-liter river water sample is 20 mg/liter. Assume this material is all organic carbon and that K_{oc}, the organic carbon–water partition coefficient for a chemical of interest, is approximately 4000 liter/kg. A laboratory filters the sample before analysis, analyzing the filtrate (water that *does* pass through the filter) and reporting the chemical concentration, C, of the filtrate.

   If the sample had *not* been filtered, how much of this chemical would have been measured? Express in terms of the measured concentration C (mg/liter) of the filtered sample.

5. In the fall of 1992, a tanker truck overturned on I-93 just north of Boston, spilling approximately 10,000 gal of gasoline and causing a major traffic jam that lasted all day. Assume that gasoline consists of primarily 2,2,4-trimethylpentane (note that this is a very crude assumption) and estimate the loss rate (in g/cm² · sec) from the resulting pool of gasoline. Note that the average wind speed that day was ≈ 12 mph; neglect any temperature effects.

Some useful data follow:

2,2,4-trimethylpentane: $H_3C-C(CH_3)_2-CH_2-CH(CH_3)-CH_3$

MW = 114

Solubility = 2.44 mg/liter $\qquad K_H = 3.1 \times 10^3$ atm · liter/mol.

6. Gasoline from a tank truck spill has contaminated a small stream that enters a lake. Although the surface slick (NAPL) has evaporated, some gasoline is now dissolved in the lake water. The lake is stratified, and the epilimnion is 4 m thick. Consider the fate of two specific compounds in the fuel: pentane (C_5H_{12}) and octane (C_8H_{18}), both initially present at concentrations of 100 ppb.
   a. Plot the expected concentration of pentane versus time, if the piston velocity for this compound is 1 cm/hr. (Neglect removal mechanisms other than volatilization.)
   b. On the same axis, plot the concentration of octane.

c. How does the composition of the dissolved gasoline qualitatively change over time?
d. What other processes do you expect would remove gasoline from the lake?

7. Trichloroethene drips from a tank truck onto a flat, impermeable parking lot at a rate of 1 liter every 5 min.
   a. If wind speed at 10-m elevation is 1.5 m/sec, estimate the flux density of TCE to the atmosphere over the resulting puddle.
   b. When the puddle has grown to its maximum extent, what is its area?
8. A manhole over a sewer is tightly sealed to the road. The sewer flows through the bottom of the cavity formed by the manhole. Because of a leaky tank upstream, the wastewater contains 2 ppm of gasoline (model as 2,2,4-trimethylpentane). Trimethylpentane in air is explosive at concentrations between approximately 1 and 6% by volume (mole percent). Is there a possibility of an explosion? Make reference to an idealized model you might use to calculate the trimethylpentane concentration in air, as well as the several possible complicating factors influencing the concentration of trimethylpentane in an actual situation.
9. A leaking tank drips 1,1,1-trichloroethane (TCA) into a small stream at a rate of 10 mg/min. Stream discharge is 10 liter/min, and depth is 25 cm.
   a. What is the maximum concentration of 1,1,1-TCA in water downstream of the tank?
   b. Does any NAPL accumulate on the stream bottom? Discuss your answer if necessary.
   c. What is the maximum flux density of 1,1,1-TCA from water to air? Assume a reaeration coefficient of 0.4/hr.
10. A slick of hexane results when a storage tank is damaged and hexane runs via a storm drain into the lake of Exercise 6.
    a. Estimate the loss rate from the lake via volatilization. Assume wind speed is 1 m/sec.
    b. Use the same conditions in (a), except chloroform runs into the lake.
11. Inspector Carol Ketchum is in pursuit of a Midnight Chemical truck that has spread waste solvents, mostly *cis*-1,2-dichloroethene, in a uniform layer 2 mm thick on the roadway. She arrives 20 min after the truck has finished emptying its load. Is she in time to collect any evidence from the road? Data follow:

    Wind speed = 3 m/sec

    Vapor pressure = 165 mm Hg at 20°C

    Molecular weight = 97

    Density = 1.28 g/cm^3 at 20°C.

12. What is the oxidation state of carbon in
    a. ethane, C_2H_6?
    b. heptane, C_7H_{16}?
    c. hexachlorobenzene, C_6Cl_6?
    d. acetic acid, CH_3COOH?
13. A 10-g (dry weight) mouse is accidentally bottled in a 25-liter jug of spring water that initially contains 10 mg/liter of oxygen. Assume that 70% of the mouse's dry weight is CH_2O.
    a. Write the relevant reaction for O_2 consumption.
    b. Does the jug become anaerobic?
    c. If 2 ppm NO_3^- and 15 ppm SO_4^{2-} are in the spring water, how much NO_3^- and SO_4^{2-} will be left when the jug reaches equilibrium?
14. Estimate the E_h and $p\epsilon$ of each of the following waters:
    a. There are bubbles and the water smells of sulfide.
    b. Small fish are swimming around.
    c. Highly chlorinated PCBs are slowly degrading.
    d. Water from this well leaves brown stains in the sink.
    e. CH_2O is converted entirely to CO_2 and water, and no soluble iron or sulfide is in evidence.
15. In an incubation performed to test for possible toxaphene degradation in organic-rich lake sediments, a jar is filled with 2 kg of dry sediment; 3 liters of water; 0.1 g of nitrate; 0.2 g of iron oxyhydroxides, $Fe(OH)_3$; and 0.2 g of sulfate. The mixture is initially bubbled with air, toxaphene is incorporated at a concentration of 0.2 ppm, and the jar is sealed. Toxaphene is a mixture of compounds formed from the chlorination of camphene, and was once commonly employed as a pesticide.
    a. Sketch, on the same set of axes, the concentrations of O_2, NO_3^-, $Fe(OH)_3$, SO_4^{2-}, methane, and toxaphene. The absolute times on the axis will be arbitrary, but the sequence of events will be predictable.
    b. Sketch the expected platinum (ORP) electrode potential as a function of time.
    c. As each oxidant described in part (a) disappears, what appears in its place?
16. The seasonally averaged sulfate concentration in the pore water of a lake sediment is 50 μM at the sediment–water interface, but only 5 μM at a depth 0.2 cm below the interface.
    a. How much sulfate can be removed each year from this lake by diffusion into the sediment? Assume an effective diffusion coefficient of 10^{-6} cm^2/sec, and a lake area of 60 ha.
    b. At what rate would this removal of sulfate cause lake alkalinity to rise if the lake has an average depth of 7 m and there were no other sources or sinks of sulfate?

c. In reality, the lake has an inflow of 5×10^6 m^3/year, and sulfate concentration in the inflow is 52 μM. Outflow is 4.5×10^6 m^3/year, with a concentration of 50 μM of SO_4^{2-}. Neglect any other possible sources or sinks of SO_4^{2-}, assume the lake to be well mixed (i.e., neglect seasonal stratification), and estimate the atmospheric input of SO_4^{2-}.

17. Agricultural drainage water in parts of the San Joaquin Valley of California contains extremely high concentrations of dissolved selenium, which originates from weathering of selenium-rich rocks in the surrounding hills. As a result of this agricultural runoff, the Kesterson National Wildlife Refuge has been heavily contaminated with this element, with disastrous results to migratory wildfowl.
    a. Two of the myriad selenium species found in natural waters are selenite, SeO_3^{2-} (and the related species $HSeO_3^-$), and selenate, SeO_4^{2-}. Write a balanced half-reaction for the conversion of $HSeO_3^-$ to SeO_4^{2-}. Is this an oxidation or a reduction reaction?
    b. Consider the following reaction:

$$\tfrac{1}{2}H_2SeO_3 + 3H^+ + 3e^- \rightleftharpoons \tfrac{1}{2}H_2Se + \tfrac{3}{2}H_2O; \ \log K = 18.3.$$

What is the pϵ of the system if $(H_2SeO_3) = (H_2Se)$? What would you predict the ratio of H_2SeO_3 to H_2Se to be if the E_h were -0.1 V? Assume the pH is 7.

18. It has been reported that some microorganisms are capable of using uranium in the (+VI) oxidation state as an oxidant with acetate (CH_3COO^-) serving as the organic carbon source. It has even been proposed that these microorganisms could be used to treat waters contaminated with uranium mine waste; although U(VI) species are highly soluble (and thus highly mobile in the environment), U(IV) tends to occur as highly insoluble species. Assuming that uranium in the form of the soluble species UO_2^{2+} is reduced to the insoluble mineral uraninite, UO_2, where would you anticipate this niche to fall in the context of an ecological redox succession? Some relevant equilibria:

$$UO_2^{2+} + 2e^- \rightleftharpoons UO_2; \qquad \Delta G^\circ = -20.44 \text{ kcal/mol}$$

$$\tfrac{1}{4}CO_2(g) + H^+ + e^- \rightleftharpoons \tfrac{1}{4}CH_2O + \tfrac{1}{4}H_2O; \qquad K = 10^{-1.2}.$$

19. The pesticide methyl parathion (Folidol-M) is leached into a stream from an abandoned disposal site. Is photolysis likely to be a significant sink for this chemical? If the travel time to the ocean is 5 days, what fraction of the chemical would you predict to reach the ocean?
20. Some chemicals, such as phosphate, are trapped in lake sediments by

sorption in an "oxidized microzone" that contains iron (+III) oxides when the overlying water contains oxygen.

a. If a chemical begins to diffuse out of the sediment after the onset of anoxia, how far up into the lake will appreciable concentrations develop if the lake remains stratified for 4 months and if mass transport is by turbulent diffusion, with $D \approx 10^{-3}$ cm^2/sec? (Transport distance can be approximated by $\sqrt{D \cdot t}$, where D is the Fickian transport coefficient and t is time.)

b. What is the lowest value D could have, and under what conditions?

21. Assume sediment oxygen consumption at a rate of 60 $mg/(m^2 \cdot hr)$ is the major sink for O_2 in a small lake 10 m deep. For simplicity, assume that the lake basin is in the shape of a rectangular box. When the lake stratifies in late spring, with a thermocline at 4-m depth, the water contains 10.8 mg/liter of oxygen. Assume the hypolimnion is fairly well mixed.
   a. Draw estimated O_2 profiles for the lake for 1 week, 1 month, and 3 months after stratification.
   b. Will this lake support a cold-water fishery?
22. A lake has become contaminated with chlorobenzene at a concentration of 2 ppb. What concentration of chlorobenzene would you expect to see in trout in this lake? How accurate is your estimate likely to be?
23. A 5-kg osprey makes a regular diet of minnows from a pesticide-contaminated farm pond. The minnows contain 1 ppm of pesticide and the bird eats 200 g of them daily. Estimate the steady-state burden of pesticide accumulated by the osprey. (Assume the birds eliminate 20% of their body burden of the pesticide per week.)
24. A 10-kg adult eagle eats 0.5 kg/day of trout containing 0.1 ppm of PCBs.
   a. Assume that the eagle excretes all PCBs with a first-order rate constant of 0.01/day. Estimate steady-state PCB concentration in the eagle.
   b. PCBs from an eagle are analyzed and found to contain a high ratio of highly chlorinated to less chlorinated congeners compared with the ratio expected on the basis of known industrial discharges. What are two plausible explanations (other than laboratory error)?
25. Quinoline is observed to degrade by photolysis under a day–night cycle in clean water in the laboratory with a half-life of about 3 weeks. A quinoline spill into a certain lake has resulted in a concentration of 1 ppm.
   a. Predict the concentration at the end of a week.
   b. Actual quinoline concentration is measured, at the end of the week, to be 0.1 ppm. Could the discrepancy between this and your prediction in (a) be due to photodegradation occurring more rapidly in the lake than in the laboratory? Explain.
26. Aniline is widely used in the manufacture of dyes and pharmaceuticals. It

undergoes a fully reversible acid–base reaction as follows:

$$\underset{\text{anilinium ion}}{C_6H_5NH_3^+} \rightleftharpoons \underset{\text{aniline}}{C_6H_5NH_2} + H^+; K = 10^{-4.6}.$$

Shown next are absorption spectra for aniline and anilinium ion; note that the ionic and nonionic forms absorb light differently.

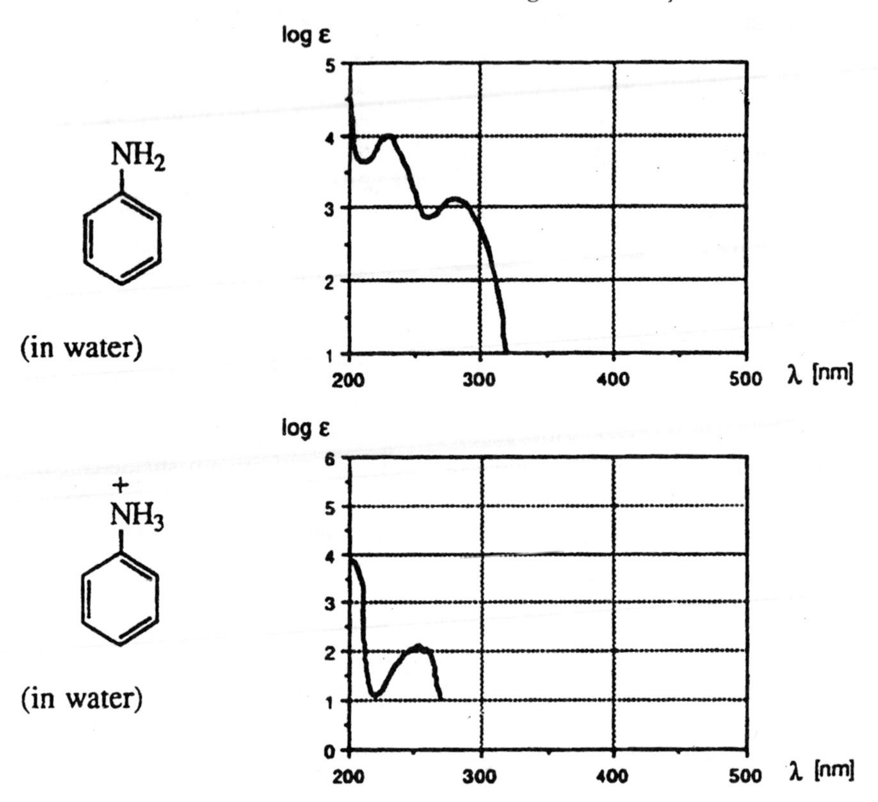

From *Environmental Organic Chemistry*, by R. P. Schwarzenbach, P. M. Gschwend, and D. M. Imboden. Copyright © 1993, John Wiley & Sons, Inc. Reprinted by permission of John Wiley & Sons, Inc.

You are hoping to develop a scheme to treat aniline-contaminated groundwater by pumping it to the surface and exposing it to sunlight, with the intent of having it degrade via direct photolysis. The pH of the groundwater is 6.5. Should you

a. acidify the water to pH 3.5 before solar treatment, with the intent of neutralizing the pH prior to discharge?
b. raise the pH of the water to 9.5 prior to solar treatment, again with the intent of neutralizing it prior to discharge?
c. leave the pH alone?

Note your logic.

27. On July 14, 1991, a train derailment resulted in the spill of approximately 13,000 gal of the soil fumigant sodium metham (Vapam) to the Sacramento River, 70 km upstream from Shasta Lake, California's largest reservoir. The spill killed almost all aquatic life in the river.

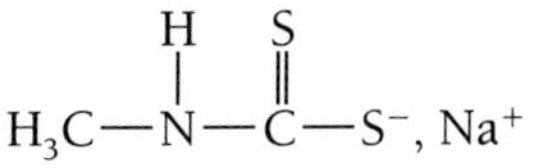

$$H_3C{-}\overset{\displaystyle H}{\overset{|}{N}}{-}\overset{\displaystyle S}{\overset{\|}{C}}{-}S^-,\ Na^+ \qquad \text{sodium salt of metham}$$

In an aqueous environment, metham (which is highly soluble) hydrolyzes to form methyl isothiocyanate and H_2S as the major products:

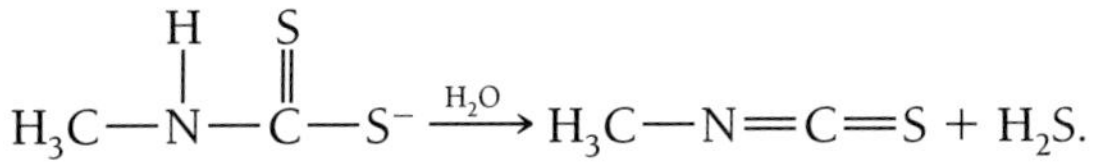

$$\underset{\text{metham}}{H_3C{-}\overset{\displaystyle H}{\overset{|}{N}}{-}\overset{\displaystyle S}{\overset{\|}{C}}{-}S^-} \xrightarrow{H_2O} \underset{\text{methyl isothiocyanate}}{H_3C{-}N{=}C{=}S} + H_2S.$$

Rate constants for this reaction are $k_a \approx 300/(M\cdot\text{sec})$, $k'_n \approx 1 \times 10^{-8}/\text{sec}$. Metham also undergoes indirect photolysis to form methyl isothiocyanate, with a pseudo-first-order rate constant of approximately $1 \times 10^{-4}/\text{sec}$.

The California Regional Water Quality Control Board urgently needs your advice, both as to the maximum concentrations of metham that will be encountered at different points along the Sacramento River, and as to the total load that will enter Shasta Lake.

a. Calculate the maximum concentration (in moles per liter) of Na^+ (which behaves as a conservative tracer) and metham that will be encountered, both at the town of Dunsmuir (10 km downstream from the spill) and at the point where the Sacramento River reaches Shasta Lake (70 km downstream from the spill).
b. How important are the processes of longitudinal dispersion, hydrolysis, and indirect photolysis in attenuating the peak concentration of metham between Dunsmuir and Shasta Lake? Quantify your answer.
c. What are the half-lives for hydrolysis and indirect photolysis?
d. How important are these two processes collectively in attenuating the total load of metham between Dunsmuir and Shasta Lake?
e. Data are currently not available as to the Henry's law constant for metham. Would you recommend measuring this parameter, or do you feel that it is reasonable to neglect air–water exchange of metham? Indicate your logic.
f. Estimate the total mass of methyl isothiocyanate (in moles) that will be transported into Shasta Lake, ignoring losses from any process (such as air–water exchange) that might remove this compound during its transport in the Sacramento River.

Consider some useful information: the Sacramento River has a discharge of 75,000 liter/min, a mean depth of 0.3 m, a pH of 7.8, a mean width of 1.2 m, and a dispersion coefficient $\approx 1.6 \times 10^2$ m²/min. Sodium metham has a commercial formulation of 33% by weight sodium salt of metham, a density of commercial formulation $\approx$ 1 g/ml, and a molecular weight of 129.2 g/mol for sodium salt of metham.

28. A soil water has an initial alkalinity of 5×10^{-4} eq/liter and initial C_T of 4×10^{-4} mol/liter.
    a. What is its pH?
    b. Denitrification occurs, and a 2×10^{-4} eq/liter amount of nitrate originally present in the water is reduced to N_2, producing additional carbon dioxide in the amount predicted by the use of "CH_2O" as the electron donor. What are the new values of Alk, C_T, and pH?
29. A helium tracer study shows that a certain stream whose depth is 1.6 m has a gas exchange coefficient for helium of 1400 cm/hr. Stream velocity is 11 cm/sec.
    a. What is the reaeration coefficient, k_r, for oxygen? Recall that dimensions of k_r are $[T^{-1}]$. Explain any assumptions.
    b. What is the gas exchange coefficient for vinyl acetate, $CH_3COOCHCH_2$? The value of H is 0.02.
    c. The stream contains 10^{-3} M vinyl acetate at a certain location. By considering only hydrolysis loss, what will be the concentration 250 m downstream? For vinyl acetate, k_a is $1.4 \times 10^{-4}/(M\cdot\text{sec})$, k'_n is 1.1×10^{-7}/sec, and k_b is $10/(M\cdot\text{sec})$. This is a hard water stream whose pH is 8.5.
    d. Use the same conditions in (c), except also consider the loss mechanism of volatilization.
30. You have been retained to investigate the characteristics of a small river that is proposed to receive water from a new wastewater treatment plant. The width of this uniform river is 10 ft (3 m), depth is 1 ft (0.3 m), slope is 0.001, and the Manning coefficient is 0.03. River velocity is thus about 0.4 m/sec. To estimate travel times and dispersion coefficients, you inject 50 g of rhodamine dye at the site of the proposed outfall. Because the river is too wide to easily distribute the dye evenly across the channel, you dump it in near the bank and figure that it will spread out soon enough. You also begin the steady bubbling of propane, a gaseous tracer, into the river at this location.
    a. Your assistant sets up a fluorometer and begins making rhodamine measurements 1 km downstream. The data, when converted to a spatial distribution at t = 2500 sec after dye release, show the following: 95% of mass lies between 800 m and 1200 m at t = 2500 sec.

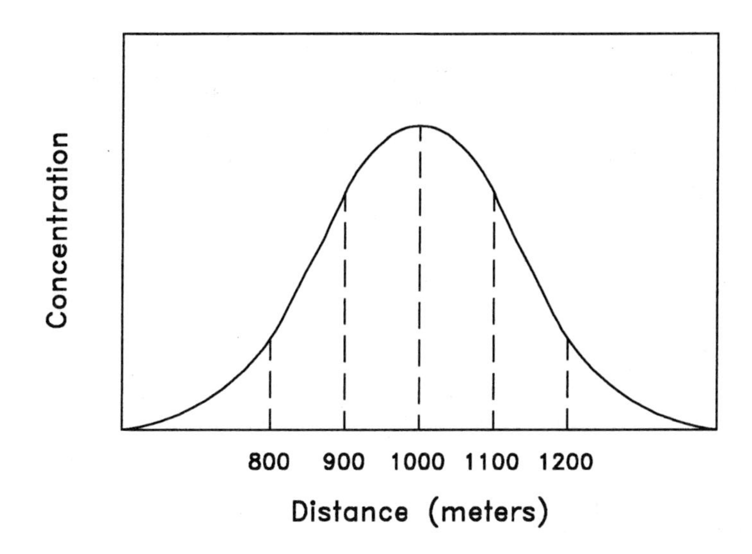

Estimate the longitudinal dispersion coefficient for the reach of stream between the proposed outfall and the measurement location.

b. Your other assistant collects samples for propane analysis. The C_3H_8 concentration is 75 μM 200 m downstream of the outfall site; it is 52 μM at 1500 m. Assume no dilution by lateral inflow. What is the reaeration coefficient?

c. How far downstream must your assistant go to collect a sample near shore that is representative of the entire channel cross section?

31. TCE from a spill accumulates in a depression of a parking lot. Estimate how fast the solvent volatilizes (in g of TCE per cm^2 per unit time) when these conditions exist:

a. The wind speed is 2 m/sec and temperature is 20°C.

b. Use the same conditions as (a), except a rainstorm covers the TCE with 1 cm of water. Assume TCE dissolves into water and quickly reaches equilibrium with the water.

32. At a certain point, toluene seeps at a rate of 9.0 mg/sec from the ground and dissolves into a stream with a discharge of 3 liter/sec and a pH of 7.5. The stream is 0.2 m deep. You have previously estimated the gas exchange coefficient (piston velocity) for propane (C_3H_8) under current discharge conditions to be 19 cm/hr, and can safely assume that both propane and toluene volatilization are controlled by the "water side."

a. What is your prediction of toluene concentration at a point B downstream if the travel time is 35 min from the seep?

b. What would be your prediction for steady-state concentration at point B if the chemical were the methyl ester of dichloroacetic acid, for which the hydrolysis rate constants are $k_n' = 1.5 \times 10^{-5}$/sec and $k_b = 2.8 \times 10^3/(M \cdot \text{sec})$? (In this case you may neglect volatilization and acid-catalyzed hydrolysis.)

33. A disgruntled postal worker dumps a liquid culture of *Anthrax* into a river 900 m upstream of the intake pipe for the municipal water works; 10^{11} organisms are introduced into the river which has a cross section of 0.6 m^2.
    a. If the river velocity is 40 cm/sec and the longitudinal dispersion coefficient is 0.1 m^2/sec, what is the maximum concentration expected at the plant?
    b. How wide (2 standard deviations) is the cloud of organisms at the time the peak concentration occurs at the intake pipe?
    c. How does the answer to (a) change if the organisms have a 10-hr half-life in the river?
34. The oxidation state of plutonium greatly affects its K_d; the more oxidized species tend to be more mobile.
    The following is a redox reaction that occurs for plutonium:

$$PuO_2^+ + 4H^+ + e^- \longrightarrow Pu^{4+} + 2H_2O \qquad \log K = 20.0.$$

    Assume the pH is 4.5 and predict which species of plutonium (PuO_2^+ or Pu^{4+}) is dominant in water under each of these conditions:
    a. The $p\epsilon$ is 7.
    b. Sulfidic conditions exist.
    c. The water is aerated.
35. A wastewater from a certain septic tank has an alkalinity of 0.2 meq/liter and pH of 7. It contains organic compounds with a total carbon concentration of 0.5 m*M* and a C to N ratio of 5 (mole to mole). (Ignore any effect this organic carbon might have on pH and alkalinity before mineralization.) The organic matter is mineralized as the water travels in a plume of contamination emanating from the septic system. The organic matter has been fully mineralized (the nitrogen is converted to NH_4^+) 5 m downgradient.
    a. What is the initial inorganic carbon content (C_T) of the water before mineralization of the organic carbon?
    b. If there are no other processes occurring (such as mixing with other groundwaters, weathering of aquifer minerals, etc.), what is the new pH?
36. A 1-liter sample of a lake water has an Alk of 0.2 meq/liter and contains 0.18 m*M* of carbonate species.
    a. What is the pH?
    b. How much does the pH change when 5×10^{-5} mol of hydrochloric acid (HCl, a strong acid) are added?

37. A small pond has a surface area of 0.15 ha, and a depth of 1.5 m, and contains 10 ppm of benzene, C_6H_6.
    a. Estimate the total benzene flux to the atmosphere if the gas exchange coefficient for methane has been determined to be 12 cm/hr.
    b. Estimate the concentration of benzene in the pond after a week, assuming that only volatilization is acting to remove benzene from the pond.
38. The total of all carbonate species in a lake water is 0.001 *M*, and the lake pH is 8.
    a. Calculate the concentration of each of the individual carbonate species.
    b. Is the lake in equilibrium with the atmosphere?
    c. If you put a piece of calcite in the lake, would it tend to dissolve? Calcium concentration is 0.1 m*M*, and the solubility product of calcite (calcium carbonate) is $10^{-8.35}$.
39. In a summer boating accident, 10 liters of gasoline is spilled into a lake. A slick, 3 m wide and 7 m long, forms on the water surface. How long will it take to evaporate? Assume the slick is of uniform thickness, and that the vapor pressure of the fuel is 25 mm Hg. The wind speed (at 10 m above the ground) is 4 m/sec. Gasoline contains approximately 7 mol/liter of liquid.
40. You are interested in producing methane from an agricultural wastewater stream. The waste stream contains dissolved sugars (model as CH_2O, in which carbon has an average oxidation state of 0), an alkalinity of 0.8 meq/liter, and a pH of 7. The initial DO is negligible; however, the water is still in equilibrium with the atmosphere with respect to dissolved nitrogen (i.e., $P_{N_2} = 0.8$ atm). You set up an experimental bioreactor (the closed tank in which you hope methanogenesis will take place) to do some experiments. Assume CO_2 and CH_4 are the only products of CH_2O fermentation.
    a. In addition to those given, what further constraints on water chemistry must be met if methane production is to begin immediately after you fill the bioreactor? You have already inoculated the batch with a suitable microbial consortium. Ionic strength, pH, and temperature (30°C) are all acceptable and there are no toxic materials in the wastewater. Mention specific chemical species.
    b. You have installed a glass window in your experimental reactor so that you can look for signs of action. After 3 days you see the first signs of bubbles. At this time, what is your best estimate of the aqueous concentration of methane, whose dimensionless Henry's law constant is 27 at the temperature of the bioreactor? (Although oxygen cannot get in, your bioreactor is not pressurized.) You may neglect the partial pressure of carbon dioxide, because CO_2 is so soluble.

c. At this exciting moment, what is your best estimate of the pH in the reactor? (It has changed, because of the CO_2 produced as part of methanogenesis.) You may neglect the possible effects of uptake of ammonium, or other biological effects on the alkalinity of the water.

41. Acid rain falls over a certain locality with a pH of 3.52. Two-thirds of its mineral acidity (negative alkalinity) is due to nitric acid; the balance is sulfuric acid. Assume base cations are negligible. As the water evolves along its path through the watershed, its chemistry is modified. Estimate its pH after each of the following steps, in sequence. *Hint:* Nitric acid and sulfuric acid are fully ionized; neglect C_T until part (b). This is an excellent assumption because carbonic acid ionizes negligibly, for most purposes, at such a low pH.
    a. Plants take up half of the nitric acid.
    b. The water acquires 2×10^{-4} mol/liter of calcium ion from mineral weathering. Simultaneously, the water acquires 0.8 m*M* of carbon dioxide from the CO_2-rich soil gas.
    c. The water flows through a wetland bordering the stream that drains the watershed. Here, the remaining nitrate is denitrified, half of the sulfate is reduced to sulfide (assume you can neglect the sulfide system in your solution procedure), and the water acquires another 0.4 m*M* of carbon dioxide from the CO_2-rich soil gas.

42. 5 kg of a chemical is instantaneously spilled into a stream whose pH is 7.5. The stream velocity is 0.3 m/sec, D_L is 50 m²/sec, and stream discharge is 1 m³/sec. Hydrolysis is dominated by base catalysis at this pH, and the second-order rate constant is $2.8/(M \cdot \text{sec})$.
    a. What is the peak concentration in the stream after an hour?
    b. Where does it occur?

43. What is the flux density of the specified chemical in each situation? State all assumptions.
    a. Nitrate ion (NO_3^-) into an anoxic lake sediment, if rate is governed by a layer of stagnant porewater in the top 0.5 cm of sediment. Above this layer, $[NO_3^-] = 6$ ppm; below, it is zero. Let D equal 10^{-6} cm²/sec.
    b. Benzene (C_6H_6) from a lake surface, where the dissolved benzene concentration in lake water is 700 ppm and the gas exchange coefficient for propane (C_3H_8) has been found to be 1.7 cm/hr.

44. A traffic accident causes a chemical lawn service tank trailer to tip over and instantly spill into a storm drain that empties into a small river. Assume the trailer holds 2 kg of a pesticide, dissolved in 4 m³ of water. The river velocity is 0.3 m/sec, the river has a uniform cross-sectional area of 2.7 m², and D_L is 300 m²/min. For a point 1.3 km downstream,

a. Estimate pesticide concentration after 1 hr.
b. Estimate concentration after 6 hr assuming that, instead of dumping its contents instantly, the trailer merely leaks pesticide solution at a rate of 0.5 m^3/hr and nobody stops the leak.

45. A certain lake becomes thermally stratified on May 15. The hypolimnion averages 4 m thick. Thereafter, O_2 in the hypolimnion begins to decrease at a rate of 0.1 ppm per day. (This is called the hypolimnetic oxygen deficit, and it is commonly used as an index of a lake's biological productivity.) When bottles of the hypolimnetic water are incubated under temperature and light mimicking natural conditions, the daily O_2 decrease in the bottles is 0.07 ppm and the rate of decrease seems to be constant for several days.
    a. Is O_2 consumption by bacteria in the bottle behaving according to zero-order, first-order, or second-order kinetics?
    b. Neglect O_2 transport across the thermocline and determine the benthic oxygen demand (rate of oxygen consumption by sediment, expressed as mg O_2/m^2-day).

    (To clarify the solution procedure, you may wish to take as your control volume a 1-m^2 piece of the lake.)

46. Methylmercury partitions strongly into the lipid-rich tissues of fish; hence, it becomes highly bioconcentrated. Assume the BCF for fish is 10^6 liters/kg (i.e., assume an equilibrium partitioning model is an acceptable approximation to the real world). What fraction of the methylmercury in a lake would actually be in fish tissue, if the lake had a volume of 10^6 m^3 and contained a metric ton (1000 kg) of fish?

47. The bottom sediment profile of a lake shows a layer of charcoal, perhaps the result of a catastrophic fire in the surrounding hills. The lead-210 activity of sediment above and below the charcoal is 0.9 DPM/g, while recent sediment registers 3.4 DPM/g. Estimate the age of the charcoal layer.

48. A spill suddenly releases 100 kg of the pesticide Malathion into a stream whose discharge is 10^4 m^3/day, and mean depth and width are 30 cm and 3 m, respectively. D is 10 m^2/min.
    a. If photolysis is the major sink ($t_{1/2}$ = 12 hr), determine the peak concentration that would be experienced at municipal water intakes both 1 and 15 km downstream.
    b. What process is more important in reducing Malathion concentration between these points (to decide which is more important, you can look at the ratio of concentrations that would occur at 15 and 1 km in the absence of dispersion, or in the absence of photolysis, and compare).

c. At a high enough pH, hydrolysis also becomes important. What second-order rate constant of hydrolysis would be required for photolysis and hydrolysis rates at pH 9 to be equal?

49. A stream has a channel area of 0.3 m^2 and a discharge of 0.07 m^3/sec. If 2 g of rhodamine dye (6.7 g/m^2) are suddenly introduced at point A in the channel at time $t = 0$,
    a. When does the maximum concentration of dye appear at point B 25 m downstream?
    b. At the instant that maximum dye concentration appears at point B, the spatial variance of the dye is 0.9 m^2. What is the 1-dimensional dispersion coefficient associated with the stream at these flow conditions?
    c. What is the dye concentration at point B at this instant?
    d. What is the advective flux of the dye at point B at this instant? What is the Fickian flux of dye at B at this instant?

50. What is the oxidation state of
    a. phosphorus in phosphoric acid, H_3PO_4?
    b. carbon in benzene?
    c. carbon in ethanol, C_2H_5OH?
    d. carbon in trichloroethene, C_2Cl_3H?

51. a. If 10^{-3} mol of $H_2CO_3^*$ are dissolved into 1 liter of pure water (by bubbling with carbon dioxide), what are the approximate final pH and alkalinity of the water?
    b. If 10^{-4} mol of calcium carbonate, $CaCO_3$, and 3×10^{-4} mol of CO_2 dissolve into 1 liter of pure water, what are C_T, Alk, and pH?
    c. If 1 liter of the water initially contained 10^{-4} mol $CaCO_3$, 3×10^{-4} mol CO_2, and 10^{-4} mol of nitric acid (which fully ionizes into NO_3^- and H^+ when in water), what are C_T, Alk, and pH?

52. a. How many moles of nitrate (NO_3^-) are needed to oxidize a mole of elemental carbon to carbon dioxide (assuming the nitrate will be reduced to nitrogen gas)?
    b. If you were making propellant (for a pyrotechnic display) out of charcoal [carbon (0)] and saltpeter (KNO_3), how much saltpeter would you need for each 12 ounces of charcoal?

53. Estimate the pϵ of a system originally containing 0.001 M glucose ($C_6H_{12}O_6$), 0.004 M nitrate, and 0.001 M sulfate, after redox equilibrium has been approached. The pH is 4. Note that

$$\tfrac{1}{5}\,NO_3^- + \tfrac{6}{5}\,H^+ + e^- \longrightarrow \tfrac{1}{10}\,N_2 + \tfrac{3}{5}\,H_2O; \qquad \log K = 21.05.$$

$$\tfrac{1}{8}\,SO_4^{2-} + \tfrac{5}{4}\,H^+ + e^- \longrightarrow \tfrac{1}{8}\,H_2S + \tfrac{1}{2}\,H_2O; \qquad \log K = 5.25.$$

54. Log K_{ow} of 1,2,2,4-tetrachlorobenzene (TeCB) is 4.5. What concentration

of this compound would you expect in rainbow trout if the measured concentration of dissolved TeCB in water is 4 ng/liter?

55. Phenyl acetate has a hydrolysis half-life of 38 days at pH 7. However, if the phenyl group has two nitro ($-NO_2$) substituents added (in *ortho-* and *para*-positions), k_n' becomes 1.1×10^{-5}/sec, and k_b becomes $9.4 \times 10^1/(M \cdot \text{sec})$. k_a can be neglected.
    a. Which compound hydrolyzes more rapidly at pH 7?
    b. A mole of the dinitrophenyl acetate and a mole of a tracer, sodium bromide, are spilled into a river, in which average velocity is 0.5 m/sec and discharge is 7 m^3/sec. If the longitudinal dispersion coefficient is 0.01 m^2/sec, determine the maximum bromide and dinitrophenyl acetate concentrations expected 8 km downstream, if river pH is 7.
56. Water from the hypolimnion of a lake has a pH of 8, and C_T is measured as 1.5 m*M*. The DO is zero.
    a. Would this water lose or gain CO_2 if it were brought into contact with the atmosphere?
    b. Is your result in (a) consistent with your understanding of the lake ecosystem? Explain *very* briefly.
    c. If 1 mmol/liter of nitrate (NO_3^-) is added to the lake, as part of an unusual treatment intended to oxidize large accumulations of organic matter in the lake, how much will this alter C_T, after denitrification (to nitrogen gas) is complete?
57. You are the environmental engineer stationed on Epsilon III, a deep space listening outpost orbiting between Earth and Mars. The 10-person crew relies on a hydroponic garden that you maintain to provide food, oxygen, and CO_2 control. Each (Earth) day, the crew, plus Brutus, the Rotweiler mascot (who was smuggled aboard as a frozen embryo), consume about 2 kg of "CH_2O". Wastes are thermally mineralized (incinerated) back to CO_2, water, and mineral salts. The station has an air volume of 400 m^3 and you seek to maintain a typical Earth P_{CO_2} of $10^{-3.5}$ atm. Your hydroponic garden contains 4 m^3 of water, continually circulated and sprayed onto plant roots. The water must stay between pH 6 and 8 or the plants will all die.
    a. What is the ratio of the total inorganic carbon in the air to the inorganic carbon contained in the water as $H_2CO_3^*$?
    b. If pH is 7.3, what is the ratio of $H_2CO_3^*$ to HCO_3^-? What is the approximate amount of inorganic carbon on board the station if P_{CO_2} is $10^{-3.5}$ atm?
    c. What is the alkalinity of the water?
    d. A tragic accident occurs in the power room, and Brutus is electrocuted when he attacks a power cable. With full ceremony, the saddened crew

puts the 6-kg (dry weight) dog into the mineralizer. Afterward you are suddenly panic-stricken by the thought that the extra inorganic carbon may kill the garden. As a quick worst-case analysis, you calculate the new pH if 1% of the extra CO_2 were to dissolve into the water of your hydroponic system. What is this worst-case pH value?

e. How much sodium hydroxide (expressed in moles added per liter of water) would you need to counteract the CO_2 effect and restore pH to 7.3?

f. What is the obvious carbon-related problem with the Epsilon III system as described?

58. A worker, performing repairs inside a large air duct, spills a gallon of contact cement, which uses toluene as a solvent. The glue spreads out to an area of 4000 cm^2. The vapor pressure of toluene in a closed headspace over the glue would be 10^{-2} atm (less than for pure toluene, because toluene is part of a mixture). The air duct is 1.5 m high by 1 m wide, and a blower is creating a 1.2 m/sec airflow past the spill.

a. What concentration of toluene (mol/m^3) do you expect to find in the air far downstream of the glue spill?

b. A coworker turns up the blower, increasing air velocity to 2 m/sec to ventilate the duct more rapidly. What is the steady-state downstream toluene concentration now?

c. The worker who spilled the glue covers it with kitty litter to a depth of 2 cm. Diffusion through the litter layer (for which you may use an effective value for $D_{molecular}$ of 0.01 cm^2/sec) now controls the toluene evaporation rate (assume the glue does not "soak up" into the litter). Estimate by what factor the concentration downstream is decreased (assume the blower is turned back to its original setting).

REFERENCES

American Public Health Association (APHA) (1960). *Standard Methods for the Examination of Water and Wastewater,* 11th ed. APHA, New York.

Aurilio, A. C. (1992). "Arsenic in the Aberjona Watershed." M. S. thesis, Massachusetts Institute of Technology, Cambridge, MA.

Ayers, J. C., Chandler, D. C., Lauff, G. H., Powers, C. F., and Henson, E. B. (1958). "Current and Water Masses of Lake Michigan," *Publication No. 3.* Great Lakes Institute, University of Michigan, Ann Arbor.

Barber, M. C., Suárez, L. A., and Lassiter, R. R. (1991). "Modelling Bioaccumulation of Organic Pollutants in Fish with an Application to PCBs in Lake Ontario Salmonids." *Can. J. Fish. Aquat. Sci.* 48(2), 318–337.

Barrio-Lage, G., Parsons, F. Z., Nassar, R. S., and Lorenzo, P. A. (1986). "Sequential Dehalogenation of Chlorinated Ethenes." *Environ. Sci. Technol.* 20(1), 96–99.

Barron, M. G. (1990). "Bioconcentration." *Environ. Sci. Technol.* 24(11), 1612–1618.

Bennett, J. P. and Rathbun, R. E. (1972). "Reaeration in Open Channel Flow." USGS Professional Paper 737.

Berner, R. A. (1980). *Early Diagenesis: A Theoretical Approach.* Princeton University Press, Princeton, NJ.

Bodek, I., Lyman, W. J., Reehl, W. F., and Rosenblatt, D. H. (Eds). (1988). *Environmental Inorganic Chemistry Properties, Processes, and Estimation Methods.* Pergamon, New York.

Bouwer, E. J. and McCarty, P. L. (1983). "Transformations of 1- and 2-Carbon Halogenated Aliphatic Organic Compounds Under Methanogenic Conditions." *Appl. Environ. Microbiol.* **45**, 1286–1294.

Brock, T. D. and Madigan, M. T. (1991). *Biology of Microorganisms.* 6th ed. Prentice-Hall, Englewood Cliffs, NJ.

Carberry, J. B. (1990). *Environmental Systems and Engineering.* Saunders, Philadelphia, PA.

Chiou, C. T., Freed, V. H., Schmedding, D. W., and Kohnert, R. L. (1977). "Partition Coefficient and Bioaccumulation of Selected Organic Chemicals." *Environ. Sci. Technol.* **11**(5), 475–478.

Cohen, B. A., Krumholz, L. R., Kim, H., and Hemond, H. F. (1995). "*In-Situ* Biodegradation of Toluene in a Contaminated Stream. 2. Laboratory Studies." *Environ. Sci. Technol.* **29**(1), 117–125.

Curtis, H. (1983). *Biology.* 4th ed. Worth, New York.

Dunne, T. and Leopold, L. B. (1978). *Water in Environmental Planning.* W. H. Freeman, New York.

Faust, B. C. and Hoigné, J. (1987). "Sensitized Photooxidation of Phenols by Fulvic Acid in Natural Waters." *Environ. Sci. Technol.* **21**(10), 957–964.

Fischer, H. B., List, E. J., Koh, R. C. Y., Imberger, J., and Brooks, N. H. (1979). *Mixing in Inland and Coastal Waters.* Academic Press, New York.

Gates, D. M. (1962). "Energy Exchange in the Biosphere." *Harper & Row Biological Monographs.* Harper & Row, New York.

Genereux, D. P. and Hemond, H. F. (1992). "Determination of Gas Exchange Rate Constants for a First Order Stream, Walker Branch Watershed, Tennessee." *Water Resources Res.* **28**(9), 2365–2374.

Genereux, D. P. (1991). "Field Studies of Streamflow Generation Using Natural and Injected Tracers on Bickford and Walker Branch Watersheds." Ph.D. thesis, Massachusetts Institute of Technology, Cambridge, MA.

Haag, W. R. and Hoigné, J. (1986). "Singlet Oxygen in Surface Waters. 3. Photochemical Formation and Steady-State Concentrations in Various Types of Waters." *Environ. Sci. Technol.* **20**(4), 341–355.

Hawker, D. W. and Connell, D. W. (1989). "A Simple Water/Octanol Partition System for Bioconcentration Investigations." *Environ. Sci. Technol.* **23**(8), 961–965.

Hemond, H. F. (1990). "ANC, Alkalinity, and the Acid-Base Status of Natural Waters Containing Organic Acids." *Environ. Sci. Technol.* **24**(10), 1486–1489.

Henderson, F. M. (1966). *Open Channel Flow.* Macmillan, New York.

Jones, J. R. E. (1949). "A Further Ecological Study of Calcareous Streams in the 'Black Mountain' District of South Wales." *J. Anim. Ecol.* **18**, 142–159.

Kim, H., Hemond, H. F., Krumholz, L. R., and Cohen, B. A. (1995). "*In-Situ* Biodegradation of Toluene in a Contaminated Stream. 1. Field Studies." *Environ. Sci. Technol.* **29**(1), 108–116.

Lapple, C. E. (1961). "The Little Things in Life." *Stanford Res. Inst. J.* Third Quarter. **5**, 95–102.

Leopold, L. B. and Maddock, T. Jr. (1953). "The Hydraulic Geometry of Stream Channels and Some Physiographic Implications." USGS Professional Paper 252.

Linsley, R. K., Jr., Kohler, M. A., and Paulhus, J. L. H. (1982). *Hydrology for Engineers.* McGraw-Hill, New York.

Liss, P. S. and Slater, P. G. (1974). "Flux of Gases across the Air–Sea Interface." *Nature (London),* **247**, 181–184.

Lyman, W. J., Reehl, W. F., and Rosenblatt, D. H. (1990). *Handbook of Chemical Property Estimation Methods,* 2nd printing. American Chemical Society, Washington, DC.

Mackay, D. (1982). "Correlation of Bioconcentration Factors." *Environ. Sci. Technol.* **16**(5), 274–278.

Malkin, J. (1992). *Photophysical and Photochemical Properties of Aromatic Compounds.* CRC Press, Boca Raton, FL.

Marcheterre, L., Choudhry, G. G., and Webster, G. R. B. (1988). "Environmental Photochemistry of Herbicides." *Rev. Environ. Contam. Toxicol.* **103**, 61–126.

Metcalf and Eddy, Inc. (1991). *Wastewater Engineering: Treatment, Disposal, and Reuse.* 3rd ed. McGraw-Hill, New York.

Mill, T. (1989). "Structure–Activity Relationships for Photooxidation Processes in the Environment." *Environ. Toxico. Chem.* **8**, 31–43.

Mitsch, W. J. and Gosselink, J. G. (1993). *Wetlands.* 2nd ed. Van Nostrand Reinhold, New York.

Morel, F. M. M. and Hering, J. G. (1993). *Principles and Applications of Aquatic Chemistry.* Wiley, New York.

Negelescu, M. and Rojanski, V. (1969). "Recent Research to Determine Reaeration Coefficients." *Water Research* **3**(3), 189–202.

Nielsen, D. R. and MacDonald, J. G. (Eds.). (1978). "Nitrogen in the Environment." *In Nitrogen Behavior in Field Soil,* Vol. I. Academic Press, New York.

O'Connor, D. J. (1983). "Wind Effects on Gas–Liquid Transfer Coefficients." *J. Environ. Eng.* **109**(3), 731–752.

Odum, E. P. (1971). *Fundamentals of Ecology,* 3rd ed. Saunders, Philadelphia, PA.

Owens, M., Edwards, R. W., and Gibbs, J. W. (1964). "Some Reaeration Studies in Streams." *Int. J. Air Water Pollut.* **8**(8/9), 469–486.

Rattray, M., Jr. (1967). "Some Aspects of the Dynamics of Circulation in Fjords." *In Estuaries* (G. H. Lauff, Ed.), Publication No. 83. American Association for the Advancement of Science, Washington, D.C.

Reid, G. K., and Wood, R. D. (1976). *Ecology of Inland Waters and Estuaries.* 2nd ed., Van Nostrand Reinhold, New York.

ReVelle, P. and ReVelle, C. (1981). *The Environment: Issues and Choices for Society.* Jones & Bartlett, Boston.

Ricklefs, R. E. (1990). *Ecology,* 3rd ed. W. H. Freeman, New York.

Rutherford, J. C. (1994). *River Mixing.* Wiley, Chichester.

Schwarzenbach, R. P., Gschwend, P. M., and Imboden, D. M. (1993). *Environmental Organic Chemistry.* Wiley, New York.

Smith, J. A., Witkowski, P. J., and Chiou, C. T. (1988). "Partition of Nonionic Organic Compounds in Aquatic Systems." *Rev. Environ. Contam. Toxicol.* **103**, 127–151.

Solo-Gabriele, H. (1995). "Metal Transport in the Aberjona River System: Monitoring, Modeling, and Mechanisms." Ph.D. thesis, Massachusetts Institute of Technology, Cambridge, MA.

Spliethoff, H. M. and Hemond, H. F. (1996). "History of Toxic Metal Discharge to Surface Waters of the Aberjona Watershed." *Environ. Sci. Technol.* **30**(1), 121–128.

Stumm, W. and Morgan, J. J. (1981). *Aquatic Chemistry: An Introduction Emphasizing Chemical Equilibria in Natural Waters.* Wiley-Interscience, New York.

Stumm, W. and Morgan, J. J. (1996). *Aquatic Chemistry: Chemical Equilibria and Rates in Natural Waters.* 3rd ed. Wiley-Interscience, New York.

Thackston, E. L. and Krenkel, P. A. (1969). "Reaeration Prediction in Natural Streams." *Am. Soc. Civil Eng. J. Sanitary Eng. Div.* **95**(SA-1), 65–94.

Thibodeaux, L. J. (1979). *Chemodynamics.* Wiley, New York.

Tinsley, I. J. (1979). *Chemical Concepts in Pollutant Behavior.* Wiley, New York.

Ulrich, H. and Stone, A. T. (1989). "Oxidation of Chlorophenols Adsorbed to Manganese Oxide Surfaces." *Environ. Sci. Technol.* **23**(4), 421–428.

Wanner, O., Egli, T., Fleischmann, T., Lanz, K., Reichert, P., and Schwarzenbach, R. P. (1989). "Behavior of the Insecticides Disulfoton and Thiometon in the Rhine River: A Chemodynamic Study." *Environ. Sci. Technol.* **23**(10), 1232–1242.

Weast, R. C. (Ed.) (1990). *CRC Handbook of Chemistry and Physics,* 70th ed. CRC Press, Boca Raton, FL.

Wetzel, R. G. (1983). *Limnology,* 2nd ed. Saunders, Philadelphia, PA.

Yu, S. L. and Hamrick, J. M. (1984). "Wind Effects on Air–Water Oxygen Transfer in a Lake." *In Gas Transfer at Water Surfaces* (W. Brutsaert and G. H. Jirka, Eds.). Kluwer, Dordrecht.

Zafiriou, O. C., Joussot-Dubien, J., Zepp, R. G., and Zika, R. G. (1984). "Photochemistry of Natural Waters." *Environ. Sci. Technol.* **18**(12), 358–371.

CHAPTER 3

The Subsurface Environment

3.1 INTRODUCTION

The environment beneath Earth's surface is composed of porous material (e.g., sand, gravel, and clay) containing water, air, and—in the uppermost layer—organic material. In most places, the uppermost layer of Earth is a thin layer of *soil,* which supplies anchorage, water, and nutrients to vegetation. Farmers and foresters have long appreciated that the characteristics of the soil partially determine the productivity and even the human habitability of entire geographic regions. Regional productivity and habitability also may depend on the availability of *groundwater,* which refers to any water derived from beneath Earth's surface. Groundwater is used by almost 50% of the public water supply systems in the United States; agriculture in the midwestern United States also depends heavily on the use of groundwater for irrigation (Epstein *et al.,* 1982).

Yet, like surface waters, the subsurface environment has been a recipient of pollutant chemicals. To effectively solve contamination problems and mitigate threats of contamination, one must have a thorough understanding of the factors that govern the transport and eventual fate of subsurface chemicals. This understanding begins with an accurate descriptive picture of the subsurface environment.

3.1.1 Nature of the Subsurface Environment

The subsurface environment is no less diverse than surface waters; its characteristics vary with depth as well as from one location to another. Figure 3-1 shows a typical subsurface environment, including an *aquifer* (water-yielding formation) containing sufficient groundwater to be a practical source of potable water. The saturated and unsaturated zones depicted usually consist of porous, granular mineral material; such unconsolidated material overlying solid rock (*bedrock*) is sometimes termed *regolith*. In the upper portion of the *unsaturated zone,* also called the *vadose* zone, both air and water are present in the pore spaces between mineral grains. The uppermost layers of the unsaturated zone, typically a few tens of centimeters in depth, are traditionally called *soil* (although the term soil is often loosely, but inaccurately,

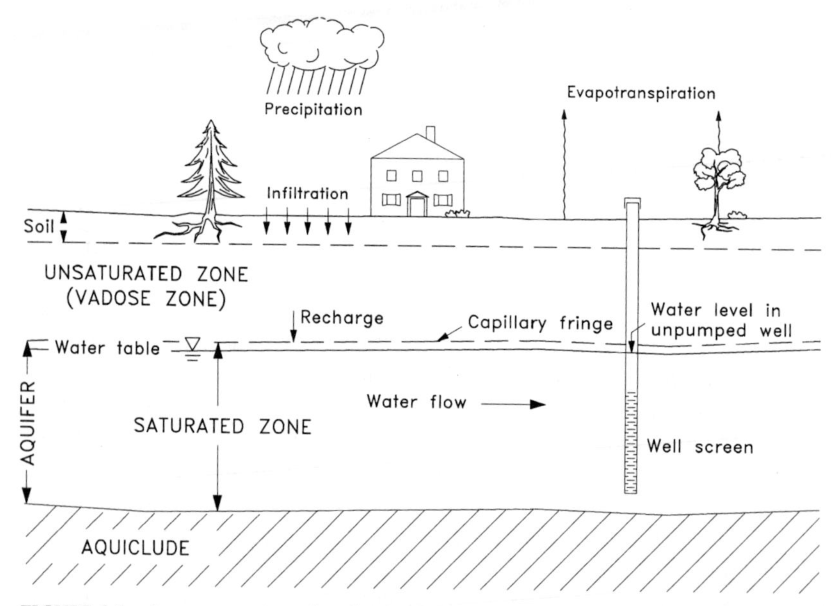

FIGURE 3-1 A representative subsurface environment, showing an upper unsaturated zone and a lower saturated zone with an aquiclude beneath. An aquiclude is nearly impermeable to water. The saturated zone above the aquiclude is a water table aquifer. A well is used to withdraw water from the aquifer.

used in reference to any granular earth material). Soil is characterized by intense biological activity; within the soil, plant roots take up water and dissolved materials; bacteria, fungi, and small soil animals are active; and a variety of natural organic acids are produced from the decay of dead roots, fallen leaves, and other organic material.

In the unsaturated zone, the water occurs as a thin film on the surface of grains or particles, and in the smallest interstices between grains; it does not completely fill the pores between them. The balance of the pore space is filled with air, the composition of which is modified by biological processes in the soil. Respiration by soil bacteria and plant roots causes the soil air to be depleted in oxygen and to be strongly enhanced in carbon dioxide (CO_2) relative to atmospheric air. (CO_2, along with organic acids, promotes the dissolution of mineral materials.) Precipitation that does not simply run off the land surface into a surface water body or storm drain enters the unsaturated zone by *infiltration*, and replaces water lost either by plant root uptake and subsequent evaporation from leaves (*transpiration*) or by direct evaporation. If the water content in the unsaturated zone becomes sufficiently high, the excess water percolates downward to the water table in a process known as *recharge*. The vadose zone, including soils, is discussed in greater detail in Section 3.3.

The *saturated zone* is formed by porous material in which all the pore spaces are filled with water. The *water table* is defined as the depth at which pore water pressure equals atmospheric pressure. If a hole is dug down to the saturated zone, the location of the water table can be easily determined; it is at the depth to which water accumulates in the hole. In coarse porous material, the location of the water table itself very nearly approximates the transition between saturated and unsaturated material; in a fine-textured porous material, enough water may move upward by capillarity to cause complete saturation of a measurable thickness above the water table (the *capillary fringe*).

In the unsaturated zone, water movement is caused by both gravity and by pore water pressure differences arising from variations in the water content from one location to another; water may even move vertically upward through the soil profile if evaporation or plant roots remove it from the near-surface soil. Water flow is impeded, however, by the fact that water can only move via the relatively thin film of water coating the particles. Such flow contrasts with water flow in the saturated zone, where water can move through the entire pore volume and occupy the full cross-sectional area of the pore spaces.

The texture of the porous material forming the saturated and unsaturated zones may range from coarse sands, through finer-textured silt, to extremely fine textured clay. Loams, consisting of roughly equal parts of sand, silt, and clay, are generally considered optimum for agricultural soils (Fig. 3-2).

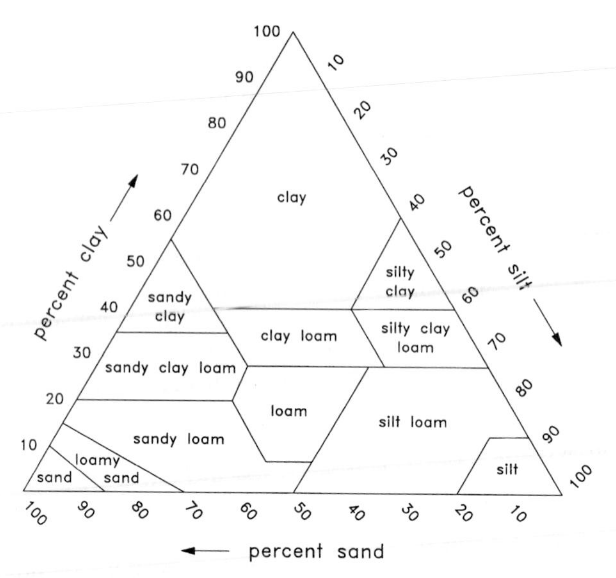

FIGURE 3-2 The soil classification system adopted by the U.S. Department of Agriculture. The components sand, clay, and silt are defined in terms of particle size (see Table 3-3 later). It should be noted that other textural classifications also are in use.

Coarse, *well-sorted* (i.e., the soil grains are all about the same size) sands and gravels are easier for water to flow through and hence form the most highly water-conductive materials (Fig. 3-3). The ease with which water is conducted through porous material is described by the parameter *hydraulic conductivity*, K, discussed further in Section 3.2.1. Hydraulic conductivity is primarily dependent on the characteristics of the porous material, but also is influenced by changes in the viscosity of water, which varies slightly with temperature.

Aquifers are regions of saturated material that are at least moderately conductive to water and may be tapped via wells for potable, industrial or agricultural water. Although many aquifers are made of coarse, unconsolidated material such as gravel, the water-bearing material also can be a porous rock such as sandstone, or even a relatively nonporous but fractured rock such as granite. Beneath an aquifer may be a nearly impermeable layer (*aquiclude*) or low-permeability layer (*aquitard*). Sometimes the aquiclude or aquitard is thin and has unsaturated material beneath it. In such a case the aquifer and water table are said to be *perched*. Perched aquifers are usually localized. A *confined aquifer* exists if the water-bearing layer is bounded on the top and bottom by an aquiclude or an aquitard; a common example occurs when a clay layer separates the water-bearing layer from overlying material. Confined aquifers transmit water much as *unconfined* aquifers do; when water is re-

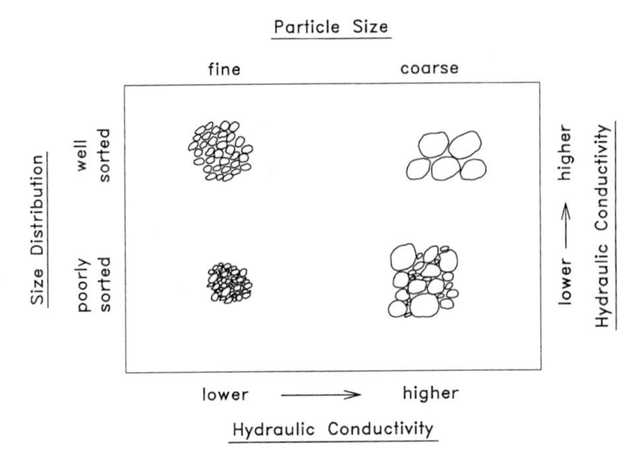

FIGURE 3-3 Particle size, size distribution, and hydraulic conductivity in soil. In a well-sorted soil, all the particles are similar in size, giving the soil a relatively high hydraulic conductivity compared with an otherwise similar but *poorly sorted* soil, in which small soil particles block the pore spaces between larger particles, thereby decreasing the hydraulic conductivity.

moved by a well, however, there can be no corresponding drainage of pores or movement of air to fill the space vacated by water, as occurs in an unconfined, water table (*phreatic*) aquifer. Instead, compression of the aquifer, as well as expansion of the water as its pressure is released at the well, accommodate the space previously occupied by water. *Flowing artesian aquifers* are confined aquifers in which water flows out of wells without any need for pumps.

3.1.2 Sources of Pollutant Chemicals to the Subsurface Environment

Public concern about subsurface contamination is relatively recent. Historically, popular views held that soil exerted a purifying effect on contaminated water passing through it and that soil could somehow cleanse chemicals dumped on the ground. It is now recognized that such filtering occurs only for moderate amounts of certain substances. As the limitations of the soil as a filter became evident, and as improved techniques of analytical chemistry revealed the extent of contamination in groundwater, public concern about subsurface contamination greatly increased.

Groundwater can become contaminated in numerous ways (Fig. 3-4). Disposal of chemicals by burying them in drums or by ponding them in *lagoons* (shallow depressions on the soil surface) has led to groundwater contamina-

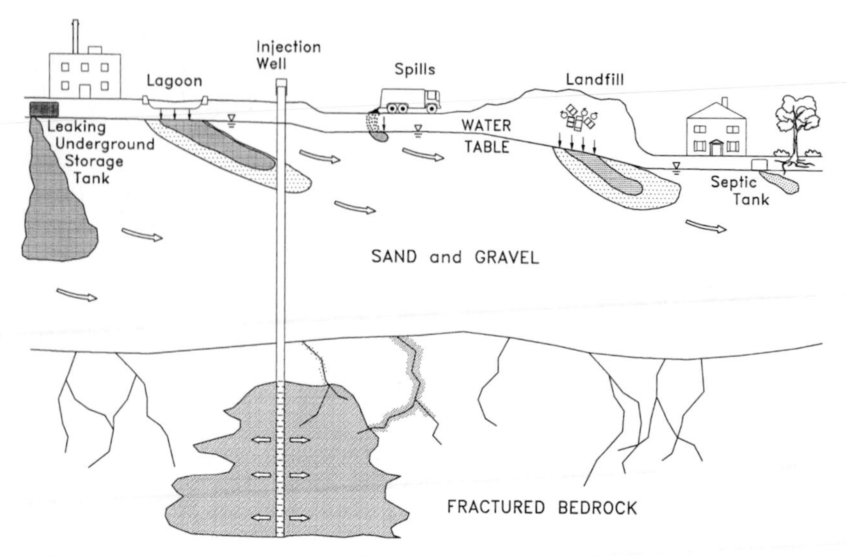

FIGURE 3-4 A representative water table aquifer of sand and gravel with fractured bedrock beneath it. Several paths by which chemicals can enter the subsurface environment are shown.

tion; chemicals can percolate through the unsatured zone to the water table during recharge. If immiscible liquid wastes such as petroleum-based solvents and fuels are released into the soil, an additional phase consisting of the water-immiscible chemical also can be established. If released in sufficient amounts, these nonaqueous phase liquids (NAPLs) also can percolate to the water table. NAPLs (such as gasoline) that are less dense than water will float on the surface of the water table. Leaking underground storage tanks (LUSTs) at gasoline stations often have NAPL on the water table beneath them. Organic liquids that are denser than water [such as the solvent trichloroethene (TCE)] can move downward through the saturated zone, partially displacing water, and settle at the bottom of an aquifer. Such a pool of dense nonaqueous phase liquid (DNAPL) can reside at the bottom of an aquifer for decades or longer, continuing to dissolve slowly into the overlying water, thereby presenting a waste stream of extremely long duration. DNAPLs tend to flow independently of water and usually are not responsive to removal by pumping, as discussed further in Section 3.4.

Subsurface chemical pollutants also come from septic tanks, which are widely used for the disposal of nonindustrial wastewater in much of suburban and rural North America. The pollutants of concern from septic tanks include pathogens as well as nutrients such as nitrogen and phosphorus. Sometimes, other toxic materials are thoughtlessly disposed of via septic systems; in earlier days chemical solvents such as methylene chloride (dichloromethane, Cl_2CH_2) were purposely added as degreasing agents. The leachate from septic systems is released into the unsaturated zone. Depending on moisture conditions, it may spread laterally for a considerable distance before reaching the water table and forming a plume of water with high concentrations of nitrogen, chloride, dissolved organic carbon (DOC), detergent residues, and sometimes phosphorus.

Other point sources of groundwater pollution include both deep *injection wells* and shallower *dry wells* used to inject chemical wastes (including radioactive waste) directly into the subsurface environment. While it is customary to make waste injections into deep aquifers that are salty or otherwise unusable for potable water, it is not unusual for injection wells to leak, or to force the flow of water from one layer of aquifer into another, resulting in contamination of an otherwise usable source of groundwater.

Agriculture is an important nonpoint source of groundwater pollution in rural areas. Widespread application of agricultural chemicals such as fertilizers and pesticides often leads to groundwater degradation due to the downward percolation of chemicals to the water table.

In all cases, an understanding of the physics of groundwater movement is necessary to adequately estimate the direction and rate at which contaminated water is moving in an aquifer. An analogy can be made with tracking a contaminant in a river, as discussed in Chapter 2, except that the underground water flow is out of sight and much more expensive to sample.

3.2 PHYSICS OF GROUNDWATER MOVEMENT

The rate at which groundwater moves through an aquifer is not usually directly measurable, and thus it must be estimated from known relationships among measurable parameters. The discussion of groundwater flow presented here is only an introduction to the topic; many of the relationships described are approximations, and most are restricted in their applicability to situations in which the flow of groundwater is steady over time. When a more detailed analysis is necessary, the reader is referred to the following texts on groundwater: Bear (1972, 1979), Freeze and Cherry (1979), Heath (1984), Strack (1989), McWhorter and Sunada (1977), and Davis and de Wiest (1966).

3.2.1 DARCY'S LAW

Hydraulic head (h) may be the most readily measurable parameter of groundwater flow. Hydraulic head is a measure of the energy per unit volume possessed by the water and is described by the sum of three terms, an elevation term, a pressure term, and a kinetic energy term,

$$h = z + \frac{p}{\rho g} + \frac{v^2}{2\,g}, \qquad [3\text{-}1]$$

where h is the hydraulic head [L], z is the height of the water above an arbitrarily chosen reference elevation [L], p is the *gauge pore water pressure* (pressure above atmospheric pressure) [M/LT^2], ρ is the water density [M/L^3], g is acceleration due to gravity [L/T^2], and v is the velocity of the water [L/T]. Within a continuous, nonmoving body of water, the third term on the right of Eq. [3-1], the kinetic energy term, is zero. In such a water body, the hydraulic head does not change with depth, because the pressure term, $p/\rho g$, increases with depth at exactly the same rate that the elevation term, z, decreases, as shown in Fig. 3-5. Note that in some fluid mechanics problems, the head may be assumed to be constant; when h is constant, Eq. [3-1] becomes the *Bernoulli equation*.

In a water-saturated porous medium, water movement occurs from areas of higher hydraulic head to areas of lower hydraulic head. The gauge water pressure at the water table (or at any other interface between free-standing water and the atmosphere, such as the water surface in a pond), is by defini-

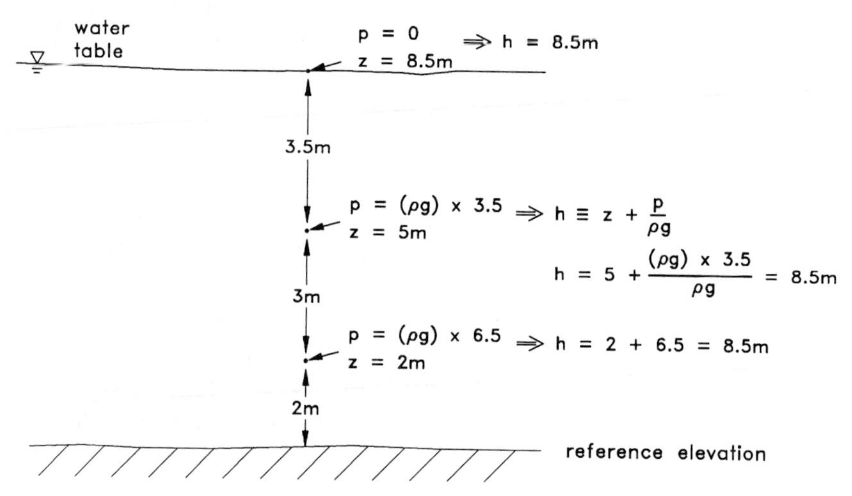

FIGURE 3-5 Under *static* conditions (no water movement), hydraulic head in a continuous volume of water is constant throughout and is equal to the sum of the elevation term and the pressure term. Decreasing the elevation relative to the reference elevation by an amount Δz increases water pressure by an amount equal to $\rho g \Delta z$; therefore, the decrease in the elevation term exactly offsets the increase in the pressure term, and the hydraulic head is unchanged.

tion equal to zero because the water pressure at such locations is equal to atmospheric pressure. The kinetic energy term is typically neglected for groundwater because velocities are so low. Therefore, the head of groundwater at the water table of an unconfined aquifer is equal to the elevation of the water table relative to an arbitrarily chosen reference elevation. Hydraulic head differences are described by the *head gradient* (dh/dx), also known as the *hydraulic gradient,* which is the rate at which hydraulic head (h) changes with distance (x). The average gradient between two points is Δh, the change in head [L], divided by Δx, the distance between the points [L]. Although the numerical values of head depend on the choice of reference elevation, the head gradient is unaffected by addition or subtraction of any arbitrary reference elevation from all points. A one-dimensional form of *Darcy's law* (Darcy, 1856), which expresses the relationship between groundwater flow and head gradient, may be written as

$$q = -K \cdot dh/dx, \qquad [3\text{-}2]$$

where q is the specific discharge [L/T], K is the hydraulic conductivity [L/T], and dh/dx is the head gradient [L/L]. *Specific discharge* (the *Darcy flux*) is equal to the amount of water flowing across a unit area perpendicular to the flow per unit time, Q/A, where discharge Q has units of [L^3/T] and A is area [L^2]. The negative sign indicates that water is moving in the direction opposite to the direction in which the head is increasing.

Darcy's law is equally applicable in three dimensions, where, assuming K is independent of direction, it takes on the form:

$$\vec{q} = -K\nabla h. \qquad [3\text{-}3]$$

(Note the similarity of this equation to the form of Fick's first law as shown in Eq. [1-4]; the difference is that Darcy's law is for water flow while Fick's first law is for mass transport by Fickian processes.)

The kinetic energy term of Eq. [3-1] may need to be included in certain calculations of groundwater flow. For example, if in a hydraulic test a sudden addition of water is made to a well in an aquifer with very high hydraulic conductivity (greater than approximately 10^{-2} cm/sec), the oscillatory response of the water level in the well to the sudden change in elevation can only be described by including a kinetic energy term (Kipp, 1985).

Hydraulic conductivity, K, is usually determined experimentally. It varies over many orders of magnitude in the subsurface environment, from values as high as 1 cm/sec for coarse gravels, to 10^{-3} cm/sec for fine sands, to 10^{-8} cm/sec or even lower for nearly impermeable clays.

To determine groundwater flow at a site, hydraulic conductivity must be measured, either in a *permeameter* test in a laboratory or *in situ*. For the permeameter, or column, test (Fig. 3-6), aquifer material is placed in a labo-

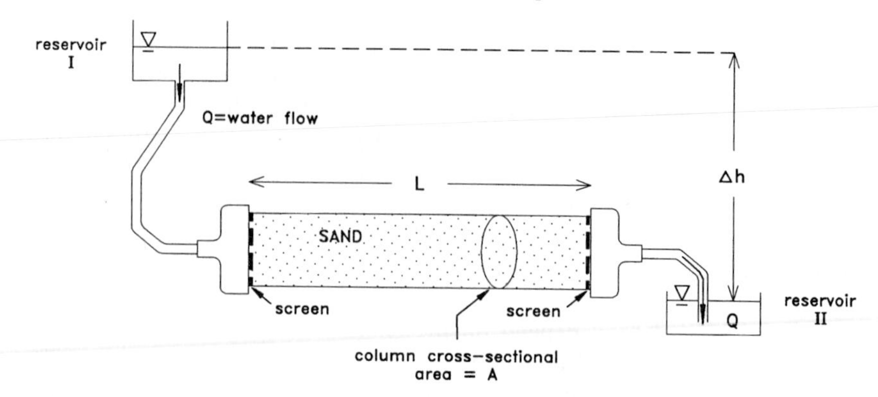

FIGURE 3-6 A simple *permeameter,* or *Darcy apparatus,* used to measure hydraulic conductivity. The column, a pipe filled with the soil sample under test, has cross-sectional area A and length L; specific discharge q is therefore equal to total column flow Q divided by A. Hydraulic conductivity K is estimated using Darcy's law: $K = -q/(\Delta h/L)$. In this apparatus, it is assumed that negligible head loss occurs in the hoses connecting the soil column to the constant head water reservoirs I and II and in the screens used to hold the soil in the column.

ratory column (usually a cylinder with an inlet at one end and an outlet at the other) across which a known head difference is imposed and through which the specific discharge is measured; K is then calculated from Eq. [3-2]. K can be measured *in situ* by observing the response of wells to pumping (*pumping tests*) or to instantaneous additions or removals of water (*slug tests*). Pumping tests average K over a substantial volume of aquifer, whereas slug tests average K over a smaller volume surrounding the well screen; permeameter tests yield K for a small sample of porous medium. Due to natural variability, as well as to the disturbance that occurs when wells are installed and porous media are sampled, none of the tests provide an extremely accurate estimate of hydraulic conductivity. Within relatively homogeneous aquifers, results may be expected to agree within an order of magnitude (factor of 10).

Table 3-1 presents approximate values of hydraulic conductivity that can be expected for different types of porous material. These values are assumed to be the same no matter in which direction the flow is occurring. In reality, the hydraulic conductivity of aquifer materials may not be *isotropic* (the same in all directions); *anisotropic* media have hydraulic conductivities that vary with direction (Fig. 3-7). Anisotropy is often a result of the way in which nonspherical particles are deposited, with their long axes oriented in a horizontal direction. Anisotropy can occur in small samples, but is especially common in sand and gravel aquifers containing horizontal clay *lenses* that

TABLE 3-1 Representative Hydraulic Conductivity Estimates for Several Types of Porous Media[a]

Unweathered marine clay	—	—	—	—									
Glacial till		—	—	—	—	—	—						
Silt, loess				—	—	—	—	—	—				
Silty sand						—	—	—	—	—			
Clean sand							—	—	—	—	—		
Gravel										—	—	—	—
K (cm/sec)	10^{-10}	10^{-9}	10^{-8}	10^{-7}	10^{-6}	10^{-5}	10^{-4}	10^{-3}	10^{-2}	10^{-1}	1	10	10^{2}
K (gal/day/ft²)	10^{-6}	10^{-5}	10^{-4}	10^{-3}	10^{-2}	10^{-1}	1	10	10^{2}	10^{3}	10^{4}	10^{5}	10^{6}

[a] Adapted from Freeze and Cherry (1979).

impede the vertical flow of water but have less of an effect on horizontal flow. For the three-dimensional anisotropic case, Darcy's law needs to be further modified to represent the directional characteristics of K, resulting in a tensor equation. For further information on anisotropy and the hydraulic conductivity tensor, the reader is referred to Freeze and Cherry (1979) or Bear (1979).

At a field site, head gradients and K both can be measured, allowing one to estimate specific discharge by using Darcy's law. To fully describe the

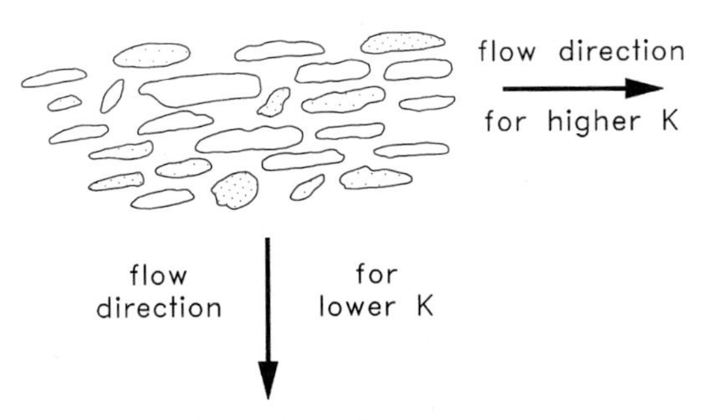

FIGURE 3-7 Anisotropy in a porous material. Hydraulic conductivity is higher in the horizontal direction than in the vertical direction because the long axes of the nonspherical particles become oriented in a horizontal direction during deposition.

advective transport of chemicals in the subsurface, however, it is also necessary to know the *porosity* of the porous media.

Porosity is defined as

$$n = \frac{V_{\text{voids}}}{V_{\text{total}}} = \frac{V_{\text{voids}}}{V_{\text{solids}} + V_{\text{voids}}}, \tag{3-4}$$

where n is the porosity [L^3/L^3], V_{voids} is the volume of the voids (spaces between grains filled with either air or water) [L^3], V_{total} is the total bulk volume of a sample of a porous medium [L^3], and V_{solids} is the volume of the grains [L^3]. Porosity is commonly measured in the laboratory by fully saturating a sample of the porous medium, weighing it, drying the sample, reweighing it, and dividing the difference in mass by the density of water to obtain the volume of the voids.

Typical aquifer porosities are in the range of 0.2 to 0.4. The rate at which nonsorbing chemicals move in the groundwater is equal to the *seepage velocity,* defined as

$$v = q/n, \tag{3-5}$$

where v is the seepage velocity [L/T], q is the specific discharge [L/T], and n is the porosity [L^3/L^3]. Note that v is numerically larger than q, although both have units of [L/T]. Seepage velocity is greater than specific discharge because flow in porous media is confined to a fraction of the total cross-sectional area; to achieve a given value of q, the water must flow through constrictions in the soil at a speed greater than q.

EXAMPLE 3-1

Two monitoring wells are located 150 m apart in an unconfined sandy aquifer. The second well is directly downgradient of the first well. The hydraulic head in the first well is 20 m and in the second well is 19 m; the hydraulic conductivity is estimated using a pump test as 10^{-3} cm/sec. What is the specific discharge in the aquifer? Given that the porosity is 0.2, what is the rate at which nonsorbing chemicals dissolved in the groundwater will move between the wells?

Use Darcy's law in Eq. [3-2] to estimate the specific discharge between the wells:

$$q = -\left(\frac{10^{-3}\text{ cm}}{\text{sec}}\right)\cdot\left(\frac{19\text{ m} - 20\text{ m}}{150\text{ m}}\right) = 6.7 \times 10^{-6}\text{ cm/sec.}$$

Then use Eq. [3-5] to estimate the seepage velocity:

$$v = \frac{q}{n} = \frac{6.7 \times 10^{-6}\ \text{cm/sec}}{0.2} = 3.3 \times 10^{-5}\ \text{cm/sec}.$$

3.2.2 Flow Nets

A *flow net* is a convenient pictorial device for visualizing groundwater flow and a powerful tool with which to perform graphic calculations of specific discharge and seepage velocity in locations where groundwater flow is reasonably steady over time and is predominantly two dimensional (usually in a horizontal plane, although sometimes in a vertical plane). The following discussion is restricted to isotropic, homogeneous porous media. For the purpose of flow net construction, anisotropic aquifers can be transformed into isotropic flow systems; the reader is referred to other texts on groundwater for a discussion of such transformations (e.g., Freeze and Cherry, 1979).

Flow nets consist of two types of lines: *streamlines* and *isopotentials*. Streamlines are lines that follow the path of representative parcels of water; water always flows parallel to streamlines. Isopotentials, drawn perpendicular to streamlines, are lines along which the hydraulic head, h, is constant. Therefore, water always flows perpendicular to isopotentials. Flow nets are often drawn to represent the horizontal movement of groundwater and associated chemicals in an aquifer; the plane of the flow net then represents the horizontal aquifer surface, and it is assumed that underneath each point on the surface, flow is essentially the same at all depths in the aquifer. An example of such a flow net is shown in Fig. 3-8.

Rules for Drawing a Flow Net

Although many groundwater hydrologists customarily employ computerized techniques for analyzing groundwater flow problems, it is valuable for the student of groundwater transport to be able to draw and interpret flow nets. To draw a flow net, the following rules must be observed:

1. Streamlines are always perpendicular to isopotentials in an isotropic aquifer.
2. *Flow boundaries*—impermeable surfaces across which flow cannot occur, such as the surfaces of clay lenses, concrete, or buried tanks—should be thought of as streamlines because water flows parallel to them. Like streamlines, flow boundaries are perpendicular to isopotentials.
3. Boundaries that can act as either *sources* or *sinks* for water (such as rivers and lakes) sometimes can be treated as lines of known hydraulic head.

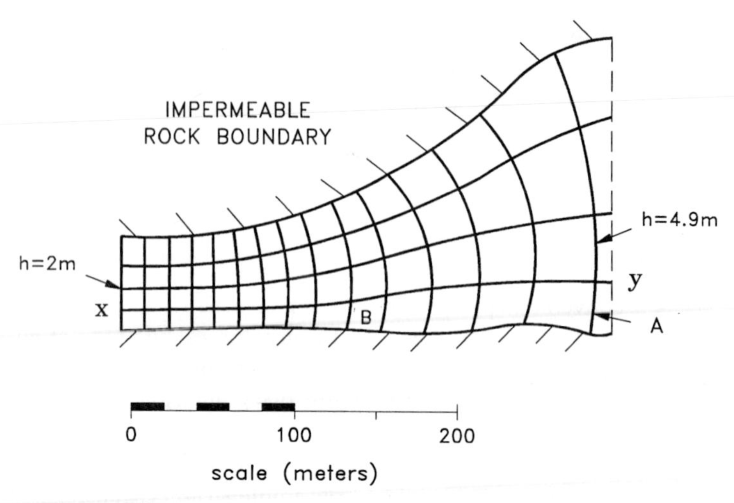

FIGURE 3-8 A flow net for a hypothetical aquifer formed in a drowned river valley, which is assumed to be of uniform thickness over its area and located between two water-impermeable walls of bedrock. The hydraulic head varies between 2 m and 4.9 m. Squares A and B and points *x* and *y* are used in Example 3-2.

Any line along which hydraulic head is constant (e.g., the shore of a lake) is by definition an isopotential.

4. Flow nets are drawn with a constant ratio of spacing between isopotentials to spacing between streamlines everywhere in the flow net. It is usually helpful to make this ratio equal to one, so that the areas enclosed by isopotentials and associated streamlines are approximately square. It can be shown that in a properly drawn flow net in a homogeneous medium, if any of the enclosed regions are square, all must be square, although the size of the squares can vary throughout the region of water flow.

5. Under conditions of no recharge and steady flow, the water table is a no-flow boundary and is perpendicular to isopotentials.

The procedure for drawing a flow net is iterative. In practice, an initial trial flow net is drawn; on inspection, it usually becomes evident that some regions of the flow net violate one or more of the preceding rules. For example, one region of the net may have squares, while other regions are filled with elongated rectangles. The regions where the squares cannot be drawn without violating additional rules must then be modified by redrawing isopotentials and streamlines. It is helpful to work in pencil and have a large eraser handy! Redrawing is continued until a satisfactory flow net, meeting all the rules, is complete. If the boundaries have been specified correctly, such a flow net contains all the information necessary to determine both specific discharge and hydraulic head at any point.

Calculating Specific Discharge from a Flow Net

The procedure for taking a properly drawn flow net and computing specific discharge is straightforward. First, the number of isopotentials between two boundaries of known head is counted. Because difference in head between any two adjacent isopotentials is the same, the head drop from one isopotential to the next is equal to the known total head difference between constant head boundaries divided by the number of intervals demarcated by the isopotentials. The hydraulic head at each isopotential then can be labeled. Next, specific discharge is computed for any convenient point on the flow net, using the known head difference between the two adjacent isopotentials, the spacing between the isopotentials, the hydraulic conductivity, and Darcy's law. Once specific discharge is known, it can be multiplied by the spacing between the corresponding streamlines to get the total water discharge in the *stream tube* (the region bordered by two streamlines). Discharge along a stream tube is both constant along its length and equal to the discharge in any other stream tube in the flow net.

By knowing specific discharge at all points in the diagram, one can calculate seepage velocity by dividing the specific discharge at each point by the porosity. The amount of time it will take a parcel of groundwater or a nonsorbing chemical to travel from one isopotential to the next is equal to the spacing between the two isopotentials divided by the seepage velocity in the square. Travel time along an entire stream tube is the sum of the travel times across the squares along the flow path. The advective flux density of a chemical in a stream tube is the product of the chemical concentration in the tube and the specific discharge in the tube.

EXAMPLE 3-2

Assume the flow net shown in Fig. 3-8 was drawn for an aquifer 3 m thick with a hydraulic conductivity of 10^{-3} cm/sec and a porosity of 0.24. Calculate specific discharge in square B, discharge in each stream tube, and total aquifer discharge. What is the travel time for a dissolved chemical across square B? From point x to point y?

In the stream tube labeled A, there are 14 intervals defined by isopotentials between the two boundaries of known head (2 and 4.9 m). In square B, the head gradient is the head change, 0.2 m, divided by the spacing between isopotentials, 20 m. By using Eq. [3-2], the specific discharge in square B can be estimated as

$$q = -(10^{-3}\ \text{cm/sec}) \cdot (-0.2\ \text{m}/20\ \text{m}) = 10^{-5}\ \text{cm/sec}.$$

Discharge in the stream tube at square B then can be estimated using the width of square B (streamline to streamline) and the thickness of the aquifer:

$$Q_{\text{tube}} = \frac{10^{-5}\ \text{cm}}{\text{sec}} \cdot \frac{10^{-2}\ \text{m}}{\text{cm}} \cdot 20\ \text{m wide} \cdot 3\ \text{m thick} = 6 \times 10^{-6}\ \text{m}^3/\text{sec}.$$

According to the rules of flow nets, discharge is equal in each stream tube. Thus, total aquifer discharge is equal to the number of stream tubes multiplied by the discharge in each tube:

$$Q_{\text{aquifer}} = 4 \cdot 6 \times 10^{-6}\ \text{m}^3/\text{sec} = 2.4 \times 10^{-5}\ \text{m}^3/\text{sec}.$$

To estimate the travel time for a dissolved chemical across square B, the groundwater seepage velocity in square B must first be estimated using Eq. [3-5]:

$$v = \frac{10^{-5}\ \text{cm/sec}}{0.24} = 4.3 \times 10^{-5}\ \text{cm/sec} = 14\ \text{m/year}.$$

Travel time across square B then can be estimated as

$$\frac{20\ \text{m}}{14\ \text{m/year}} = 1.5\ \text{year}.$$

Travel time from point x to point y is the sum of travel times along the flow tube, [i.e., the sum of the travel times through each square (isopotential to isopotential)]. Given that the travel time for square B already has been calculated, estimate the travel time, τ_{square}, for each of the other squares in the stream tube as follows:

$$\tau_{\text{square}} = \tau_{\text{square B}} \cdot \frac{(\text{length of square})^2}{(\text{length of square B})^2}.$$

Add the travel times over all the squares to estimate the travel time from point x to point y as approximately 22 years.

3.2.3 Groundwater Wells

To characterize flow in the subsurface environment, groundwater wells often are installed. Wells can be used to measure hydraulic head gradients, and thus to determine the direction of groundwater flow; to conduct aquifer tests; to monitor groundwater quality through periodic sampling; and to alter groundwater flow or remove groundwater in order to *remediate* (clean up) a contaminated aquifer. Of course, the primary purpose of groundwater wells is the extraction of water for human use. This is precisely the reason why groundwater contamination is of such concern, because the drinking of groundwater provides a direct route for human exposure to subsurface pollutant chemicals.

Construction of Wells

Regardless of their purpose, most wells consist of a vertical hole drilled from the ground surface into the aquifer, with a slotted screen or similar construction through which water may flow from the surrounding porous media into the hole. Often much of the well is lined, or *cased,* with steel or polyvinyl chloride (PVC) plastic, with a *screen* at the desired depth to provide a flow path into the well from the porous aquifer material (see Fig. 3-9). A packing of gravel may be placed between the screen and the surrounding aquifer to maximize ease of water movement to the screen and prevent plugging by fine particles; often, a deliberate attempt is made to wash fine particles away from the vicinity of the screen to create a *natural gravel pack.* To minimize the vertical flow of water along the casing and into the screened area, some type of sealing material, such as bentonite or concrete, may be placed between the casing and the porous media above the screen. A seal also may be placed around the well casing at the ground surface to prevent surface water from seeping downward along the casing into the well. The construction of wells is an art, relying on the skill and experience of the drilling personnel and their "feel" for soil and rock conditions. The size and construction of a well depend on its intended use, with production wells used for water supply being larger than wells used for monitoring groundwater quality or for measuring hydraulic heads. A *piezometer* is a type of well whose primary purpose is the measurement of head at a given location and depth; diameters of at most a few centimeters are commonly used, and the interval of screen is short and centered at the depth where the head measurement is desired.

When water is pumped from a production or remediation well, water from the surrounding aquifer enters the well in response to the head gradient created by the water removal. This leads to a lowering of the hydraulic head around the well, forming a *cone of depression* (Fig. 3-10). It takes some time for this cone of depression to fully develop; while it is developing, the flow of water must be analyzed by *transient* techniques that account for flow changes over time (see Section 3.2.4). Ultimately, the water removed by a well must be replaced—in an unconfined aquifer, this is usually by rainwater percolation or by inflow from a river—or else the well will go dry. Wells often are

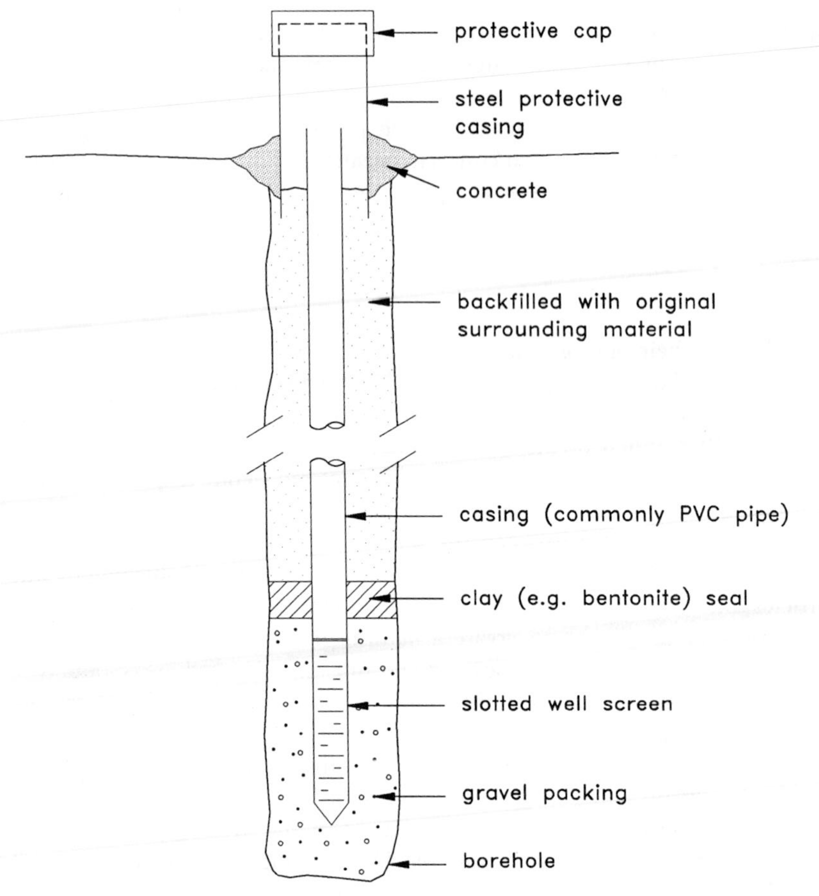

FIGURE 3-9 Cross section of a typical groundwater monitoring well. Groundwater production wells are similar but may be much larger in diameter.

operated at a *safe yield* less than their maximum flow capability to prevent running dry.

Darcy's Law Applied to Wells

At steady state, the principal of conservation of mass dictates that the rate at which water crosses an imaginary cylindrical boundary at a radius r from the well must be the same as the rate at which water is pumped from the well. The area of the cylindrical boundary is equal to its circumference ($2\pi r$) multiplied by its height (b), which is approximated by the saturated thickness

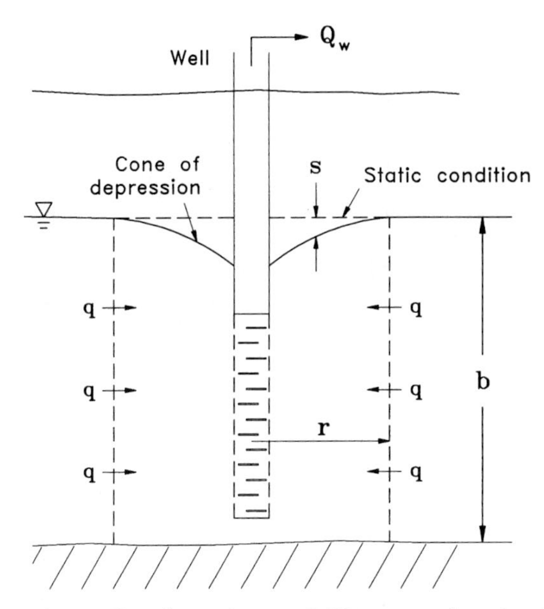

FIGURE 3-10 Steady-state flow of water into a well. Water passes through an imaginary cylinder of radius r and height equal to the aquifer thickness (b). Conservation of mass requires that water flow out of the imaginary cylinder (which equals Q_w) must equal flow into it $(2\pi rb) \cdot (q)$. Because, by Darcy's law, q is given by the product of hydraulic conductivity (K) and gradient (ds/dr), this flow also is equal to $(2\pi rb) \cdot (K) \cdot (ds/dr)$.

of the aquifer. The specific discharge at any point is equal to the product of hydraulic conductivity (K) and hydraulic gradient (dh/dr). Therefore, the rate at which water is pumped from the well is given by

$$Q_w = -K(dh/dr) \cdot 2\pi rb, \qquad [3\text{-}6]$$

where Q_w is the rate at which water is pumped from a well [L^3/T], K is the hydraulic conductivity [L/T], dh/dr is the hydraulic gradient in the radial direction (change in hydraulic head divided by change in radial distance) [L/L], r is the radial distance from the well [L], and b is the thickness of aquifer from which water is being drawn [L]. For convenience, changes in water table height (h) also can be expressed in terms of the distance the water table lies below the height it had in the absence of pumping (the *static condition*); this distance is defined as the *drawdown* (s), as shown in Fig. 3-10. Therefore, the quantity ($-dh/dr$) in Eq. [3-6] can be replaced with ds/dr:

$$Q_w = K(ds/dr) \cdot 2\pi rb, \qquad [3\text{-}7a]$$

or

$$ds/dr = Q_w/2\pi Krb. \qquad [3\text{-}7b]$$

Note that the slope of the water table in the radial direction, given by the gradient *ds/dr*, is proportional to the rate at which the well is pumped; the slope is inversely proportional to hydraulic conductivity, distance from the well, and aquifer thickness.

Equations [3-6], [3-7a], and [3-7b] assume that the saturated aquifer thickness b is constant (as in a confined aquifer). In unconfined (phreatic) aquifers, which are somewhat more susceptible to subsurface contamination, the saturated thickness varies as the hydraulic head changes; thus b is not, strictly speaking, constant. Unless otherwise stated, however, it is assumed in the following discussions that changes in the water table height of a phreatic aquifer are relatively small compared with the saturated thickness. When this is not the case, more complex expressions are needed to describe the hydrodynamics, and the reader is referred to Bear (1979).

Steady-State Drawdown in an Aquifer or Well

In designing a production or remediation well for an aquifer, it is important to be able to predict total drawdown in the aquifer. Theoretically, in an idealized aquifer having unlimited extent and no recharge, the cone of depression advances outward to infinity. Thus, steady-state analysis is approximate and useful only after pumping has occurred for some time. Steady-state drawdown at any given radius r_1 from the well, relative to drawdown at another radius r_2, can be determined by integrating Eq. [3-7b],

$$s_1 - s_2 = (Q_w/2\pi Kb) \cdot \ln(r_2/r_1), \qquad [3\text{-}8a]$$

where s_1 is the drawdown at a radial distance r_1 from the well [L], s_2 is the drawdown at a radial distance r_2 from the well [L], Q_w is the rate at which water is pumped from a well [L^3/T], K is the hydraulic conductivity [L/T], and b is the aquifer thickness [L].

In practice, absolute drawdown may be estimated if it is possible to define a *radius of influence*, *R*, which represents the horizontal distance beyond which pumping of the well has little influence on the aquifer; that is, beyond R, no significant drawdown due to pumping is assumed to exist (see also Section 3.2.4). A variety of formulae for estimating R are given by Lembke (1886 and 1887); one formula is

$$R = b\sqrt{K/2N},$$

where N is annual recharge by precipitation. Sometimes R can be based on hydrogeologic conditions, such as the presence of a known constant-head boundary. Given an appropriate value for R and substituting it in Eq. [3-8a],

$$s = (Q_w/2\pi Kb) \cdot \ln(R/r), \qquad [3\text{-}8b]$$

where s is the drawdown at a radial distance r from the well [L], Q_w is the rate at which water is pumped from a well [L^3/T], K is the hydraulic conductivity [L/T], b is the aquifer thickness [L], and R is the radius of influence [L].

One special case of Eq. [3-8b] is the drawdown in the well casing itself; Eq. [3-8b] gives the drawdown, s_w, in the well if r is set equal to r_w, the radius of the well:

$$s_w = (Q_w/2\pi\, Kb) \cdot \ln(R/r_w) \qquad [3\text{-}8c]$$

Equations [3-8a] to [3-8c] are different applications of the *Thiem equation,* which estimates drawdown in an aquifer or well under steady-state conditions. As previously mentioned, it is assumed that the changes in saturated aquifer thickness are small compared with the total saturated depth. This is necessarily true in a confined aquifer, but not always in an unconfined (phreatic) aquifer. If drawdown becomes a significant fraction of the saturated aquifer thickness, more complicated expressions for drawdown are obtained; see Bear (1979). For an unconfined aquifer in which drawdown is a significant fraction of the saturated thickness, Eq. [3-8a] must be expressed in terms of head instead of drawdown:

$$h_2^2 - h_1^2 = (Q_w/K\,\pi) \cdot \ln(r_2/r_1), \qquad [3\text{-}9]$$

where h_2 is the hydraulic head at a radial distance r_2 from the well [L], and h_1 is the hydraulic head at a radial distance r_1 from the well [L].

In Eqs. [3-8a] and [3-8b], the quantity $(K \cdot b)$ appears. This quantity measures the ability of an aquifer to deliver water to a well and is called *transmissivity*—abbreviated T, with units [L^2/T].

The groundwater flow into a well also can be analyzed by using a flow net. Because the well is symmetrical, lines of constant drawdown (isopotentials) are circles centered at the center of the well (Fig. 3-11). Streamlines are arranged radially around the well if there is no regional flow in the aquifer.

For a confined aquifer, even if there is flow in the aquifer in the absence of the well, the resulting flow field due to both the well pumping and the preexisting flow can be determined by applying a simple but powerful concept called *superposition.* Given a flow net showing head values for preexisting flow in a confined aquifer, one can subtract the drawdown at each point in the aquifer caused by well pumping (as shown in Fig. 3-11) to create the hydraulic head distribution that would actually result from placement and operation

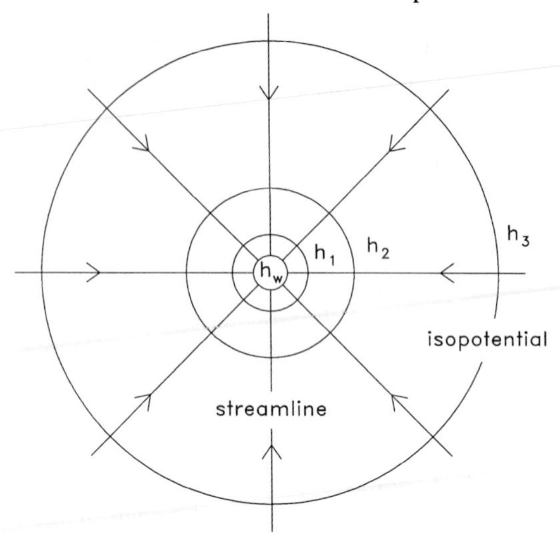

FIGURE 3-11 The flow net resulting from pumping a well in a spatially uniform aquifer. Hydraulic head in the well, h_w, is lower than in the surrounding aquifer, and water flow is radially inward. The hydraulic gradient is greatest near the well, and decreases radially outward. Note that, by virtue of the rules for flow nets, the change in head between any two adjacent isopotentials (which are circular here because of the symmetry of the situation) is constant throughout the diagram.

of a well in the aquifer. When the resultant isopotentials are drawn, and streamlines are extended through these new isopotentials following the rules for constructing flow nets, an accurate flow net that includes the effect of the pumping well results. This simple technique of superposition allows quantitative flow solutions to be obtained in even complex situations.

Superposition in phreatic aquifers is more complicated than superposition in confined aquifers, because drawdown and hydraulic head are not additive. However, for unconfined aquifers in which drawdown is a sufficiently small fraction of aquifer thickness, the technique of superposition can be used as an approximation. The reader is referred to Strack (1989) for further details on superposition in unconfined aquifers.

EXAMPLE 3-3

The hydraulic conductivity in a confined aquifer 3 m thick is estimated to be 10^{-2} cm/sec. The hydraulic gradient is approximately 0.0012, and the porosity is 0.25. A well 20 cm in diameter is installed in the aquifer and is pumped at an average rate of 10 liter/min.

a. What is the drawdown in this well if the radius of influence is 75 ft?
b. Estimate the hydraulic gradient in the aquifer at distances of 5 and 20 m, both directly upstream and downstream of the well.

a. Drawdown in the well at steady state can be estimated using the Thiem equation in the form of Eq. [3-8c]:

$$s_w = \frac{10 \text{ liter/min}}{(2\,\pi)(10^{-2} \text{ cm/sec} \cdot 3 \text{ m})} \cdot \frac{1 \text{ min}}{60 \text{ sec}} \cdot \frac{1000 \text{ cm}^3}{1 \text{ liter}} \cdot \frac{1 \text{ m}}{100 \text{ cm}}$$

$$\cdot \ln\left(\frac{(75 \text{ ft} \cdot 0.3048 \text{ m/ft})}{0.1 \text{ m}}\right) = 48 \text{ cm}.$$

b. Gradients due to the well being pumped can be superimposed on the aquifer's hydraulic gradient in the absence of pumping. Use Eq. [3-7b] to estimate the gradient, ds/dr, created by the well.

At 5 m:

$$\frac{ds}{dr} = \frac{10 \text{ liter/min}}{(2\,\pi)(3 \text{ cm}^2\text{/sec})(500 \text{ cm})} \cdot \left(\frac{1 \text{ min}}{60 \text{ sec}}\right) \cdot \left(\frac{1000 \text{ cm}^3}{1 \text{ liter}}\right) = 0.018.$$

At 20 m, using Eq. [3-7b], ds/dr is equal to 0.0044.

Upstream of the well, the hydraulic gradient will be increased due to pumping:

at 5 m: gradient = 0.0012 + 0.018 = 0.019

at 20 m: gradient = 0.0012 + 0.0044 = 0.0056.

Within the radius of influence downstream of the well, the hydraulic gradient of the aquifer is sloping away from the well, but maintains the same sign; in contrast, the pumping of the well tends to cause water to flow in the opposite direction, toward the well, and thus the gradient sign is reversed:

at 5 m: gradient = 0.0012 − 0.018 = −0.017

at 20 m: gradient = 0.0012 − 0.0044 = −0.0032.

At all four points, the hydraulic gradient is toward the well.

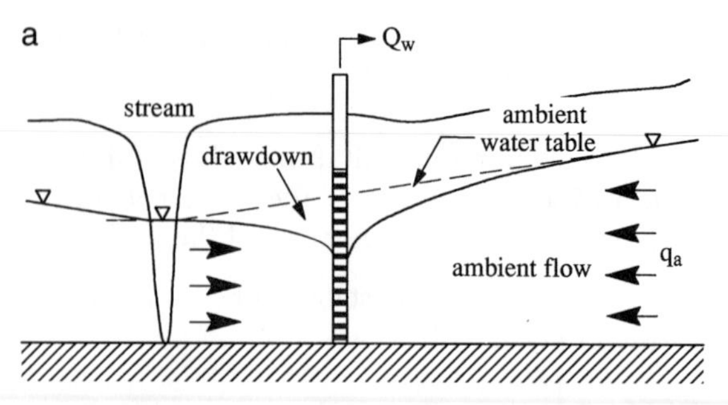

FIGURE 3-12 (a) A well located in an aquifer that is hydraulically connected to a stream. In this idealized situation, stream depth is equal to aquifer thickness b. q_a is the specific discharge that would occur in the aquifer in the absence of pumping. (b) The rate of well pumpage (Q_w) determines whether water from the stream enters the well. In the top two panels, the well does not draw in stream water; $Q_w \leq \pi dbq_a$. In the bottom panel, the streamlines show the section of stream from which water is drawn into the well. (c) The fraction of water in the well that comes from the stream (Q_s/Q_w) is plotted as a function of the dimensionless quantity $Q_w/\pi dbq_a$. Induced recharge from the stream is zero if Q_w is less than or equal to πdbq_a (adapted from Wilson, 1993).

Capture Curves

The flow net is a helpful tool for depicting the area of an aquifer from which a well captures water. It is essential to be able to delineate this area to avoid pulling contaminated groundwater into a well used for drinking water production; areas of known contamination must lie outside the well's *capture zone*. Alternatively, if a well is to be used for groundwater remediation (removal of contaminated groundwater), the capture zone must enclose the contaminated areas. *Capture curves*, which are the boundaries of capture zones, may be drawn readily by inspection if a flow net is available; the capture zone includes all stream tubes that terminate in the well. The following are two examples of capture curves.

Capture of River Water

Figure 3-12a depicts a well located in an aquifer that is hydraulically connected to a stream. Figure 3-12b shows the streamlines associated with three different rates of well pumpage (Q_w). In the absence of pumping from the well, water flow in the aquifer would be perpendicular to the stream, and isopotentials would be parallel to the stream; the stream is assumed to be a constant head boundary. (For this to be a good approximation, the stream depth must be a large fraction of the aquifer thickness, as shown.) When the

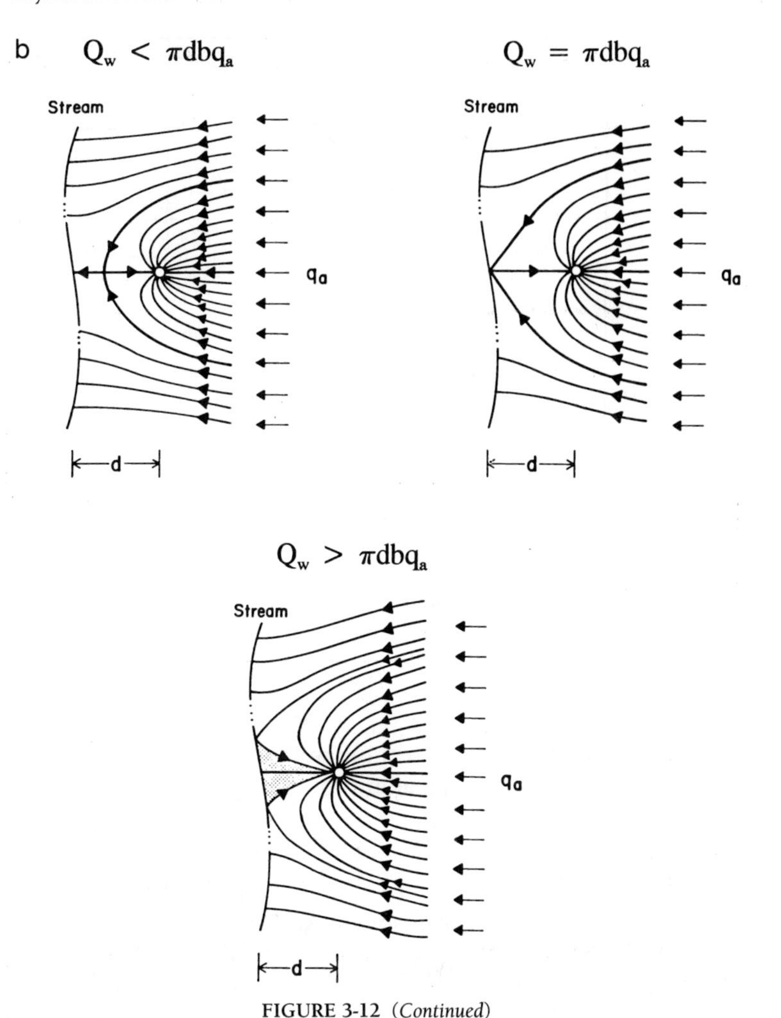

FIGURE 3-12 (*Continued*)

well is pumped, water is drawn from the stream into the well (bottom panel of Fig. 3-12b) if the pumpage rate exceeds the quantity πdbq_a, where d is the distance between the stream and the well [L], b is the aquifer thickness [L], and q_a is the specific discharge that would occur in the aquifer in the absence of pumping [L/T]. Figure 3-12c shows the fraction of water in the well that comes from the stream (Q_s/Q_w) as a function of the dimensionless quantity $Q_w/\pi dbq_a$. When Q_w is less than or equal to πdbq_a (top two panels of Fig. 3-12b), no water is drawn from the stream into the well. Regardless of

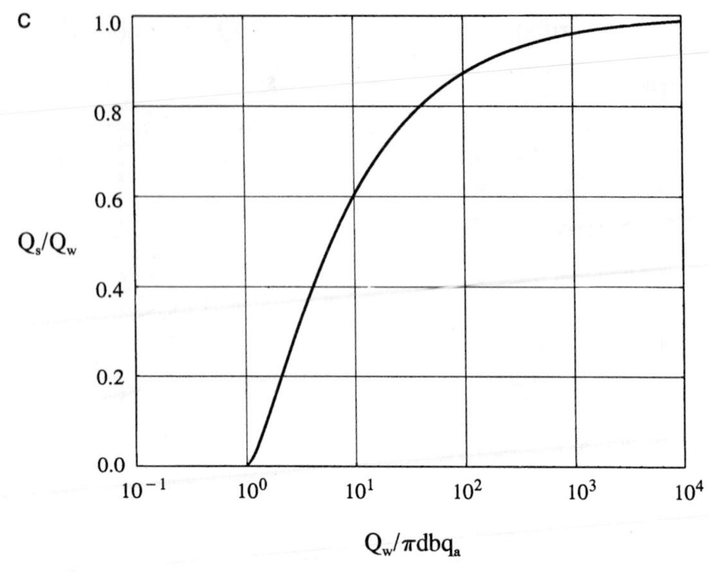

FIGURE 3-12 *(Continued)*

whether water from the stream enters the well, when the well is pumping at steady state, the capture zone is that area of the aquifer traversed by streamlines that enter the well.

Capture Zone of One or More Wells in an Aquifer

Figure 3-13 shows a set of curves representing the zone from which water is pulled into a well located in an aquifer having a uniform background flow. The width of the region from which water is intercepted increases as the pumping rate increases and decreases as the regional specific discharge of the aquifer increases. To use these curves, the quantity Q_w/bq_a is calculated (Q_w is the well pumpage rate, b is the aquifer thickness, and q_a is the specific discharge that occurs in the aquifer in the absence of pumping of the well). The corresponding capture curve is then taken from the figure. Capture curves can be overlaid on a map of a site and used to determine whether a well will draw in water from a contaminated region. Such a curve can be used either to minimize the likelihood of drawing in contaminated groundwater or to locate wells and maximize the extent to which they capture contaminated groundwater in a remediation scheme.

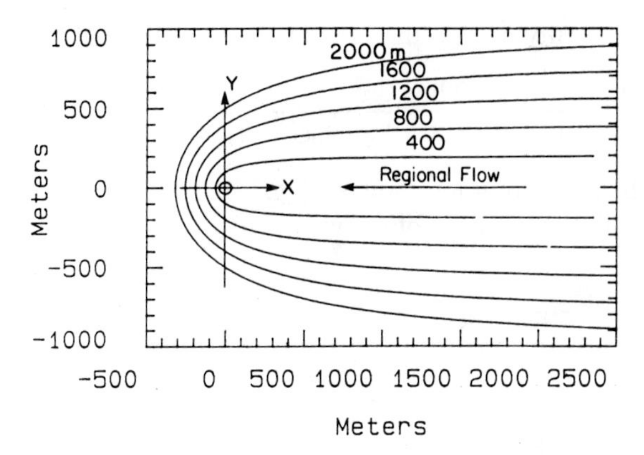

FIGURE 3-13 Type curves for the capture zone of a single pumping well located at the point (0,0), for several values of Q_w/bq_a. In the absence of dispersion, all water lying within the capture zone, along with any contaminants it may be carrying, will eventually end up in the well. Note that these curves do *not* form a flow net (adapted from Javandel and Tsang, 1986).

EXAMPLE 3-4

A well serving a small brewery is pumped at an average rate of 10 liter/sec. Customer complaints about the beer led to the discovery of hydrocarbons, possibly from gasoline, in the well water. The brewery has filed suit against every gasoline service station within a 2.5-km radius of the well. Station A is 2000 m north and 300 m east; station B is 300 m due east; and station C is 200 m southwest. Hydrogeologists determine that the phreatic aquifer is in an infinite flow domain, is 15 m thick, has a transmissivity of 0.003 m^2/sec, and has a regional hydraulic gradient of 0.004 from north to south. Which of the lawyers will have a difficult time defending his or her client?

First, calculate the hydraulic conductivity from the transmissivity and aquifer thickness:

$$K = T/b = (0.003\ \text{m}^2/\text{sec})/15\ \text{m} = 2 \times 10^{-4}\ \text{m/sec}.$$

Then use Eq. [3-2] to estimate the specific discharge of the aquifer:

$$q_a = -K\,dh/dx = -(2 \times 10^{-4}\ \text{m/sec}) \cdot -(0.004) = 8 \times 10^{-7}\ \text{m/sec}.$$

Calculate the quantity used in the capture curves of Fig. 3-13:

$$\frac{Q_w}{bq_a} = \frac{10 \text{ liter/sec}}{(15 \text{ m})(8 \times 10^{-7} \text{ m/sec})} \cdot \frac{1 \text{ m}^3}{1000 \text{ liter}} = 830 \text{ m}.$$

By plotting the gas station locations on Fig. 3-13, it appears that only station A lies within the capture zone of the well (use the 800-m curve).

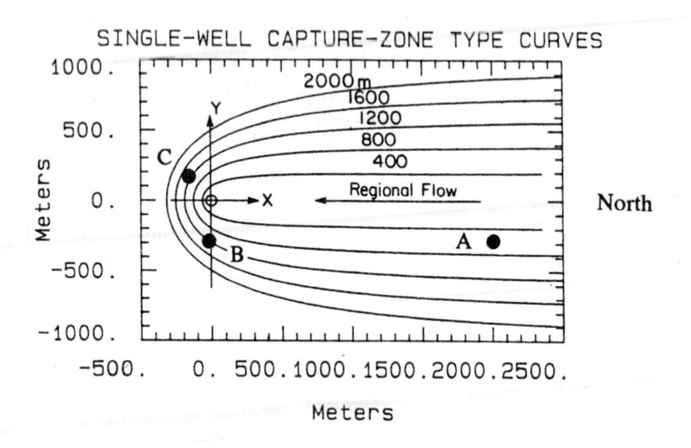

3.2.4 Unsteady (Transient) Storage and Flow

In the preceding discussions, the flow of water in an aquifer was assumed to be at steady state (i.e., unchanging over time), which is a reasonable approximation in many situations. For example, if the movement of a plume of contaminated groundwater over a period of several years is of concern, average seepage velocities, which neglect fluctuations throughout any given year, often can be used. Likewise, if a remediation well is expected to be pumping out contaminated water for many months, the fact that aquifer flows are changing throughout the first few days of well operation probably can be ignored.

In other situations, however, rapidly changing flow velocities and hydraulic heads must be taken into account. For example, a municipal drinking water well operator may need to know how the spread of a recently spilled chemical will be influenced by short-term changes in the operation of the well field. A hydrogeologist may analyze the magnitude and the rate of head change that occur in the seconds following implacement of a slug into a well to give information on the hydraulic conductivity and water storage characteristics of an aquifer. *Unsteady,* or *transient,* flow varies with time; in unsteady flow, the amount of water stored in the aquifer, and the head at various points in an aquifer, also change with time. For example, before a well is pumped, regional flow may be steady, and the water table may be quite stable in the vicinity of the well. After pumping has occurred for a long time, a steady state may be achieved with a stable cone of depression established. Between these times, an ever-growing drawdown zone is expanding outward from the location of the well, and water is being taken from storage in the aquifer. Analysis of this dynamic situation requires understanding the relationship between the change in water storage and the associated change in hydraulic head.

Specific Yield and Storativity

In an unconfined aquifer (as shown in Fig. 3-1), the amount of water stored can be related to the hydraulic head, because both are proportional to the height of the water table. As water table height changes, the hydraulic head is raised or lowered by the same amount. The ratio of the change in *depth* of water stored (change in volume of water per unit surface area) divided by a unit decline in hydraulic head is called the *specific yield,* S_y. In a coarse porous medium, S_y is approximately equal to the porosity. Sometimes S_y can be estimated from the change in head accompanying a known amount of recharge by a rainfall event.

In a confined aquifer, the quantity corresponding to specific yield is called *storativity,* S, and is typically much smaller than S_y. Storativity is a function of the compressibility of water and aquifer material and is proportional to aquifer thickness. A related term is *specific storage,* S_s, which is storativity divided by aquifer thickness. Storativity increases as thickness of the aquifer increases, whereas specific yield is not a function of aquifer thickness. Removal of large volumes of water from confined aquifers can result in measurable ground subsidence.

Transient Flow in an Aquifer

A two-dimensional differential equation expressing the relationship between water flow and water storage in an aquifer can be written by equating the *net* groundwater flow (as given by Darcy's law) entering a point in the aquifer to the rate of change of storage in the aquifer at that same point. If q_x is specific discharge in the horizontal x direction and q_y is specific discharge in the horizontal y direction, a gradient in q_x or q_y at a point means that a net flow of water is occurring either into or out of the point. In an unconfined aquifer, the rate at which water is entering or leaving the point also is equal to the specific yield multiplied by the rate of change of hydraulic head (or water table height) and divided by aquifer thickness. Thus, by conservation of mass,

$$S_y \cdot dh/dt = b(-dq_x/dx - dq_y/dy). \quad [3\text{-}10a]$$

Because q_x is equal to $-K \cdot dh/dx$ and q_y is equal to $-K \cdot dh/dy$, mass conservation of water at the point also can be written as

$$S_y \cdot dh/dt = Kb[(d^2h/dx^2) + (d^2h/dy^2)]. \quad [3\text{-}10b]$$

Eq. [3-10b] assumes that the aquifer is horizontally isotropic (i.e., K is the same in both x and y directions). Solutions to the preceding differential equations under appropriate boundary and initial conditions describe the time-varying hydraulic head, h, in two dimensions. Several solutions are given by Carslaw and Jaeger (1959). More complicated boundary and initial conditions are treated by Hantush (1964), Reed (1980), and Wang and Anderson (1982). Note that in Eq. [3-10b] the quantity (Kb) could be replaced by T, the aquifer transmissivity.

Theis Equation

A common application of a solution to Eq. [3-10b] is the prediction of time-varying drawdown in the vicinity of a well that is pumped at a constant rate. The *Theis equation* (Theis, 1935) was derived for confined aquifers and applies to unconfined aquifers for some period of time after a well begins to be pumped. The Theis equation can be written as

$$s(r, t) = [Q_w/4\pi T]W(u), \quad [3\text{-}11]$$

where $s(r, t)$ is the drawdown [L] at time t and radius r from the well, Q_w is the rate of well pumpage [L^3/T], T is the aquifer transmissivity [L^2/T], and $W(u)$ is the well function—the exponential integral—which is widely available in tabulated form (Abramowitz and Stegan, 1965). Table 3-2 lists values of $W(u)$ for various values of u, which also is a dimensionless quantity:

$$u = (r^2 \cdot S)/(4\ T \cdot t). \quad [3\text{-}12]$$

As shown by Cooper and Jacob (1946), Eq. [3-11] can be written for small values of u (less than 0.01) as

$$s(r, t) = \left(\frac{0.183\ Q_w}{T}\right)\log\left(\frac{2.25\ Tt}{r^2S}\right), \quad [3\text{-}13]$$

where $s(r, t)$ is the drawdown [L] at time t and radius r from the well, Q_w is the rate of well pumpage [L^3/T], T is the aquifer transmissivity [L^2/T], t is time [T], r is the radius from the well [L], and S is the storativity [dimensionless]. In Eq. [3-13], the drawdown is proportional to $\log(t)$, indicating that

TABLE 3-2 The Well Function, $W(u)$[a,b]

u	1.0	2.0	3.0	4.0	5.0	6.0	7.0	8.0	9.0
$\times 1$	0.219	0.049	0.013	0.0038	0.0011	0.00036	0.00012	0.000038	0.000012
$\times 10^{-1}$	1.82	1.22	0.91	0.70	0.56	0.45	0.37	0.31	0.26
$\times 10^{-2}$	4.04	3.35	2.96	2.68	2.47	2.30	2.15	2.03	1.92
$\times 10^{-3}$	6.33	5.64	5.23	4.95	4.73	4.54	4.39	4.26	4.14
$\times 10^{-4}$	8.63	7.94	7.53	7.25	7.02	6.84	6.69	6.55	6.44
$\times 10^{-5}$	10.94	10.24	9.84	9.55	9.33	9.14	8.99	8.86	8.74
$\times 10^{-6}$	13.24	12.55	12.14	11.85	11.63	11.45	11.29	11.16	11.04
$\times 10^{-7}$	15.54	14.85	14.44	14.15	13.93	13.75	13.60	13.46	13.34
$\times 10^{-8}$	17.84	17.15	16.74	16.46	16.23	16.05	15.90	15.76	15.65
$\times 10^{-9}$	20.15	19.45	19.05	18.76	18.54	18.35	18.20	18.07	17.95
$\times 10^{-10}$	22.45	21.76	21.35	21.06	20.84	20.66	20.50	20.37	20.25
$\times 10^{-11}$	24.75	24.06	23.65	23.36	23.14	22.96	22.81	22.67	22.55
$\times 10^{-12}$	27.05	26.36	25.96	25.67	25.44	25.26	25.11	24.97	24.86
$\times 10^{-13}$	29.36	28.66	28.26	27.97	27.75	27.56	27.41	27.28	27.16
$\times 10^{-14}$	31.66	30.97	30.56	30.27	30.05	29.87	29.71	29.58	29.46
$\times 10^{-15}$	33.96	33.27	32.86	32.58	32.35	32.17	32.02	31.88	31.76

[a] Linsley *et al.* (1982).

[b] Each entry in the table represents the value of $W(u)$ for a value of u equal to the product of the top row and the left column of the table. For values of u less than 10^{-2}, $W(u)$ may be approximated as $W(u) = -0.5772 - \ln(u)$.

pumping data may be plotted on semilog paper as straight lines (Fig. 3-14), and aquifer transmissivity T can be determined by noting Δs_{10}, the difference in drawdown between any two values of time differing by a factor of 10:

$$T = 0.183\, Q_w/\Delta s_{10}. \qquad [3\text{-}14]$$

Much can be learned about an aquifer by comparing the theoretical drawdown calculated by the Theis equation with actual drawdown data obtained over a long time period from a pumping test. For example, Fig. 3-15 shows evidence of aquifer recharge, perhaps from a river or a lake, because drawdown levels off after prolonged pumping instead of continuing to follow the theoretical curve. In contrast, drawdown that increases more rapidly at long times than predicted by the theoretical curve extrapolated from early pumping times (Fig. 3-16) may indicate limited geographic extent of an aquifer. Other problems also may be solved with the Theis equation; for example, fluctuating drawdown in an aquifer surrounding an intermittently pumped well may be calculated using *superposition* of several Theis solutions, each containing a time lag corresponding to a time interval between pump startup and shutdown. For further information, the reader is referred to Driscoll (1986) and Kruseman and de Ridder (1983).

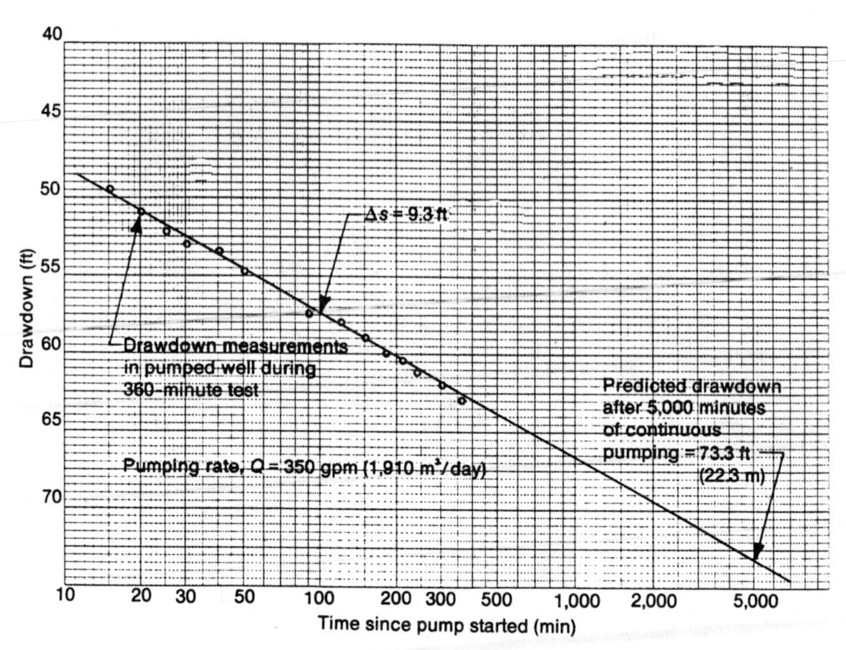

FIGURE 3-14 Plot of drawdown as a function of time for a well in an infinite aquifer with no recharge. Theoretically, as described by the Theis equation, drawdown increases indefinitely after pumping has begun. The graph can be extended beyond the time of the test to predict the amount of drawdown that would occur after a longer period of continuous pumping (Driscoll, 1986).

3.2.5 Dispersion

Mechanical Dispersion

Chemicals dissolved in groundwater tend to spread out as the groundwater moves, just as chemicals spread out in a lake or a stream. This mixing causes dilution of a mass of chemical into an increasingly larger volume of water, and also may cause a chemical to appear earlier at a downgradient point than would be predicted based only on groundwater velocity. In the groundwater environment, mixing is normally not dominated by turbulence, as it is in most surface waters, because the flow of groundwater is much slower. Most mixing is due to the tortuous, winding paths that water must follow as it travels through porous media. As illustrated previously in Fig. 1-7, some parcels of water follow wide, direct routes, while others follow narrow paths that zigzag back and forth at substantial angles to the average direction of flow. This process is called *mechanical dispersion.*

Groundwater dispersion can be treated mathematically in the same way as

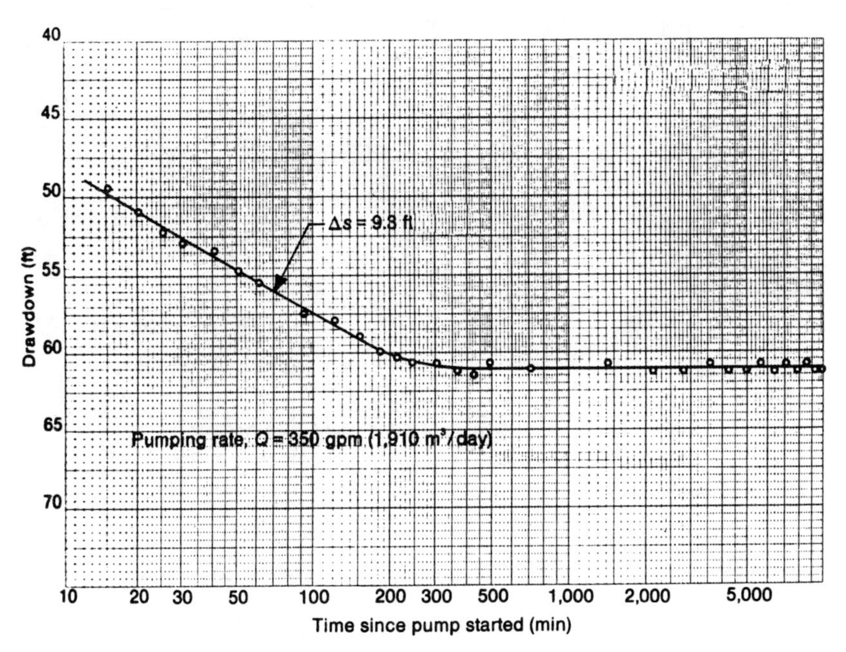

FIGURE 3-15 Plot of drawdown as a function of time for a well in an aquifer that receives recharge from a source such as a river or another aquifer; note that drawdown theoretically can reach a steady state in this case (Driscoll, 1986).

turbulent diffusion and dispersion in surface water are treated, by applying Fick's first law. In one dimension the dispersion coefficient, D, is often approximated by

$$D = \alpha \cdot v, \qquad [3\text{-}15]$$

where D is the mechanical dispersion coefficient [L^2/T]; α is the *dispersivity* of the aquifer, approximately equal to the median grain diameter of the aquifer solids [L]; and v is the seepage velocity [L/T]. Table 3-3 shows ranges of grain diameters for different porous media; see also Klotz *et al.* (1980) and Fried and Gombarnous (1971).

In two- or three-dimensional flow, mixing occurs not only along the axis of flow but also along axes perpendicular to the flow. Longitudinal dispersion, which occurs in the direction of seepage velocity, is normally larger than dispersion perpendicular to flow. In an anisotropic aquifer, the components of D have a different relationship to water velocity depending on the direction of seepage velocity relative to the axes of anisotropy. For analysis of D as a tensor quantity, the reader is referred to Gelhar and Axness (1983).

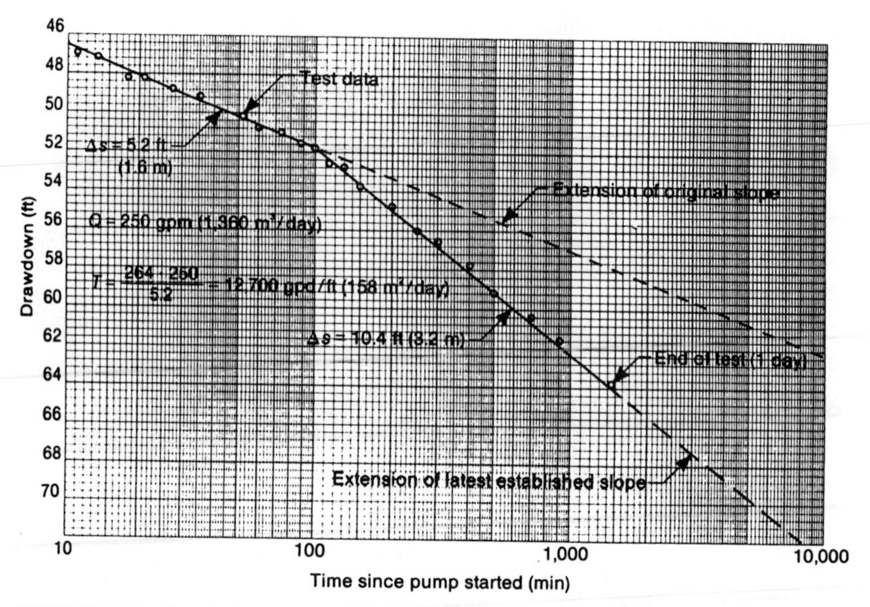

FIGURE 3-16 Plot of drawdown as a function of time for a well in an aquifer having a finite extent. Note that drawdown increases more rapidly at longer times than it does in the case of the infinite aquifer (Fig. 3-14). Comparison of actual well pumping data with theoretical curves allows estimation of aquifer properties, including storativity, transmissivity, extent, and connection to sources of recharge (Driscoll, 1986).

TABLE 3-3 Comparison of Particle Sizes (Diameters) of Various Porous Media (USDA System versus USCS)[a]

Particle	Size range (mm)	
	USDA	USCS
Cobbles	76.2–254	>76.2
Gravel	2.0–76.2	4.76–76.2
Coarse	12.7–76.2	19.1–76.2
Fine	2.0–12.7	4.76–19.1
Sand	0.05–2.0	0.074–4.76
Very coarse	1.0–2.0	NA
Coarse	0.5–1.0	2.0–4.76
Medium	0.25–0.5	0.42–2.0
Fine	0.1–0.25	0.074–0.42
Very fine	0.05–0.1	NA
Fines	NA	<0.074
Silt	0.002–0.05	[b]
Clay	<0.002	[b]

[a] Fuller (1978). NA, not available.

[b] USCS silt and clay designations are determined by response of the soil to manipulation at various water contents instead of by particle size.

In an aquifer, the total Fickian transport coefficient of a chemical is the sum of the dispersion coefficient and the *effective molecular diffusion coefficient*. For use in the groundwater regime, the molecular diffusion coefficient of a chemical in free water must be corrected to account for tortuosity and porosity. Commonly, the free-water molecular diffusion coefficient is divided by an estimate of tortuosity (sometimes taken as the square root of two) and multiplied by porosity to estimate an effective molecular diffusion coefficient in groundwater. Millington (1959) and Millington and Quirk (1961) provide a review of several approaches to the estimation of effective molecular diffusion coefficients in porous media. Note that mixing by molecular diffusion of chemicals dissolved in pore waters always occurs, even if mechanical dispersion becomes zero as a consequence of no seepage velocity.

Field Scale Dispersion

Often the value of D that must be used in a model to make a predicted contaminant plume correspond to actual measurements in the field is appreciably larger than the value of D given by the product of a seepage velocity and median grain size, Eq. [3-15]. This effect is due at least in part to variations in the properties of the porous media in the aquifer; for example, the presence of volumes of clay amid expanses of coarse gravel will cause portions of contaminated water to move either much more slowly or much more rapidly than average, resulting in mixing. This phenomenon is called *macrodispersion*. For a more thorough discussion of macrodispersion, and methods for estimating its effects, the reader is referred to Gelhar and Axness (1983), Güven *et al.* (1984), and Schwartz (1977). The value of D used in numerical models of groundwater transport also has a strong effect on the performance of many numerical solution techniques and sometimes is made artificially large to minimize computational errors; in this case, D begins to lose its connection to physical reality and may be better thought of as a model-fitting parameter.

One-Dimensional Dispersion in a Column

Dispersion can be demonstrated in laboratory studies in which water moves at a known seepage velocity, v, through porous media. A *conservative* chemical tracer, such as salt or dye that does not stick to particles or decay, is initially placed at the upstream end of a laboratory column (ideally as a thin slice across the column perpendicular to the direction of the seepage velocity). After the pulse injection, the tracer moves by advection along the column at a speed equal to the seepage velocity. Because there is a high concentration gradient near the edges of the tracer mass, Fickian mass transport results in

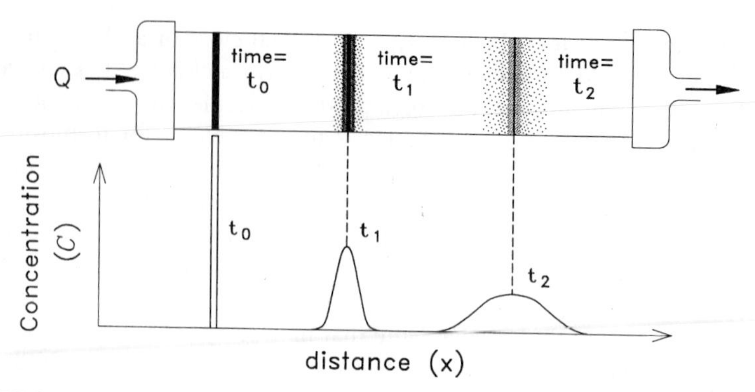

FIGURE 3-17 Dispersion of a *pulse* of a tracer substance in a sand column experiment. Note the parallel between this and the corresponding dispersion of a tracer in a flowing river (Fig. 2-4). The same equation, with a correction factor for porosity in the case of the sand column, describes both situations. However, the physical processes responsible for the Fickian transport differ; mechanical dispersion dominates in the sand column, while turbulent diffusion and the dispersion associated with a nonuniform velocity profile dominate in the river.

the spreading out of the tracer as it moves along the column, as shown in Fig. 3-17.

If the Fickian transport coefficient is known, it is possible to predict the distribution of the tracer at any time and location after it is introduced into the column. At the time of injection of the tracer ($t = 0$), the concentration is high over a short length of column. At a later time t_1, the center of the mass of tracer has moved a distance equivalent to the seepage velocity multiplied by t_1, and the mass has a broader Gaussian, or normal, distribution; see Eq. [2-6]. The solution to Eq. [1-5] for this one-dimensional situation gives the concentration of the tracer as a function of time and distance,

$$C(x, t) = \frac{M}{n\sqrt{4\pi D_L t}} e^{-(x - vt)^2/(4D_L t)}, \qquad [3\text{-}16]$$

where C is the concentration of the tracer [M/L^3], M is the mass of tracer injected per unit area of the column [M/L^2], x is the distance downgradient from the point the tracer was originally injected [L], D_L is the longitudinal Fickian transport coefficient [L^2/T], n is the porosity [L^3/L^3], t is the time since injection [T], and v is the seepage velocity [L/T]. Equation [3-16] shows that the maximum concentration decreases with both increasing time and increasing distance along the column. Note that this equation is essentially the same as Eq. [2-10], with a correction for porosity because only part of the volume of the column is filled with water available for dilution of the tracer.

If the chemical is not conservative but decays with a first-order decay rate k, Eq. [3-16] is multiplied by e^{-kt}, as in Eq. [2-11].

Another analysis of practical importance is the prediction of a *breakthrough curve*. The breakthrough curve in a one-dimensional column is the plot of chemical concentration at a *fixed* location (usually the outlet of the column) as a function of time. The breakthrough curve at a point in a column can be predicted from Eq. [3-16] by substituting in a fixed distance and by solving for concentration at various times. In an aquifer, a breakthrough curve might describe the concentration of a contaminant observed in a downstream well over time. In general, however, two- or three-dimensional analysis is required in an aquifer.

If seepage velocity changes in the laboratory column, the rate of mixing by mechanical dispersion changes proportionally. For example, if seepage velocity is halved, the dispersion coefficient is halved. However, it then takes twice as long for the tracer to reach a given location downgradient, so the time available for mixing is doubled. The effects cancel, and the amount of spreading that the tracer undergoes in traveling to that point is unchanged. This can be demonstrated by replacing D_L in Eq. [3-16] by the equivalent quantity, $\alpha x/t$, based on Eq. [3-15]. The resulting Eq. [3-17] gives concentration at the center of mass when the center of mass is a distance x from the origin. Note that the concentration is not a function of time:

$$C(x) = \frac{M}{n\sqrt{4\pi\alpha x}}. \qquad [3\text{-}17]$$

Equation [3-17] does not hold at very low seepage velocities because mechanical dispersion no longer dominates Fickian mass transport. When the mechanical dispersion coefficient becomes less than the effective molecular diffusion coefficient, the longer travel times associated with lower velocities do not result in further decreases in Fickian mass transport.

EXAMPLE 3-5

A column experiment is set up in the laboratory. Sand with a mean grain size of approximately 0.5 mm is packed into a cylindrical column, 1.5 m in length and 10 cm in diameter; water flows through the column with a seepage velocity of 1 m/hr. Porosity is 0.3. Five milligrams of salt are injected into the column (a pulse injection).

a. What will be the concentration of salt after an hour at a distance 0.9 m down the column?

b. When the tracer mass is centered 1.3 m down the column, what is the concentration of tracer at this location?

The mechanical dispersion coefficient D in the longitudinal direction can be approximated by Eq. [3-15]:

$$(0.5 \text{ mm}) \cdot (1 \text{ m}/1000 \text{ mm}) \cdot (1 \text{ m/hr}) = 5 \times 10^{-4} \text{ m}^2/\text{hr}.$$

a. Use Eq. [3-16] to obtain:

$$C(0.9 \text{ m}, 1 \text{ hr})$$

$$= \frac{(5 \text{ mg}/(\pi \cdot (0.05 \text{ m})^2)) \cdot \exp {}^{-}(0.9 \text{ m} - 1 \text{ m/hr} \cdot 1 \text{ hr})^2/(4 \cdot 5 \times 10^{-4} \text{ m}^2/\text{hr} \cdot 1}{(0.3)\sqrt{(4\pi)(5 \times 10^{-4} \text{ m}^2/\text{hr})(1 \text{ hr})}}$$

$$= 180 \text{ mg/m}^3.$$

At 0.9 m, after 1 hr, C is 0.18 mg/liter. Note that the center of mass has passed the 0.9-m mark.

b. Use Eq. [3-17] to obtain:

$$C(1.3 \text{ m}) = \frac{(5 \text{ mg}/(\pi \cdot (0.05 \text{ m})^2))}{(0.3)\sqrt{(4\pi)(0.0005 \text{ m})(1.3 \text{ m})}} = 23{,}000 \text{ mg/m}^3.$$

At 1.3 m, C is 23 mg/liter at the center of mass.

Now consider the case of a continuous input of a chemical, as opposed to a pulse injection. Spreading of the contaminant *front* can be described by another solution to Eq. [1-5], based on the initial condition that at time $t = 0$, all the water downgradient of a point $x = 0$ contains no chemical, while all the water upgradient contains a concentration C_0. [Note that there is initially a very steep (theoretically infinite) concentration gradient of chemical at location $x = 0$.] As fluid flow proceeds, the boundary between the water containing no chemical and the upstream water containing chemical at a concentration C_0 blurs as mixing occurs in response to the steep concentration gradient. The concentration of chemical at location x at some time t after

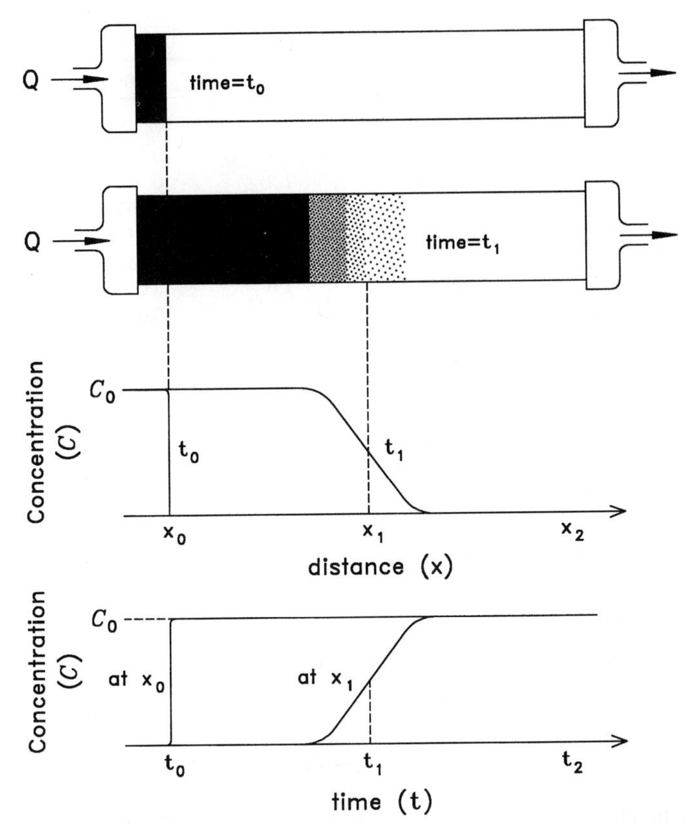

FIGURE 3-18 Dispersion of a *continuous* tracer injection in a sand column experiment. The behavior of a *front* of the tracer is shown in the next to last panel; tracer concentration is presented as a function of distance at fixed times t_0 and t_1. A *breakthrough curve,* a plot of concentration as a function of time at a fixed point, is shown in the bottom panel. (Compare with Fig. 3-28, which shows breakthrough curves for pulse inputs.)

the initial condition may be predicted by

$$C(x, t) = (C_0/2) \cdot \text{erfc}[(x - vt)/(2\sqrt{Dt})] \cdot e^{-kt}, \qquad [3\text{-}18]$$

where C_0 is the initial concentration of chemical [M/L^3], erfc(x) is the complementary error function, and k is a first-order decay rate [T^{-1}].

The corresponding tracer experiment in a sand column is shown in Fig. 3-18. The *complementary error function,* or *erfc,* is a tabulated function; selected values are given in Table 3-4. Note that erfc is equal to 1 − erf, where erf,

TABLE 3-4 The Complementary Error Function[a]

x	erfc(x)	x	erfc(x)
0	1.0		
0.05	0.943628	1.1	0.119795
0.1	0.887537	1.2	0.089686
0.15	0.832004	1.3	0.065992
0.2	0.777297	1.4	0.047715
0.25	0.723674	1.5	0.033895
0.3	0.671373	1.6	0.023652
0.35	0.620618	1.7	0.016210
0.4	0.571608	1.8	0.010909
0.45	0.524518	1.9	0.007210
0.5	0.479500	2.0	0.004678
0.55	0.436677	2.1	0.002979
0.6	0.396144	2.2	0.001863
0.65	0.357971	2.3	0.001143
0.7	0.322199	2.4	0.000689
0.75	0.288844	2.5	0.000407
0.8	0.257899	2.6	0.000236
0.85	0.229332	2.7	0.000134
0.9	0.203092	2.8	0.000075
0.95	0.179109	2.9	0.000041
1.0	0.157299	3.0	0.000022

$$\mathrm{erfc}(x) = 1 - (2/\sqrt{\pi})\int_0^x e^{-\epsilon^2} d\epsilon$$
$$\mathrm{erfc}(-x) = 2 - \mathrm{erfc}(x)$$

[a]Adapted from Freeze and Cherry (1979).

the *error function,* is obtained by integration of the normal (Gaussian) curve. It also can be obtained from the equation

$$\mathrm{erfc}(x) = 1 - \frac{2}{\sqrt{\pi}} \sum_{n=0}^{\infty} \frac{(-1)^n x^{2n+1}}{n!(2n+1)}. \qquad [3\text{-}19]$$

Figure 3-19 presents equations for the transport of a conservative tracer from pulse and continuous sources in one, two, and three dimensions. Note that the one-dimensional solution for a continuous input starting at $t = 0$ uses a slightly different boundary condition at $x = 0$ than is used in the statement of Eq. [3-18]; instead of letting $C = C_0$ for all $x < 0$ at $t = 0$, a mass input at $x = 0$ is turned on at $t = 0$. The results are not distinguishable except for small times and locations very close to $x = 0$. (Note that unlike Eq. [3-18], porosity must be taken into account in Fig. 3-19 because the expression uses mass divided by the column area rather than concentration in the pore water.) Also note, again, that first order decay can be accomodated in these equations by multiplying by e^{-kt}.

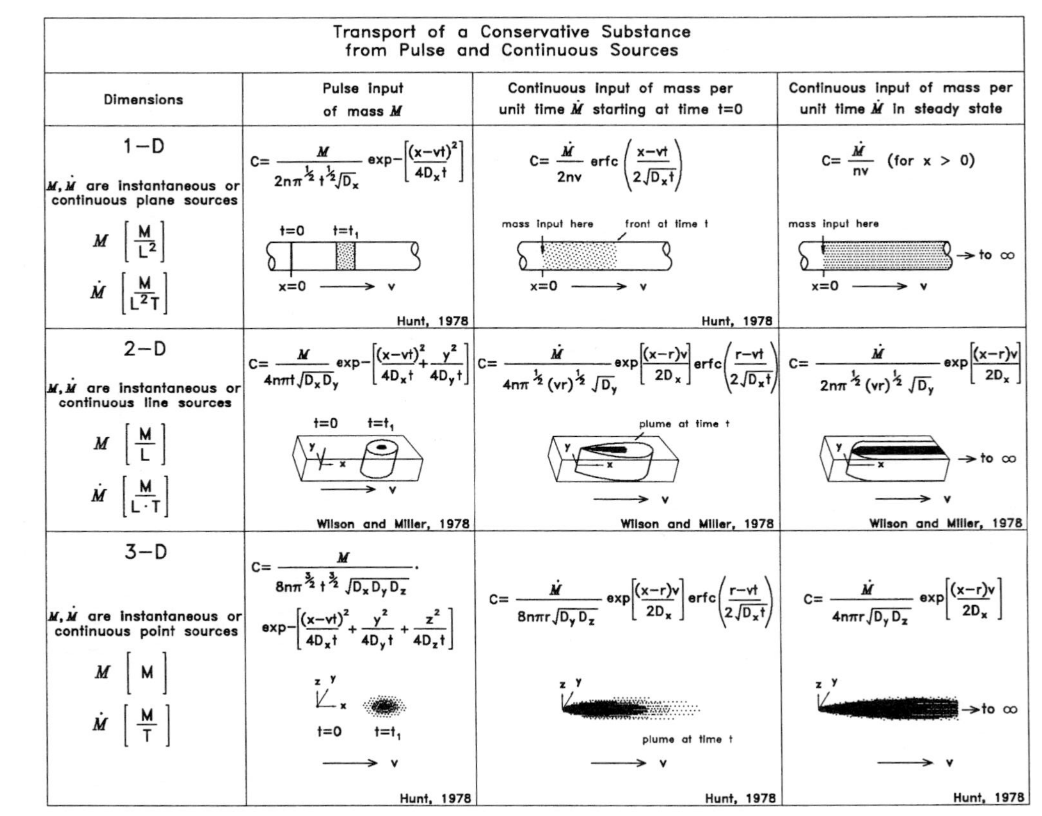

FIGURE 3-19 Solutions to the advection–dispersion equation (Eq. [1-5]) for a conservative solute. Cases for continuous input of mass beginning at time $t = 0$ are adapted from references cited, assuming x and/or r are much larger than D/v; r equals $(x^2 + y^2 D_x/D_y)^{1/2}$ in two dimensions or $(x^2 + y^2 D_x/D_y + z^2 D_x/D_z)^{1/2}$ in three dimensions. Note that the definitions of M and $\dot{M}$ vary with the number of dimensions.

EXAMPLE 3-6

Consider the same experimental apparatus described in Example 3-5, in which the inflow of pure water is replaced by inflow of a solution with a salt concentration of 5 mg/liter beginning at $t = 0$. What will be the concentration of salt a distance 0.9 m down the column after 1 hr?

Use Eq. [3-18] to obtain

$$C(0.9 \text{ m}, 1 \text{ hr}) = \frac{5 \text{ mg/liter}}{2} \cdot \text{erfc}\left(\frac{0.9 \text{ m} - (1 \text{ m/hr})(1 \text{ hr})}{2\sqrt{(5 \times 10^{-4} \text{ m}^2/\text{hr})(1 \text{ hr})}}\right)$$

$$C = 2.5 \text{ mg/liter} \cdot \text{erfc}\,(-2.2) = 2.5 \text{ mg/liter} \cdot (2 - \text{erfc}\,(2.2))$$

$$C = 2.5 \text{ mg/liter} \cdot (1.998) = 5 \text{ mg/liter}.$$

3.3 FLOW IN THE UNSATURATED ZONE

The saturated zone is the region within which chemical pollution is generally of most concern because the saturated zone is the source of drinking water. Before pollutant chemicals spilled or disposed of on the land surface can reach the saturated zone, however, they must first move through the unsaturated zone (vadose zone). While vadose zone contamination is of concern per se, the fate and transport of chemicals in the vadose zone also are of interest because they affect the transmission of chemicals to the saturated zone.

3.3.1 THE NATURE OF THE UNSATURATED ZONE

The upper layer of the vadose zone is usually soil, a zone occupied and significantly modified by the biota, including plant roots, microbes, and animals (Beven and Germann, 1982). The physical and chemical nature of a soil is determined by the climate, the type of vegetation occurring in the area, and the nature of the parent geologic material (e.g., bedrock, glacial till, outwash sands, and gravels) from which the soil is developed. The study of soils is itself an entire field of science (Bohn *et al.*, 1985; Brady, 1990; Birkeland, 1984; Buol *et al.*, 1973).

Two important forces in soil formation (soil *diagenesis*) are the deposition and decomposition of organic material from plant leaves and roots and the movement of water through the soil. Figure 3-20 shows the typical layers, or *horizons*, that develop in a soil in a humid, temperature climate. The uppermost layer of soil is often formed of undecomposed and partially decomposed

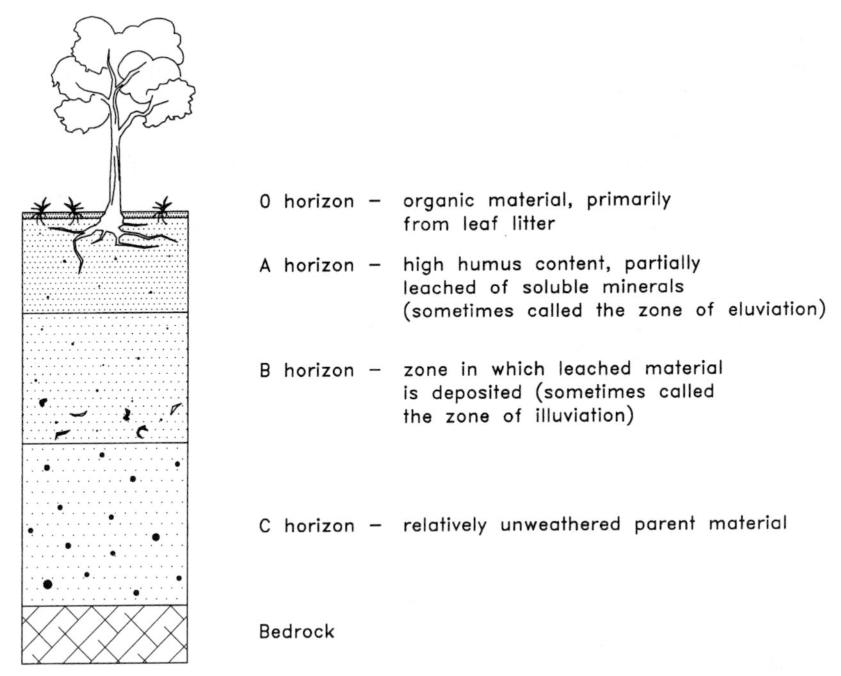

FIGURE 3-20 A generalized soil profile typical of a humid temperature climate, showing O, A, B, and C horizons.

plant material. This uppermost layer forms the *O horizon*. Below this is a zone, sometimes called the *A horizon*, or the *zone of eluviation*, from which mineral and organic components have been leached by the passage of water. This *weathering* process yields pore water enriched in elements such as calcium, potassium, sodium, silica, iron, and aluminum. Weathering is aided by the acidity imparted to the pore water by carbon dioxide (from root respiration and microbial decomposition of organic matter), and by organic acids, produced during decomposition of organic material, which help solubilize elements such as iron and aluminum. The mineral materials may completely dissolve (as is the case with limestone, $CaCO_3$) or may *weather incongruently*, leaving a secondary material. Incongruent weathering occurs when only some of the atoms forming a mineral enter solution, thereby leaving a mineral "residue" behind. If the original mineral material contains large amounts of aluminum and silica (e.g., as in feldspars), the mineral may be incongruently weathered into a *clay*. For example, the feldspar mineral albite ($Na_2O \cdot Al_2O_3 \cdot 6SiO_2$) can weather to the clay Na-montmorillonite ($Na_2O \cdot 7Al_2O_3 \cdot 22SiO_2 \cdot nH_2O$) by losing Na^+ and H_4SiO_4 (dissolved silica). On further weathering, the remaining sodium and additional silica can dissolve in pore water, leaving behind kaolinite ($Al_2O_3 \cdot 2SiO_2$), a fairly stable clay. Further weathering of kaolinite produces gibbsite, $Al(OH)_3$, a pure aluminum oxyhydroxide of low solubility.

Below the zone of leaching is a *zone of illuviation,* or zone of deposition, sometimes called the *B horizon,* in which dissolved organic matter and previously solubilized iron and aluminum are deposited. Deposition occurs when the organic acids and associated complexed metals are sorbed onto soil minerals or when the organic acid molecules themselves are mineralized by bacterial action, causing the previously complexed metals to precipitate. Beneath this depositional soil horizon is a relatively unweathered mineral material, often called the *C horizon,* or *parent material,* because it is the original material from which the soil profile developed.

In all but extreme climates, the upper portion of the soil profile is extensively occupied by plant roots, which remove both water and mineral nutrients. Plants and other biota (such as insects and small mammals) create extensive networks of voids often referred to as *macropores,* which result in a heterogeneous, *biporous* (i.e., there are two porosity values, for micro- and macropores), and structurally very complex material. Macropores (and pipes, which are larger, continuous macropores) can play a significant role in water transport, although the exact role of flow through macropores versus flow through the rest of the soil matrix is not completely understood. For an overview of the types and mechanisms of formation of macropores, the reader is referred to Beven and Germann (1982).

The preceding description of soil formation is extremely generalized and does not include many soil types. For example, peat soils are predominantly composed of partially decomposed plant remains. In arid areas, upward water flow and evaporation at the soil surface can produce a hard *caliche* soil. In very warm and humid areas, leaching and organic matter decomposition occur quite rapidly and soil minerals can become highly weathered, resulting in *saprolite* soils. In ecosystems in these areas, biomass, instead of soil particles, may become the major repository of plant nutrients.

3.3.2 Water Transport in the Unsaturated Zone

Beneath the soil profile, the vadose zone may consist of unconsolidated or consolidated materials; in this text, only unconsolidated materials are considered. The movement of water in the vadose zone is more difficult to describe than flow in the saturated zone, because water transport occurs only via water-filled pores and the fraction of the pore spaces filled with water (the *percent saturation*) is highly variable. The water content θ refers to the fraction of the bulk soil volume that contains water; θ can range between zero

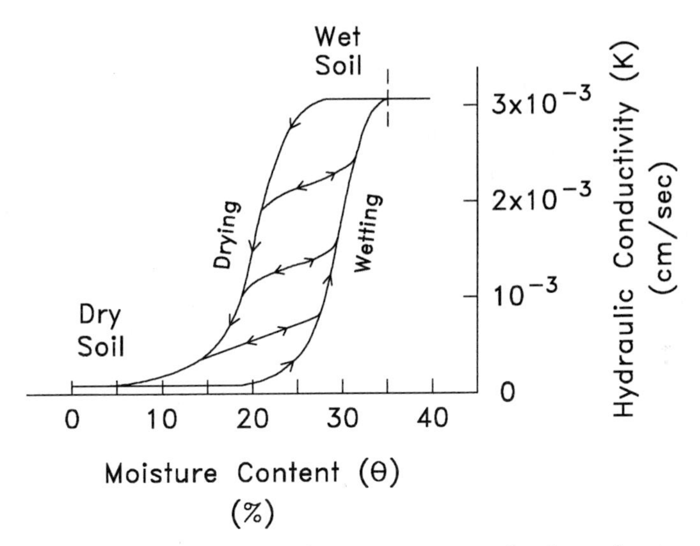

FIGURE 3-21 Hydraulic conductivity of a certain unsaturated soil as a function of water (moisture) content (θ). Higher conductivity occurs when θ is larger and the pore water is under less suction. Note that the relationship between K and θ depends on whether the soil most recently was dried or wetted.

and n, the soil porosity. The relationship between the hydraulic conductivity, K, and the water content, θ, is complex and difficult to predict; therefore, it is usually measured experimentally and expressed in the form of a $K - \theta$ curve, as shown in Fig. 3-21. One major complication in the $K - \theta$ relationship is *hysteresis;* hydraulic conductivity differs depending on whether the porous media most recently have been dried or wetted to obtain a given moisture content. For dry material, regardless of its texture, hydraulic conductivity is low.

In the unsaturated zone, the pressure of the pore water is *less* than atmospheric pressure. The symbol ψ is often used to represent the pressure head of pore water,

$$\psi = p/\rho g, \qquad [3\text{-}20]$$

where ψ is the pressure head of pore water [L], p is the pressure of the pore water [M/LT^2], ρ is the density of water [M/L^3], and g is the acceleration due to gravity [L/T^2]. Confusion can arise because some scientists regard ψ as positive and others as negative. When ψ is thought of as a pressure head, it is negative in the unsaturated zone. Sometimes ψ is called *soil suction,* however, and is presented as a positive quantity. To further

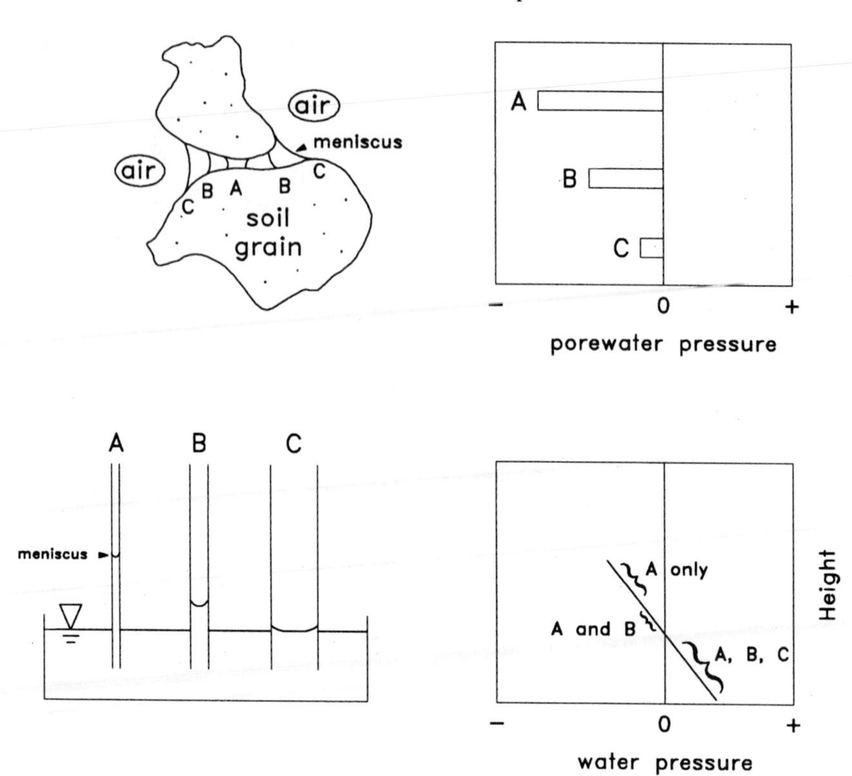

FIGURE 3-22 Soil water under three different values of water content (upper half of figure). Whenever both air and water are present in soil pore spaces (conditions A, B, and C), pore water pressure is rendered negative by the force of surface tension acting over a meniscus. The meniscus may be thought of as a flexible diaphragm that is under tension, thus pulling on the water on its convex side. As water content decreases and the meniscus retreats into smaller pore spaces (B, then A), surface tension forces act over a smaller area of water, and the resulting water pressure is more negative. The same effect occurs in capillary tubes (lower half of figure), where the most suction (most negative pressure, and thus most capillary rise) is developed in the tube of smallest diameter. The lower right panel shows that hydrostatic pressure applies in all three tubes; pressures are negative in any water that has risen above the water table.

confuse the issue, ψ is often presented in units of pressure [M/LT^2] instead of head [L].

Soil suction is a function of capillary forces acting between water and soil grains. Other factors being equal, soil suction is greatest in fine textured media, in which the *menisci* between soil water and soil gas have the smallest radii and thus produce the greatest pressure difference between water and air. Large pores require larger menisci between air and water; these create smaller pressure differences. This phenomenon is easily demonstrated with water in

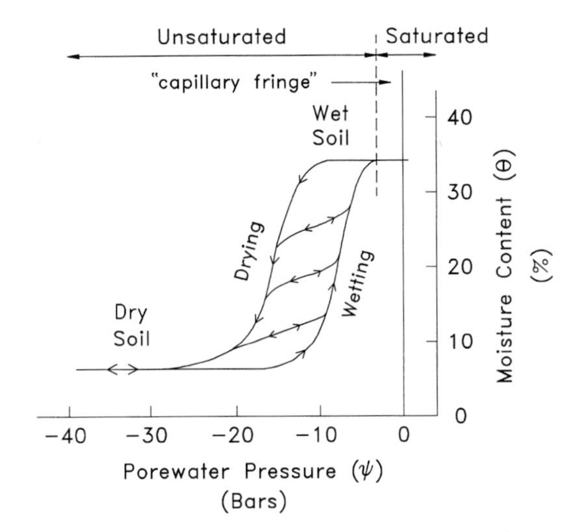

FIGURE 3-23 Relationship between water (moisture) content (θ) and pore water pressure (ψ) in a certain unsaturated soil. Pore water suction is the negative of pore water pressure. As in the case of the $K - \theta$ relationship (Fig. 3-21), this relationship is influenced by whether the soil most recently was wetted or dried.

glass capillary tubes of various diameters; water rises the most in the smallest tubes (Fig. 3-22).

The relationship between ψ and the water content of a porous medium is also determined empirically; Fig. 3-23 shows typical $\psi - \theta$ curves. When ψ is zero, the porous medium is at its saturated moisture content, which is equal to porosity (assuming that there is negligible trapped gas). The pressure head can become somewhat negative before air enters (and water leaves) some soils; the value ψ_a at which air begins to enter is called the *air entry* value. Water leaves the matrix as ψ (defined as pressure head) becomes more negative than ψ_a; as water leaves the soil, the hydraulic conductivity also decreases.

The reversal of relative hydraulic conductivities of unsaturated fine and coarse textured porous media is an interesting consequence of the greater suction required to remove water from fine-textured material. As shown in Fig. 3-24, a coarse-textured "sandy" soil has a higher hydraulic conductivity than a fine-textured "clayey" soil at full saturation, when ψ is zero. As pore water pressure decreases, however, the soil water content decreases to a lesser extent for the fine-textured soil; a point is reached at which the fine-textured soil holds appreciably more water than the coarse-textured soil and has a higher hydraulic conductivity because many of its pore spaces, despite their smaller size, remain filled with water.

Flow is more difficult to model in the unsaturated zone because hydraulic

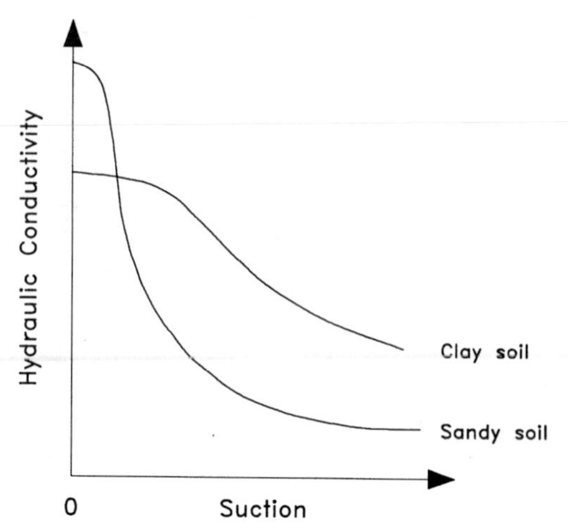

FIGURE 3-24 Curves of hydraulic conductivity versus pore water suction. The relative hydraulic conductivities of clayey and sandy soils can become reversed at low pore water pressures (high pore water suction), when the water content of the sandy soil becomes much lower than that of the clayey soil, resulting in fewer water-filled pathways in the sandy soil.

conductivity changes as water content changes. In unsaturated flow models, the Darcy–Richards equation, which is a more general form of Darcy's law, is used. For isotropic porous media in three dimensions,

$$\vec{q} = -K(\psi)\nabla(\psi + z), \qquad [3\text{-}21]$$

where $\vec{q}$ is the specific discharge [L/T], $K(\psi)$ is the hydraulic conductivity (a function of pore water pressure) [L/T], ∇ is the gradient operator, ψ is pressure head [L], and z is the height above an arbitrarily chosen reference elevation [L].

Water flow in the vadose zone also is much more difficult to measure than is water flow in the saturated zone. Piezometers placed in unsaturated porous media do not fill with water, so head is less readily measured than it is in saturated media, and thus the direction and magnitude of the hydraulic gradient are more difficult to determine. For values of ψ between 0 and -1 atm, *tensiometers* may be employed to directly measure pore water pressure. For very low values of ψ, other techniques may be used, such as the indirect assessment of ψ from measurements of the humidity of the air in the pore spaces.

In the absence of other forces in the vadose zone, there still is a net tendency for water to move downward in response to gravity, as indicated by the elevation term (z) in Eq. [3-21]. In arid zones, however, the pressure term in Eq. [3-21] can overwhelm the elevation term, and the fate of water in the vadose zone may be determined almost entirely by spatial variability in the properties of the porous media, with water often moving horizontally along regions that are most conductive and/or have the highest suction under the prevailing moisture conditions.

3.4 THE FLOW OF NONAQUEOUS PHASE LIQUIDS

Transport in unsaturated porous media is sufficiently complex when only two fluid phases, air and water, are present; flow becomes even more complicated when a third fluid phase, such as an immiscible organic fluid, is involved. This third fluid phase (NAPL) arises when liquid hydrocarbon fuels or solvents are spilled accidentally on the ground surface or when they leak from underground storage tanks. The resulting subsurface flow problem then involves three fluids, air, water, and NAPL, each having different interfacial tensions with each other, different viscosities, and different capillary interactions with the soil. The adequate description of three-phase flow is still a topic of active research, but a few qualitative generalizations can be drawn.

Small quantities of NAPL can move into porous media under both gravity and capillary effects and can become essentially immobilized, due to discontinuities that develop in the NAPL as it spreads out; these discontinuities are very much like the ones formed in water films in very dry porous media. Discontinuities prevent flow of NAPL from one region to another; the amount of NAPL present when flow stops is called the *residual saturation*. The effect can be likened to that of water breaking into discontinuous, discrete droplets on the bottom of a greasy kitchen sink, thereby preventing flow of the last drops of water down the drain. (In this domestic example the water, not the grease film, is the discontinuous phase.) Residual saturation depends on soil texture and on the surface tension (σ) between the NAPL and water; this surface tension is approximately 30 to 50 dyn/cm for many immiscible solvents and fuels. Residual saturation also depends on the initial water content of the porous media when the NAPL is introduced.

NAPL, especially when distributed discontinuously, may show little or no tendency to move unless the applied hydraulic gradients are extremely high, beyond what is normally achieved in the field (Fig. 3-25). The same problem of residual saturation occurs during the recovery of oil from a petroleum reservoir; a significant fraction of the petroleum cannot be recovered even with the most advanced techniques. In coarse soils, NAPL with sufficient vapor pressure can sometimes be recovered below residual saturation by volatilization into the air in pore spaces, and subsequent removal of the air

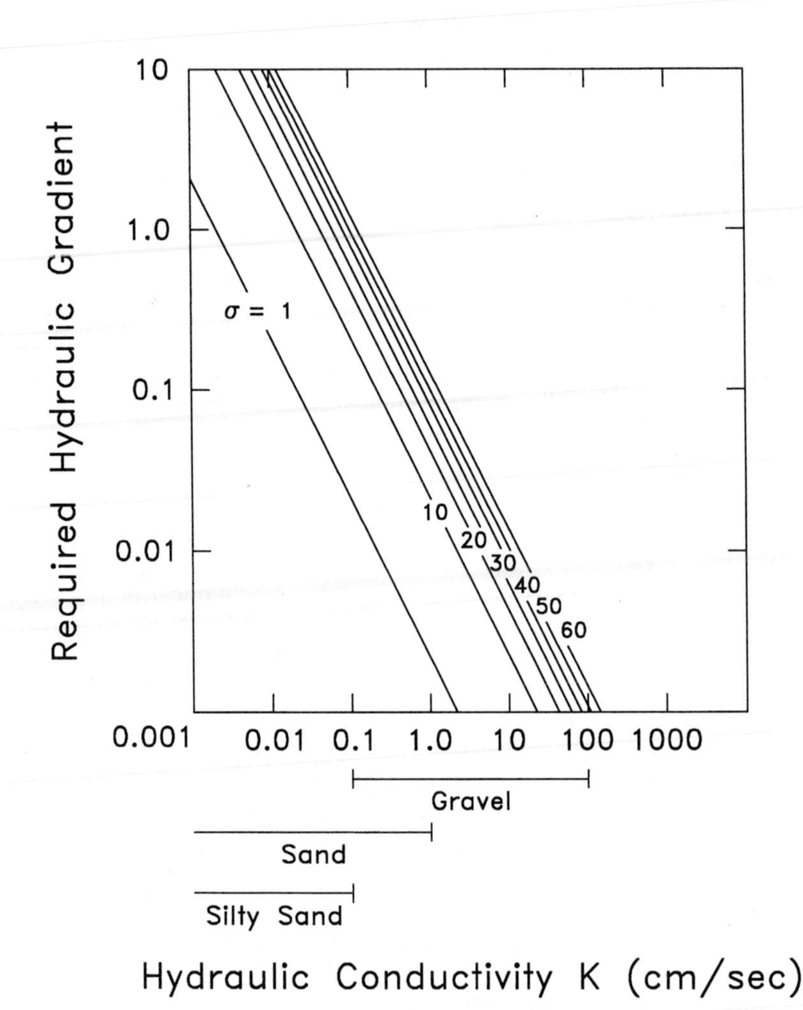

FIGURE 3-25 Hydraulic gradient required to move NAPLs with varying surface tensions (σ, in dyn/cm) through different porous media with varying hydraulic conductivities. In the case of a sand having a hydraulic conductivity of 10^{-2} cm/sec and a NAPL having a surface tension of 50 dyn/cm with respect to water, the required gradient to move the NAPL would be greater than 10. In a groundwater system, typical hydraulic gradients are on the order of 0.01 or less (adapted from Wilson and Conrad, 1984).

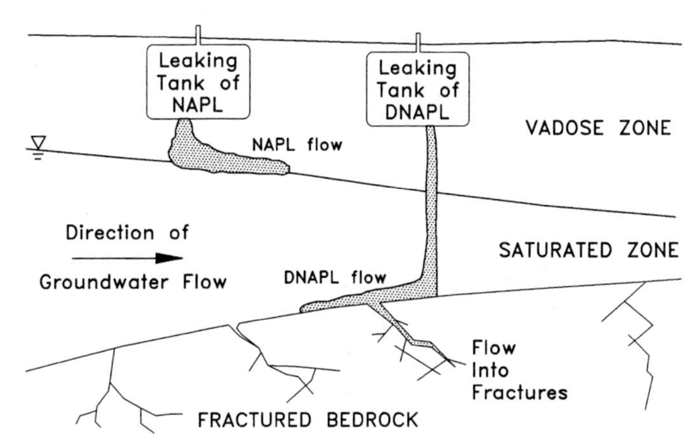

FIGURE 3-26 The behavior of NAPL and DNAPL in the subsurface environment. NAPL floats on the water table, slowly dissolving into the groundwater flowing by, thus creating a plume in the downgradient direction. DNAPL tends to sink to the bottom of the aquifer, where it may flow contrary to the groundwater flow direction, following instead the slope of an aquiclude such as bedrock. DNAPL may also flow into bedrock fractures, and become exceedingly difficult to either locate or remove.

phase. This remedial technique is known as *vacuum extraction*. NAPL also may be removed if large quantities of water are added to the vadose zone at one location and removed at another; the NAPL is mainly removed by dissolution into the water flowing past. This can be a very slow process, due to the low solubility of most NAPLs. The use of cosolvents to increase the solubility of NAPL has been proposed to accelerate dissolution into the moving phase.

If NAPL exceeds residual saturation over a large volume of porous media, bulk downward flow of NAPL—perhaps to the water table—may occur under the force of gravity. If the NAPL is less dense than water (which is generally the case for hydrocarbon fuels and many solvents), it floats on the water table, spreading out to form the equivalent of a thick underground oil slick. This geometry, often called a *bolus*, may be a fraction of a meter thick and tens of meters in horizontal extent (Fig. 3-26). Recovery efforts commonly involve creating a cone of depression in the underlying water, thus encouraging bulk flow of NAPL into a collection well (Fig. 3-27). Again, this recovery technique is only adequate to recover a fraction of the NAPL; when the NAPL distribution becomes discontinuous, bulk flow stops and residual saturation is reached.

In the case of a dense NAPL (DNAPL), whose density exceeds 1 g/cm^3 (Table 1-3), downward flow takes place, until a condition of residual saturation occurs or an aquiclude or aquitard is reached (see Fig. 3-26; see also Schwille (1988)). Such a DNAPL is exceedingly difficult to recover or even to

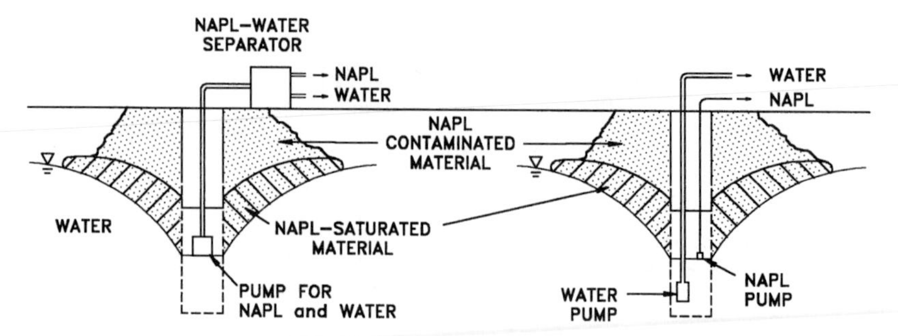

FIGURE 3-27 Typical schemes to recover a floating NAPL, such as gasoline, from a phreatic aquifer. In each case a cone of depression is created by pumping to encourage the NAPL to flow toward the collection point. Either a NAPL/water separator or two pumps are required.

locate. In principle, a DNAPL may be partially recovered if a subsurface pool can be located and a suitably screened well placed in the zone of DNAPL-saturated porous media. Disturbing the subsurface environment, however, may only make the DNAPL move deeper.

DNAPL located at the bottom of an aquifer serves as a source of pollution to groundwater flowing past. Considering the low solubility of many chlorinated solvents and the unacceptability of their presence at concentrations higher than a few parts per billion, it is apparent that even a relatively small volume of DNAPL can contaminate an immense volume of groundwater; unfortunately, centuries or even millennia may be required for the ultimate removal of DNAPL from an aquifer under natural conditions. DNAPL also commonly will enter channels, fractures, holes, and cracks in an underlying aquitard or aquiclude, thereby sinking even farther (see Fig. 3-26). No satisfactory technique has yet been developed to locate DNAPL in bedrock fractures, much less to predict its movement.

3.5 RETARDATION

The previous sections have described the movement of groundwater in both the saturated and the unsaturated zones. The equations presented are sufficient to predict the transport of conservative chemicals that neither decay with time nor sorb onto particles.

Many chemicals, however, do sorb onto the grains of aquifers, or onto iron oxyhydroxide or organic coatings on the grains. As discussed in Section 1.8.3, the term "sorb" includes both *adsorptive* processes, in which a chemical sticks to the two-dimensional surface of a solid, and *absorptive* processes, in which a chemical diffuses into a three-dimensional solid. In groundwater, the transport of chemicals which sorb is slowed, or *retarded.* Such chemical behavior can be described by using a partition (K_p) or distribution (K_d) coefficient (as presented in Section 1.8.3) that relates the amount of chemical sorbed to the soil to the amount dissolved in the water. If the equilibrium between sorbed and dissolved chemicals occurs sufficiently rapidly, and if the concentration on the solid material is proportional to aqueous concentration (i.e., a *linear* sorption isotherm exists), the equations for the transport of a conservative tracer can be readily modified by a *retardation factor* so that they describe the behavior of a sorbing chemical.

The retardation factor R [$(M/L^3)/(M/L^3)$] is defined as

$$R \equiv \frac{\text{mobile chemical concentration} + \text{sorbed chemical concentration}}{\text{mobile chemical concentration}}.$$

Therefore,

$$R = 1 + \frac{\text{sorbed concentration}}{\text{mobile concentration}}. \quad [3\text{-}22]$$

The rate at which the center of mass of a sorbing chemical in equilibrium with aquifer material moves through the aquifer is equal to the seepage velocity, v, divided by R. Likewise the Fickian transport coefficient D must be adjusted by dividing by R. As an example, if the retardation factor for a chemical is five, a plume of dissolved chemical will advance only one-fifth as fast as a parcel of water. Transport with retardation produces the same amount of dispersion per distance traveled for a sorbing chemical as for a nonsorbing chemical, however, because the effect of the smaller dispersion coefficient of the sorbing chemical is offset by the longer travel time.

In Eq. [3-22], concentration is expressed as mass of chemical per volume of porous media. The volume of porous media, also termed aquifer volume, is defined to include both particle grains and pore water. Equation [3-22] can be rewritten in terms of the aqueous chemical concentration (C_{aq}), the sorbed chemical concentration (C_s), the water-filled porosity, n, the distribution coefficient K_d, and the *bulk density* of the porous material ρ_b. Bulk density is defined as the weight of dry solids divided by the volume of aquifer from which they were taken.

Bulk density is related to the density of individual soil particles, ρ_s, by the equation

$$\rho_b = (1 - n) \cdot \rho_s, \quad [3\text{-}23]$$

where ρ_b is bulk density [M/L^3], n is porosity [L^3/L^3], and ρ_s is soil particle density [M/L^3]. [Soil particle density is often approximately 2.65 g/cm^3, the density of silica (SiO_2).]

Expressed on the basis of aquifer volume, the mobile concentration of a chemical in an aquifer is the product of the aqueous chemical concentration and the porosity ($C_{aq} \cdot n$), while the sorbed concentration is the product of the sorbed chemical concentration and the bulk density ($C_s \cdot \rho_b$). Substitute these products into Eq. [3-22] to get:

$$R = 1 + \frac{(C_s \cdot \rho_b)}{(C_a \cdot n)}. \qquad [3\text{-}24a]$$

Because by definition $K_d = C_s/C_{aq}$, Eq. [3-24a] reduces to the commonly used expression:

$$R = 1 + K_d \cdot \rho_b / n. \qquad [3\text{-}24b]$$

It must be noted that a retardation coefficient is strictly applicable only when a linear relationship exists between C_s and C_{aq} and the partitioning equilibrium is rapidly established. In some nonlinear cases, or when the kinetics of sorption and desorption are slow, mixing is increased as chemicals are released from aquifer solids after some time has passed, resulting in a "smearing" or tailing of a pollutant peak (Fig. 3-28).

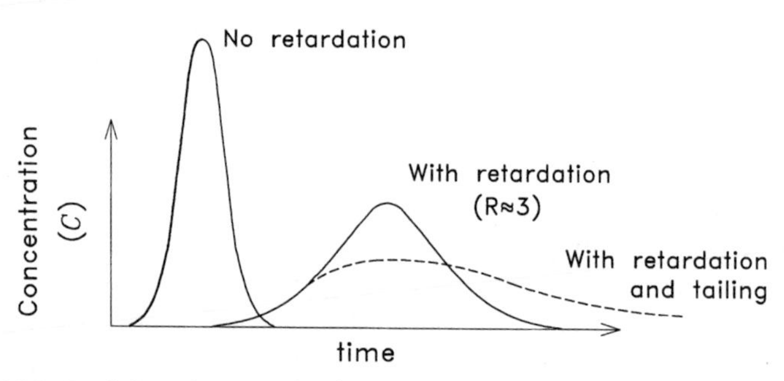

FIGURE 3-28 Breakthrough curves for three cases: (a) a pulse input of a conservative tracer substance that is not retarded; (b) a pulse input of a substance that moves through the porous medium with a retardation factor of three; and (c) a pulse input of a substance that is retarded, but for which sorption to the solid phase is either not a linear function of dissolved concentration or not at local equilibrium. Case (c) exhibits "tailing" (i.e., the breakthrough curve is asymmetrical—skewed toward longer times).

3.5.1 Chemical Sorption by Organic Carbon

A large number of organic chemical pollutants are *hydrophobic,* literally "afraid of water." These chemicals have limited solubility in water but do tend to dissolve easily into oils, fats, nonpolar organic solvents, and organic carbon in the soil. To a first approximation, the partition coefficient for many hydrophobic chemicals in soil is not especially sensitive to the exact source or nature of the organic carbon. Accordingly, K_{oc}, the organic carbon–water partition coefficient, can be used to estimate the extent of sorption. K_{oc} can be expressed as

$$K_{oc} = \frac{\text{chemical concentration sorbed to organic carbon (mg/g)}}{\text{chemical concentration in water (mg/ml)}}. \qquad [3\text{-}25]$$

Because the fraction of organic material in porous media is rarely 100% and is typically less than 1% (notable exceptions exist in wetland sediments and peatlands), the partitioning of a hydrophobic organic compound between water and bulk soil can be estimated by the equation

$$K_d = f_{oc} \cdot K_{oc}, \qquad [3\text{-}26]$$

where f_{oc} is the fraction of soil that is organic carbon [M/M].

TABLE 3-5 **Relationships to Calculate K_{oc} from K_{ow}[a]**

Equation[b]	No.[c]	r^2[d]	Chemical classes represented
$\log K_{oc} = 0.544 \log K_{ow} + 1.377$	45	0.74	Wide variety, mostly pesticides
$\log K_{oc} = 0.937 \log K_{ow} - 0.006$	19	0.95	Aromatics, polynuclear aromatics, triazines, and dinitroaniline herbicides
$\log K_{oc} = 1.00 \log K_{ow} - 0.21$	10	1.00	Mostly aromatic or polynuclear aromatics; two chlorinated
$\log K_{oc} = 0.94 \log K_{ow} + 0.02$	9	NA	*s*-Triazines and dinitroaniline herbicides
$\log K_{oc} = 1.029 \log K_{ow} - 0.18$	13	0.91	Variety of insecticides, herbicides, and fungicides
$\log K_{oc} = 0.524 \log K_{ow} + 0.855$	30	0.84	Substituted phenylureas and alkyl-*N*-phenylcarbamates

[a] Lyman *et al.* (1990). NA, not available.
[b] K_{oc}, organic carbon-water partition coefficient; K_{ow}, octanol–water partition coefficient.
[c] Number of chemicals used to obtain regression equation.
[d] Correlation coefficient for regression equation.

The preceding expression is useful for soils in which f_{oc} is greater than approximately 0.001; in these soils, sorption to organic carbon dominates. For lower values of f_{oc} (values of 10^{-4} may occur in some aquifer materials), direct sorption onto mineral phases of the soil can become important, and K_{oc} is no longer a good predictor of sorption.

K_{oc} can be estimated from K_{ow}, the octanol–water partition coefficient (defined in Section 1.8.3). Table 3-5 shows some correlations between K_{oc} and K_{ow} for different classes of hydrophobic organic compounds. For further descriptions of the hydrophobic behavior of chemicals, the reader is referred to Schwarzenbach *et al.* (1993).

EXAMPLE 3-7

For an aquifer solid with a bulk density of 2 g/cm³ containing 0.5% organic carbon, estimate the retardation factor for the common polycyclic aromatic hydrocarbon (PAH) naphthalene ($C_{10}H_8$; see Fig. 1-11), used in mothballs. If the porosity of the aquifer is 0.24, the hydraulic conductivity is 10^{-3} cm/sec, and the hydraulic gradient is 0.001, how fast will a plume of naphthalene travel?

From Table 1-3, $\log K_{ow} = 3.36$. By using Table 3-5, an estimate of K_{oc} for PAHs is

$$\log K_{oc} = 0.937 \log K_{ow} - 0.006.$$

Therefore,

$$\log K_{oc} = (0.937) \cdot (3.36) - 0.006$$

$$K_{oc} \approx 1400 \text{ ml water/g organic carbon.}$$

Use Eq. [3-26] to estimate K_d:

$$K_d = (0.005 \text{ g carbon/g soil}) \cdot (1400 \text{ ml water/g carbon}) = 7 \text{ ml water/g soil.}$$

Use Eq. [3-24b] to estimate the retardation factor:

$$R = 1 + (7 \text{ ml/g}) \cdot (2 \text{ g/cm}^3)/0.24 = 59.$$

Use Eq. [3-2] to estimate specific discharge:

$$q = 10^{-3} \text{ cm/sec} \cdot 0.001 = \frac{10^{-6} \text{ cm}}{\text{sec}}.$$

Use Eq. [3-5] to estimate seepage velocity:

$$v = \frac{10^{-6}\ \text{cm}}{\text{sec}}/0.24 = 4.2 \times 10^{-6}\ \text{cm/sec}.$$

Then divide by the retardation factor to estimate how fast a plume of naphthalene will travel:

$$\frac{v}{R} = \frac{4.2 \times 10^{-6}\ \text{cm/sec}}{59} = 7.1 \times 10^{-8}\ \text{cm/sec}.$$

3.5.2 Sorption by Ion Exchange

In contrast to hydrophobic organic compounds, which are uncharged, many inorganic pollutants in water occur as ions, either as positively charged cations or negatively charged anions. (Certain organic compounds also can occur as ions; such compounds are invariably more soluble in water than their uncharged counterparts.) Unlike hydrophobic organic compounds, which essentially dissolve into solid organic material in the aquifer, ionic substances may sorb onto some aquifer solids by *ion exchange.*

According to the electroneutrality principle, the sum of all positive charges in a droplet of water must equal the sum of all negatively charged ions in that droplet. The same is true for aquifer solids (although in certain cases, such as surface complexation, discussed in Section 3.5.3, the volume over which the sum of charges is zero can include a thin film of water next to the particles). Ion exchangers have a structure containing an excess of fixed negative or positive charge and take up ions from solution as needed to neutralize the charge. Common ion exchangers in aquifers are clays, in which an excess of negative charge arises when tri- and divalent cations in the crystal structures are replaced by di- or monovalent ions in a process termed *isomorphous substitution.* To satisfy electroneutrality, such clay minerals take up positively charged *counterions* (in this case, cations) from the water. A typical clay might have a cation exchange capacity of 2 meq/g; for each gram of clay, 2 meq of cations such as hydrogen ions, sodium ions, or calcium ions are held within the clay structure. (Note that 2 mmol of H^+ or Na^+ form 2 meq, whereas only 1 mmol of Ca^{2+} is needed to form 2 meq.)

Ion exchangers can preferentially absorb some types of ions relative to other ions; therefore, the ratio between different counterions on an ion exchanger is usually not the same as the ratio between those ions in solution. For example, consider a clay ion exchanger that is selective for calcium. In a groundwater regime that initially contains sodium as the only cation, the clay necessarily contains only sodium ions. If a contaminant pulse containing calcium reaches the ion exchanger, the exchanger preferentially sorbs the calcium, exchanging it with sodium and thereby leaving the exchanger enriched in calcium relative to the pore water. Conversely, the moving pore water necessarily becomes depleted in dissolved calcium. The calcium is thereby retarded; it cannot move until it is exchanged back into the water under altered conditions. As the calcium-rich water passes by the exchanger, and the aquifer solids are again surrounded by pore water containing only sodium, the exchanger must ultimately return to its sodium form (i.e., all the counterions are sodium). As it does so, the calcium is released to the pore water.

Unfortunately, cation exchange does not lend itself to simple mathematical description as does idealized sorption of hydrophobic organic compounds. For low concentrations of a contaminant ion in a constant background of other ions, however, the ion exchange process often is treated approximately as a linear partitioning process, and use of a simple retardation factor in a transport model may be justified. Distribution coefficients for ionic contaminants in an aquifer are usually determined experimentally. Typically, a *batch test* is performed in which the ionic concentration on a fixed volume of aquifer solids and the ionic concentration in the associated pore waters are analyzed; the distribution coefficient is taken as the ratio of the concentrations. Because sorption by ion exchange is affected by the concentrations of all other ions in the groundwater, the distribution coefficient estimated by a batch test can be used only as long as the background chemistry of the water stays essentially constant. For a more complete discussion of ion exchange equilibria, the reader is referred to Walton (1969) and Freeze and Cherry (1979).

EXAMPLE 3-8

Copper-rich waste disposed of in lagoons behind an electroplating operation has percolated down to the groundwater and dissolved, creating a plume. A sample of the porous medium and associated pore water is taken from the plume area and analyzed. The results are

Copper concentration in pore water = 10^{-6} mol/liter

Copper concentration on aquifer solids = 10^{-3} mol/kg

Aquifer bulk density = 2.5 g/cm^3

Aquifer porosity = 0.3.

If groundwater seepage velocity is 600 ft/year, how fast will the plume migrate?

By assuming the relationship between dissolved and sorbed copper is reasonably linear, a retardation coefficient can be meaningfully applied. The distribution coefficient, K_d, can be estimated as

$$K_d = \frac{\text{conc on solids}}{\text{conc in water}} = \frac{10^{-3}\ \text{mol/kg}}{10^{-6}\ \text{mol/liter}} = 10^3\ \text{liter/kg}.$$

From Eq. [3-24b], the retardation factor can be estimated as

$$R = 1 + \frac{(10^3\ \text{liter/kg}) \cdot (2.5\ \text{kg/liter})}{0.3}$$

$$R \approx 8300.$$

Therefore, the plume would be expected to advance at approximately:

$$\frac{v}{R} = \frac{600\ \text{ft/year}}{8300} = 0.07\ \text{ft/year}.$$

3.5.3 Surface Complexation

In addition to being subject to ion exchange, ionic chemical species are also subject to sorption onto metal oxyhydroxides, such as iron oxyhydroxide, $Fe(OH)_3$, which may be present in an aquifer. In this process of *surface complexation,* ionic chemicals form both electrostatic and chemical bonds with the metal or oxygen atoms of the oxyhydroxide surface. The net charge of the oxyhydroxide surface does not need to be completely balanced by the sorbed ions; electroneutrality is satisfied by including additional layers of ions associated with water lying within a few nanometers of the surface. In principle, it is possible to predict the equilibrium between a metal oxyhydroxide surface and pore water from a knowledge of the equilibrium constants for each chemical complex that can be formed on the metal oxhydroxide surface. Because hydrogen ions typically compete with metal ions for surface sites, the amount of sorption that occurs is pH dependent. For details on the reactions and equilibrium constants associated with surface complexation to $Fe(OH)_3$, the reader is referred to Dzombak and Morel (1990). In practice, partition coefficients are often determined experimentally, as in the case of ion exchange. In fact, the distinction between surface complexation and ion exchange is often not made; aquifer solids may exhibit some characteristics of each sorption mechanism.

3.5.4 Nonideality in Retardation

The application of a simple retardation factor is justifiable for locations where a single distribution coefficient is adequate to describe chemical sorption, and equilibrium between phases occurs rapidly. These conditions may not be met for several reasons, including (1) inherent nonlinear characteristics of sorption, especially at higher concentrations; (2) the possibility of changing background pore water chemistry, which has major effects on ion exchange; and (3) slow kinetics, especially where a substantial radius of the aquifer grains is penetrated by solute ions or molecules, resulting in long equilibration times (Wu and Gschwend, 1986). When compared with ideal tracer behavior, these nonideal situations change the shape of the breakthrough curve. *Tailing,* in which a pulse of chemical passes through an aquifer but leaves behind chemical residues sorbed to the solids, is quite common. These residual chemicals can continue to "bleed off" after the bulk of the pulse has passed, resulting in much slower removal of the chemical from the aquifer than would otherwise be predicted (see Fig. 3-28).

3.6 Biodegradation

Organic pollutant chemicals are susceptible to biodegradation in the groundwater environment, just as they are in surface waters. Recall that biodegradation refers to the transformation of an organic compound into other compounds through microbial action. If biodegradation results in the formation of inorganic species (e.g., carbon dioxide, water, and mineral salts), it may be referred to as *mineralization.* Complete mineralization typically involves oxidation using oxygen, but also can occur under anoxic conditions.

As in surface waters, the agents of biodegradation in groundwater are primarily bacteria and fungi. In groundwater, these microorganisms are very likely to be attached to grains of aquifer material, forming a biofilm. The advection of groundwater carries nutrients past the microbes, potentially exposing them to larger amounts of nutrients than if the microbes were moving with the water. This phenomenon was described in Section 2.6.3 in the context of biofilms in streams and rivers.

3.6.1 Modeling Biodegradation in Biofilms

In groundwater, as in surface waters, biofilms can be very effective in transforming pollutant chemicals. Modeling the behavior of biofilms is more complex than modeling batch or continuous cultures (Section 2.6.3), although the quantities V_{max}, K_s, y, and C that are used to describe the growth and uptake rates of cells in batch or continuous cultures are equally applicable to attached cells. The major difference between modeling biofilms and modeling batch or continuous cultures is that cells in biofilms, being embedded in a matrix of extracellular polysaccharide material (bacterial slime), may not be exposed to the full concentration of a chemical in the water flowing past the biofilm surface. Instead, chemicals must diffuse through the biofilm to the cell walls before being taken up; this diffusion implies the existence of a concentration gradient from the bulk water to the bacterial cell walls, with a lower chemical concentration at the cell walls than in the bulk water. Thicker biofilms increase this effect, because chemicals must diffuse greater distances, along smaller gradients, to cells farther from the bulk water. An upper limit on the thickness to which a biofilm can grow is determined by the thickness at which the average growth rate of cells is not high enough to replace cells that die.

Figure 3-29 illustrates the geometry of an idealized biofilm. Transport is influenced by L_f, the biofilm thickness, and by D_f, the effective molecular diffusion coefficient in the biofilm. In general (and especially when the biofilm model is to be applied to a stream or river), one also must consider the existence of a stagnant water layer, of thickness L_b, through which transport from the bulk water to the biofilm surface occurs by molecular diffusion. (This layer is analogous to the stagnant water layer invoked as a barrier to gas exchange in the thin film model discussed in Section 2.3.1.) The biofilm shown in Fig. 3-29 is a *fully penetrated biofilm*, so-called because the chemical, though depleted to concentrations much less than in the bulk water, is present at a significant concentration throughout the biofilm. A *shallow biofilm* is one in which the chemical gradients are small enough that one can consider concentrations within the film to be essentially the same as in the bulk water. A *partially penetrated biofilm* exists when a chemical is fully consumed within the outer portion of the biofilm, creating a depleted zone close to the surface of attachment, within which the concentration of the chemical is zero. Note that such a situation will exist in steady state only if some other chemical serves as nutrient to cells in the depleted zone.

A mathematical model of a biofilm is more complex than a model of a batch or a continuous culture, mainly because the pertinent mass balance equations contain diffusive mass transport terms. (Recall from Section 2.6.3 that a model of a batch culture contains no mass transport equations at all, while a continuous culture, or chemostat, model contains only simple mass inflow and mass outflow.) The following set of equations characterizes the biofilm of Fig. 3-29 at steady state:

1. Conservation of mass in the stagnant boundary layer requires that the diffusive flux into the stagnant water layer from bulk water equals the diffu-

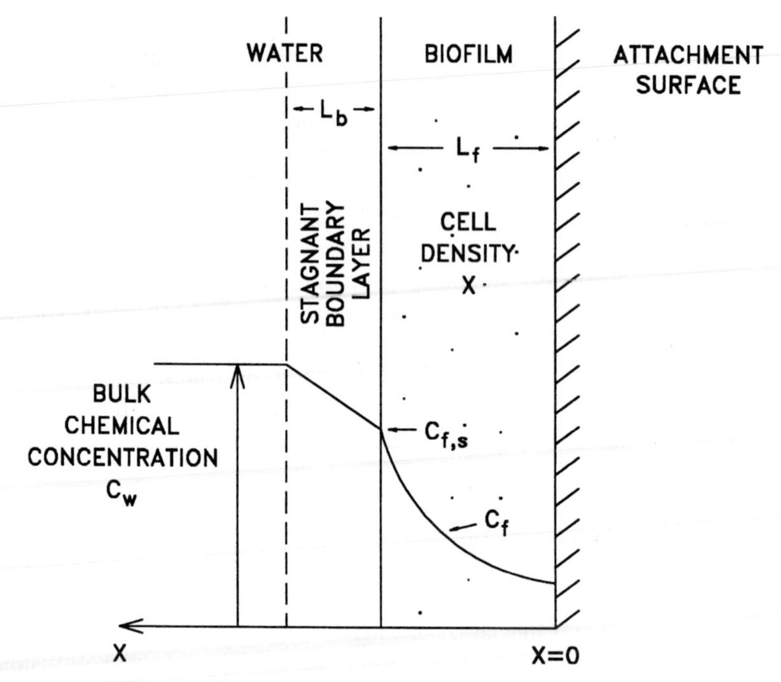

FIGURE 3-29 Diagram of a fully penetrated biofilm. The rate of uptake of a chemical is governed by the uptake capabilities of the microbial cells composing the biofilm as well as by Fickian transport limitations in both the biofilm and the stagnant boundary layer. In a partially penetrated biofilm, C_f would decrease to zero at some point in the biofilm. In a shallow biofilm, C_f would be approximately equal to $C_{f,s}$ throughout the biofilm.

sive flux from the boundary layer into the biofilm,

$$D_b \cdot \frac{C_w - C_{f,s}}{L_b} = D_f \cdot \left(\frac{\partial C_f}{\partial x}\right)_{\text{at biofilm surface}}, \qquad [3\text{-}27]$$

where D_b is the molecular diffusion coefficient in the bulk water [L^2/T], C_w is the chemical concentration in the bulk water [M/L^3], $C_{f,s}$ is the chemical concentration in the biofilm immediately adjacent to the stagnant water layer [M/L^3], L_b is the thickness of the stagnant water layer [L], D_f is the effective molecular diffusion coefficient in the biofilm [L^2/T], C_f is the chemical concentration in the biofilm [M/L^3], and x is the distance perpendicular to the biofilm surface [L].

2. Mass balance within an arbitrarily chosen biofilm section, or "slice," taken parallel to the surface of attachment, is described by the one-dimensional, advection–dispersion–reaction equation, Eq. [1-5], with steady-state conditions and no advection. The sink term is microbial uptake, modeled using the parameters discussed in Section 2.6.3; see Eqs. [2-71a] and [2-72],

$$\frac{\partial}{\partial x}\left(D_f \cdot \frac{\partial C_f}{\partial x}\right) = V_{max} \cdot \frac{C_f}{C_f + K_s} \cdot X, \qquad [3\text{-}28]$$

where V_{max} is the maximum possible chemical uptake rate per cell [M/(cell·T)], K_s is the half-saturation constant [M/L^3], and X is the cell density [cells/L^3].

3. The cell density X at any point within this layer can grow according to Monod kinetics. In addition, a first-order rate constant for cell death, d, is usually introduced; net cell growth, which equals zero at steady state, is equal to the difference ($\mu - d$). (In other than shallow biofilms, μ is not independent of location as implied here, but varies with location in the biofilm.)

$$\frac{dX}{dt} = (\mu - d) \cdot X = 0, \qquad [3\text{-}29]$$

where dX/dt is the change in cell density with time [M/(L^3·T)], μ is the specific growth rate [T^{-1}], and d is the cell death rate [T^{-1}]. In Eq. [3-29], μ is a function of the chemical concentration C_f, according to Monod kinetics,

$$\mu = (V_{max} \cdot y) \cdot \frac{C_f}{C_f + K_s}, \qquad [3\text{-}30]$$

where y is the cell yield [cells/M]. Note that Eq. [3-30] is identical in form to Eq. [2-73].

4. If the film is growing on a solid surface, the chemical flux at the surface must be zero:

$$\left(\frac{\partial C_f}{\partial x}\right)_{\text{at solid surface}} = 0. \qquad [3\text{-}31]$$

5. At steady state, the chemical flux into the biofilm is entirely consumed by bacteria in the film. The diffusive flux into the control volume of the entire film equals the sink strength (chemical uptake rate by microorganisms) integrated over the biofilm,

$$D_f \cdot \left(\frac{\partial C_f}{\partial x}\right)_{\text{at biofilm surface}} = \int_{x=0}^{x=L_f} V \cdot X \cdot dx, \qquad [3\text{-}32]$$

where V is the rate of chemical uptake per cell [M/(cell·T)], a function of the chemical concentration C_f.

Any number of simplifying assumptions may be made to yield more mathematically tractable models. For example, in some cases it may be appropriate to assume first-order uptake by cells (as when C_f is small compared with K_s) or zero-order uptake (if C_f is expected to be much greater than K_s).

It is possible to estimate the chemical concentration that would be just barely sufficient to maintain a shallow biofilm. Such a condition exists at the downgradient end of a zone in which biodegradation is active. When nutrient concentrations decrease below this critical value, the biofilm withers. Setting μ equal to d in Eq. [3-30] and solving for C_f gives C_{min}, the lowest chemical concentration at which a shallow biofilm can exist in steady state:

$$C_{min} = K_s \cdot \frac{d}{(V_{max} \cdot y) - d}. \qquad [3\text{-}33]$$

C_{min} also can be thought of as the lowest steady-state concentration at which a biofilm can be expected to degrade a chemical, unless cometabolism occurs. (Recall from Section 2.6 that during cometabolism, cell enzymes transform a chemical even though the transformation yields no energy to the cell.) Note that there can be a substantial lag time in the response of a biofilm to its environment. For example, a thick biofilm may do a good job of degrading even very low concentrations of pollutants for a short time, after which its effectiveness will decay if microbial cells do not grow rapidly enough at the lower pollutant concentration to replace cells that die.

Microbial transformation by biofilm kinetics may be incorporated in computer simulations of chemical fate and transport in groundwater. Modeled biodegradation rates are combined with advective and dispersive transport simulations of known chemical inputs to an aquifer to obtain concentration estimates for a chemical as a function of time and location in the aquifer. There are many complicating factors that arise, however, when simulating processes in an actual aquifer, including incomplete characterization of aquifer properties and of heterogeneity. Furthermore, the microbial community is more complex than can be described even by the full set of equations given previously. One reason is that a microbial community usually contains many different species, each with different uptake and growth kinetics. (An exception might be a bioreactor that is first sterilized and then inoculated with a pure culture of a bacterium.) Competitive or symbiotic interactions among the species add further complications. The practical manifestation of these numerous uncertainties is that models are typically "calibrated" by adjusting transport and microbial kinetic parameters until model results approximate some set of actual measurements in the field. Thus, although biofilm models are useful conceptually, in practice the prediction of biodegradation by attached bacteria in the environment still relies on a great deal of empiricism. In fact, simple models that use empirical first-order rate constants are often applied to estimate chemical lifetimes in the subsurface environment. Some empirical constants are shown in Table 2-8 in Section 2.6.3. For an example of biodegradation of specific chemicals in anoxic biofilms, the reader is referred to Bouwer and Wright (1988).

EXAMPLE 3-9

Glyphosate is a broad-spectrum herbicide widely used in agriculture. Soil incubation studies have shown a biodegradation rate constant for glyphosate of 0.1/day (Lyman *et al.*, 1990) based on the rate of disappearance from incubated soil. The rate as inferred from studies using ^{14}C-labeled glyphosate is 0.0086/day, an order of magnitude smaller.

The glyphosate concentration in the pore water below an agricultural field is found to average 1000 ppb. What might be the glyphosate concentration in water that recharges an unconfined aquifer beneath the cropland? Depth to the water table is 2 m, and annual recharge is 35 cm of water; assume θ averages 0.1, and neglect possible sorption of this very water-soluble chemical.

First, an estimate of the travel time is needed. Estimate the seepage velocity using Eq. [3-5], but use the water-filled porosity θ instead of porosity, n:

$$v = (35 \text{ cm/year})/0.1 = 350 \text{ cm/year}.$$

Travel time to the water table then can be calculated:

$$\tau = \frac{200 \text{ cm}}{350 \text{ cm/year}} = 0.57 \text{ year} = 209 \text{ days}.$$

The concentration of glyphosate in the water reaching the water table then can be calculated using the expression for first-order decay, Eq. [1-19]:

$$C_{\text{recharge water}} = (1000 \text{ ppb}) \cdot e^{-(0.1/\text{day}) \cdot (209 \text{ day})} = 9 \times 10^{-7} \text{ ppb}.$$

This calculation suggests that the amount of glyphosate reaching the water table should be minuscule. A different conclusion would be reached using the rate constant determined from the measurement of $^{14}CO_2$ evolved from the incubated soil:

$$C_{\text{recharge water}} = (1000 \text{ ppb}) \cdot e^{-(0.0086/\text{day}) \cdot (209 \text{ day})} = 170 \text{ ppb}.$$

Another factor to consider is the possibility of more rapid downwater movement through preferred flow pathways such as macropores created by plant roots in the soil.

3.6.2 Natural and Enhanced Biodegradation in the Field

When modeling chemical concentrations and distributions in groundwater, biodegradation frequently must be considered, because it is a significant fate process for many pollutant chemicals. From a practical standpoint, biodegradation of pollutants can be a very important tool in the task of cleaning up groundwater contamination. In some cases, allowing biodegradation to proceed in a contaminated aquifer without human intervention may be deemed the most feasible process for removing contaminants. Although this remedial action amounts to doing nothing and letting natural processes take their course, it is an appropriate response to contamination in some cases. Such a remedial action is an example of cleanup by *natural attenuation.*

Natural attenuation by itself, however, often is not sufficient to achieve a desired extent or rate of contaminant removal from an aquifer. In these instances, one remedial option may be to enhance the natural rate of biodegradation of pollutant chemicals in the aquifer. This strategy, called *in situ bioremediation,* is considered to be one of the most attractive remedial techniques from a cost perspective, because many of the high costs associated with pumping and treating groundwater or excavating contaminated aquifer material are avoided. Furthermore, the potential exposure of cleanup workers to pollutant chemicals is reduced if many of the contaminants are mineralized while still in the aquifer.

From the preceding discussion of microbial kinetics, several possible methods for enhancing the rate of biodegradation of a chemical should be evident. Because the biodegradation rate depends on V_{max}, K_s, and cell density (X), one can, in principle, increase the degradation rate by increasing V_{max}, decreasing K_s, or increasing microbial cell density within the aquifer. The first two strategies require changing the microbial species present, either by changing conditions within the aquifer to better suit a native population having more favorable kinetic parameters or by introducing additional microbial species. The introduction of additional genetic elements, such as plasmids, to convey new degradation capabilities to existing microbial species also has been proposed.

Of course, no such changes or enhancements will increase the biodegradation rate if some chemical other than the pollutant of interest is limiting to the microbial community. The most frequent such limitation is the availability of oxygen (or another suitable oxidant). Thus, a necessary and sometimes sufficient strategy for enhancing biodegradation involves the introduction of an oxidant to an aquifer. Injecting gaseous oxygen, in a technique known as *sparging,* is a direct means by which an oxygen limitation is mitigated. Alternatives to injection of gaseous oxygen (either pure oxygen injection or air injection) include the introduction of hydrogen peroxide (H_2O_2), which decomposes to release oxygen, or the injection of nitrate, which is also a favorable oxidant (see Section 2.4.3). One advantage of nitrate injection is that more equivalents of oxidizing capacity can be dissolved in a liter of water as nitrate than can be dissolved as gaseous oxygen at ordinary pressure. In addition, cleanup strategies such as vacuum extraction may, as a secondary effect, increase the availability of oxygen to microorganisms in an aquifer.

Supplying oxygen to an oxygen-limited microbial community increases the metabolic activity of any existing population of aerobic, biodegrading organisms; it also leads to their growth, and thus an increase in cell numbers. In some instances, it has been suggested that cell density also can be limited by the availability of certain mineral nutrients, such as phosphate (recall the discussion of the Redfield ratio in Section 2.4.2). Thus, another strategy sometimes employed to enhance *in situ* biodegradation is to inject a mineral nutrient solution, alone or in conjunction with an oxidant.

Altering nutrient and redox conditions also is likely to cause changes in the relative proportions of different microbial species present in an aquifer, and thus bring about changes in the microbial kinetic parameters. These changes occur even if no new species of microorganisms are added; however, it also has been proposed that new microbes having the specific metabolic capability to degrade a pollutant of concern could be injected into a contaminated aquifer. Although this remedial action sounds reasonable, such a strategy often fails for ecological reasons, because the newly introduced microbes are not able to compete with the well-adapted indigenous microbial communities. An interesting variant on this strategy is to introduce carriers for genes that code for the manufacture of key enzymes and, if incorporated into indigenous bacteria, will lead to the production of enzymes required for the degradation of a pollutant. Introducing new genes frequently may not be necessary, however, given the enormous diversity of natural microbial populations and the fact that novel degradation abilities often are found to appear in contaminated areas.

A related approach is to develop a dense community of microorganisms that, although they may be unable to derive energy from and grow upon a particular pollutant may have the ability to degrade the chemical via cometabolism. To induce the growth of such a bacterial community it is necessary to introduce a growth substrate that will favor the development of the desired bacteria. One approach is to foster the growth of *methane-oxidizing bacteria (methanotrophs),* which produce enzymes (*methane monooxygenases*) that have a very broad range of oxidation capabilities. Methanotrophs are widely present in the environment, and can be stimulated to grow by introducing methane as a growth substrate (and assuring that sufficient oxygen is available).

In the case of oxidized pollutants, one could consider the introduction of reductants to promote microbial transformation. Such a technique can, in principle, promote biodegradation of certain highly chlorinated organic compounds (some of which are known to degrade by microbial reduction in highly reducing soils and sediments). Microbial reduction, requiring the establishment of reducing environmental conditions, already is used as a strategy in certain treatment processes to remove nitrate from water by means of denitrification.

3.7 CONCLUSION

Although the subsurface environment may be somewhat simpler than surface waters, it is less well understood, perhaps because it cannot be observed directly and is accessible only at considerable expense through excavation and drilling.

Transport in the subsurface environment is slow compared with the other environmental media. Contaminants may move only tens of meters per year by advection, contrasting sharply with surface waters, which travel this far in minutes or hours, and air, which may travel this far in seconds (as discussed in the next chapter). Similarly, Fickian transport coefficients are rarely higher than thousandths of a square centimeter per second and are often no larger than a fraction of the molecular diffusion coefficient in free water. Many organic compounds that would rapidly volatilize into the atmosphere from surface waters may reside in groundwaters for decades or longer.

The subsurface environment also shares many processes with surface waters. Microbially mediated redox reactions and biodegradation processes are significant in each medium, and much of what was discussed in Chapter 2 on these topics applies directly to the subsurface as well. The presence of particles, and their potential to absorb chemicals, also is common to both surface waters and groundwaters; when modeling subsurface transport, the high ratio of solid material to water requires special recognition of even moderate sorbing tendencies.

In Chapter 4, the atmosphere is discussed. Again, both contrasts and similarities are seen among the three environmental media.

EXERCISES

1. The hydraulic conductivity of an aquifer is 5×10^{-3} cm/sec, the hydraulic gradient is 0.02, and the porosity is 0.2.
   a. What is the specific discharge, q_a, in this aquifer?
   b. Calculate the maximum velocity at which dissolved salt could move through this aquifer.
   c. Assume the aquifer is 5 m thick. What is its transmissivity? What is the discharge per unit of distance perpendicular to the water flow (bq_a)?
2. A well of 0.1-m radius is installed in the aquifer of the preceding exercise and is pumped at a rate averaging 80 liter/min.
   a. Estimate drawdown in this well if the radius of influence is 100 ft.
   b. Estimate the hydraulic gradient in the aquifer at distances of 20 and 75 ft, both directly upstream and downstream of the well.
3. In a sand having a median grain size of 1 mm and porosity of 0.25, how high must specific discharge be to make the mechanical dispersion coefficient equal to the effective molecular diffusion coefficient?
4. A domestic well is located 30 m upstream from a neighbor's septic system. If domestic use of water is 1000 liter/day, what is the minimum regional flow (bq_a) that will just suffice to keep septic effluent from entering the well (easiest to do analytically)?
5. A municipal well pumping 1200 m³/day is located 100 m from a major river. Aquifer transmissivity is 800 m²/day, aquifer thickness is 30 m, and the background hydraulic gradient (i.e., in the absence of the well) is 0.002 toward the river.
   a. Discuss the calculation of drawdown if the well is 2 m in diameter.
   b. A large spill releases carbon tetrachloride (CCl_4) into the river upstream, resulting in the river water now containing 50 ppm dissolved CCl_4. What maximum concentration do you expect to see in the well if no corrective action is taken? What do you recommend the municipal well operator do to prevent well contamination?
6.

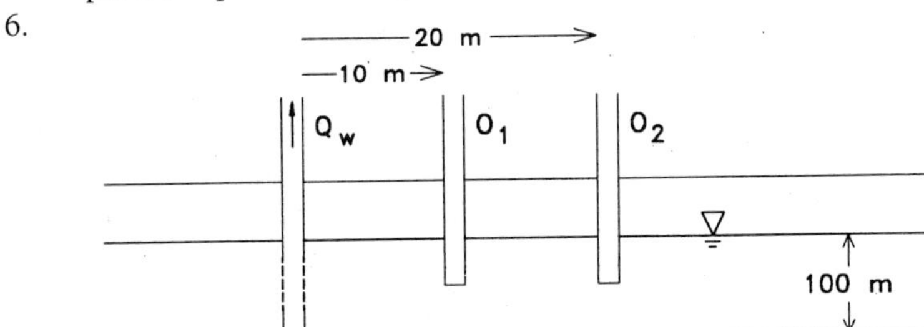

A well in a phreatic aquifer 100 meter thick supplies a small manufacturing plant, and is pumped continuously at a rate of 1200 liter/min. The difference in water levels between observation wells O_1 and O_2, located 10 and 20 m away, respectively, is seen to stabilize at 3 cm, whereas before pumping the level in O_1 was 1 cm higher than in O_2. The specific yield of the aquifer material is estimated as 0.2.

Following plant shutdown for 6 months (during which time the well was not pumped), the well again is pumped, this time at a rate of 3000 liter/min.

a. What is aquifer hydraulic conductivity?
b. Tabulate drawdown versus time at O_1 for the first week after pumping starts at the 3000 liter/min rate (e.g., tabulate for 0, 1, 2, 4, 8 days).

7. In the flow net diagram that follows, water from a leaky pressurized sewer pipe is seeping into the unpaved street above. The backfill around the pipe has a hydraulic conductivity of 10^{-3} cm/sec, much higher than the surrounding clayey soil. How much water is being discharged per meter of pipe, if head in the pipe is 1 m, and head of the groundwater at street level is 0.5 m? How accurate is the flow net shown?

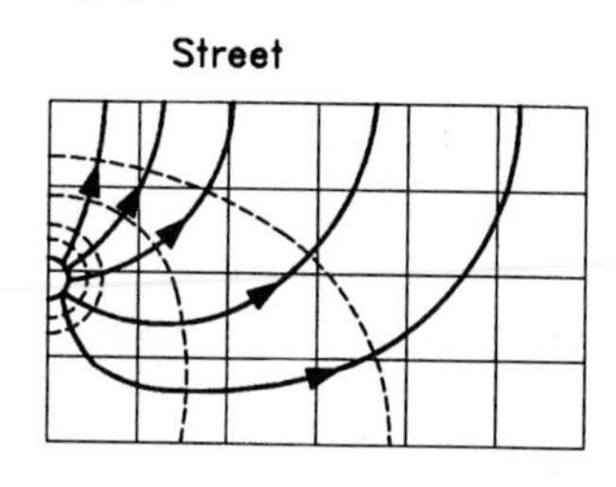

8. Water moves through a sand column 2 m long with a specific discharge of 15 cm/hr. Sketch the breakthrough curve of an initially sharp front of a tracer if porosity is 0.3 and median sand grain diameter is 2 mm. (Be sure to indicate the width of the breakthrough curve.)
9. A small shop has been dumping used solvents into an abandoned well, shown as follows. Sketch a flow net indicating where the solvents are likely to appear at the lake bed. If dissolved solvent moves with

the water, roughly how long will it take to travel from the well to the lake?

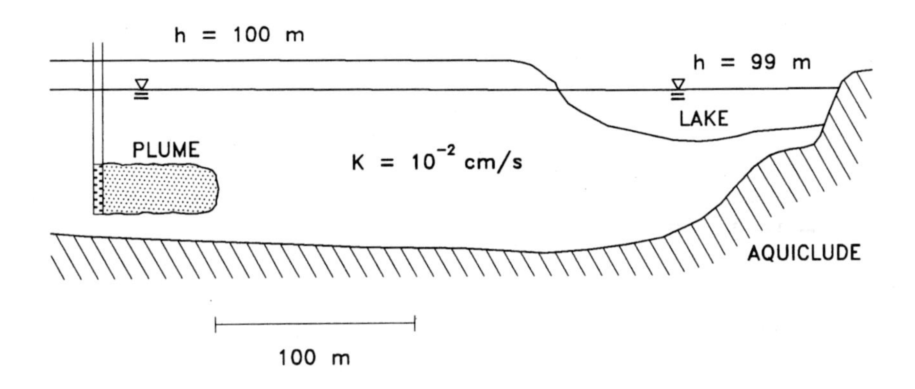

10. At what speed does a pulse of dissolved benzene move through organic-rich wetland soil (f_{oc} = 50%) if the specific discharge is 10^{-3} cm/sec and porosity is 0.4? Assume that ρ_b for these sediments is 1.5 g/cm^3.
11. Shortly after a spray application of pesticides, soil water in a potato field contains 100 ppb of the pesticide. How soon will the concentration decrease to below detection limits (0.5 ppb) for the analytical method being used, if $k'_n = 10^{-7}$/sec, $k_a = 10^{-2}$ liter/(mol·sec), and k_b = 0.3 liter/(mol·sec)? Soil water pH is 8.
12. A fully penetrating well is installed in a confined aquifer. Aquifer thickness is 12 m and conductivity is 10^{-4} m/sec. Storativity is 0.001. The well is pumped at a rate of 100 liter/min for 10 hr. The pumps are then shut down. What is the drawdown 5 m from the well right when the pumping is stopped, and 4 hr after it is stopped?
13. A laboratory column of 10 cm^2 cross-sectional area is 1 m long. Porosity is 0.25 and influent flow rate is 8 liter/day. Particle grain size is 0.5 mm. At time t = 0, injection of a nonsorbing contaminant into the influent at a constant concentration is started. Estimate contaminant concentration as a function of the time at the outflow.
14. At what rate must well A be pumped to capture all the pollutant plume from the landfill? Natural gradient is 0.001 from north to south, aquifer hydraulic conductivity is 10^{-4} m/sec, and aquifer thickness is 10 m.

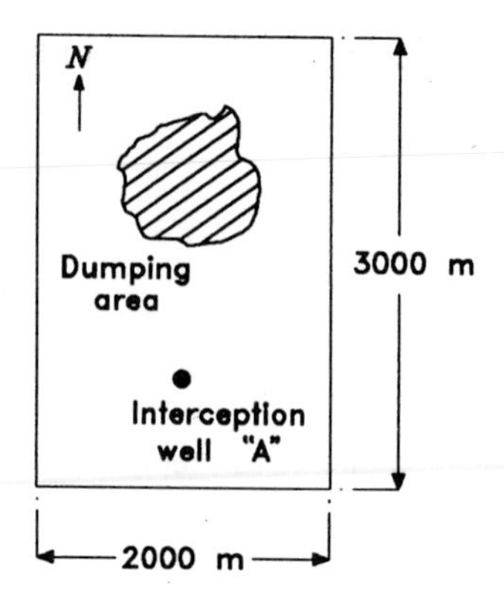

15. A very short pulse injection of tracer is made into a column. When the center of mass has traveled 2 m, the standard deviation is 0.1 m. The molecular diffusion coefficient is 10^{-5} cm²/sec.
   a. What is the longitudinal dispersion coefficient *D* if average velocity of pore water is 1 mm/sec?
   b. What is the dispersivity α?
   c. How does *D* vary with velocity? How does α vary with velocity?
   d. If the tracer were replaced by a different tracer having much lower molecular diffusivity (perhaps something like latex microspheres having an effective molecular diffusivity of 10^{-8} cm²/sec), would the tracer distribution be different at this point in the column? (Assume the spheres move freely through the porous medium and are not filtered out or stuck.)
   e. If the microspheres were 0.1 mm in diameter, would they be likely to pass through the porous medium freely (i.e., would they fit in the pores)?
16. Estimate a retardation factor for the transport of dissolved chloroform ($CHCl_3$) under the following conditions:
   a. In the saturated zone, where $n = 0.3$, $f_{oc} = 1\%$, and soil bulk density = 2.1 g/cm³.
   b. In the same soil but in the unsaturated zone, with $\theta = 0.15$. A pulse of dissolved chloroform is being carried through the soil by water flow. Assume the system is *closed* (i.e., negligible amounts of chloroform vapor escape to the atmosphere over the time period of interest).
17. Groundwater contains 10 ppm of ethyl acetate, $CH_3COOC_2H_5$, under a landfill. If the seepage velocity is 1 m/day, what is the maximum distance downgradient at which a concentration in excess of 0.2 ppm could be expected? Answer for a water pH of (a) 4 and (b) 10.

18. A total of 400 liters of benzene leaks from an underground storage tank before a leak is discovered. The water table lies a few feet below the tank.
    a. Discuss the possibility of recovering some of the chemical, and how it might be done.
    b. Discuss the maximum concentrations expected in the groundwater.
    c. Discuss retardation, assuming $f_{oc} = 0.5\%$.
    d. Discuss the rate of biodegradation.
19. At Maxey Flats, Kentucky, radioactive waste was buried in trenches. As rainwater infiltrated the soil, the trenches became saturated, and water containing dissolved radioisotopes began to move through the soil as the trenches "overflowed." Assume a disposal trench overflow scenario that looks like the following diagram:

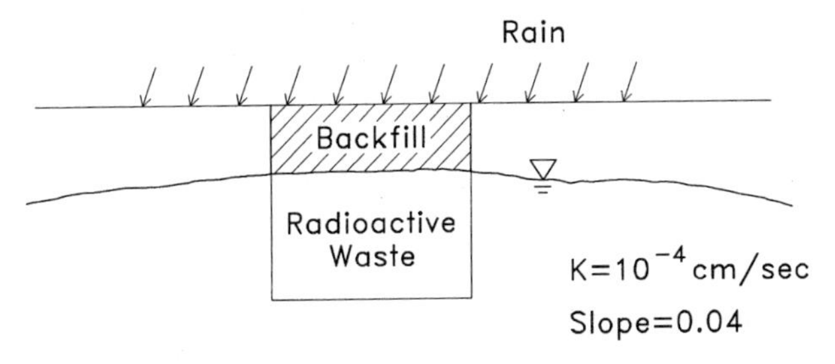

    a. How long will it take a tracer such as chloride to move 50 m from a disposal trench? Assume $n = 0.22$.
    b. How long would it take plutonium to move this far? K_d for plutonium is found to be approximately 5×10^3 ml/g.
    c. What effect would the presence of macropores have on your answer to part (b)?
20. A tanker truck full of trichloroethene has ruptured, dumping 25,000 gal of the solvent. Quick work by cleanup crews has recovered most of it, but some has reached the water table and dissolved in the groundwater.

    Aquifer characteristics include a 10-m saturated depth, a transmissivity of 2×10^{-3} m^2/sec, a regional gradient of 0.0005, and a porosity of 0.25. It is proposed to remove, treat, and reinject the water to clean up the remaining trichloroethene. Water will be removed at well A, treated, and reinjected at well B, 200 m away downgradient. The spill extent is poorly known, but the consultant has assumed that if the steady-state pumping rate is sufficient to cause any contaminant midway between A and B to enter well A, all contaminant eventually will be captured.
    a. What is the maximum concentration of trichloroethene expected in the aquifer?
    b. Discuss the proposed pumping and reinjection scheme with regard to its feasibility.

c. One treatment alternative being considered is to use columns of peat from a local wetland as an adsorbent. Columns 5 m long and 1 m^2 in cross section are proposed. If peat has porosity of 0.5, f_{oc} of 50%, density of 1.2 g/cm^3, and dispersivity of 1 cm, how much water will each peat column treat until breakthrough?

d. Another treatment alternative being considered is air bubbling. In this scheme, air bubbles would pass through a column of contaminated water, reaching equilibrium with the water while they rise through the column. How many volumes of gas per unit volume of water are necessary to reduce the trichloroethene concentration below 1 μg/liter?

21. A confined aquifer receives recharge in an agricultural area where the soil water often contains appreciable levels of nitrate. Potential electron acceptors in the recharge water include NO_3 at 5 ppm, O_2 at 7 ppm, and SO_4^{2-} at 12 ppm. Organic matter in the aquifer consumes these electron acceptors via microbial activity at a rate equivalent to the oxidation of 1 μM of CH_2O (oxidized to CO_2) per liter of water per day. Average seepage velocity is 1 m/day.

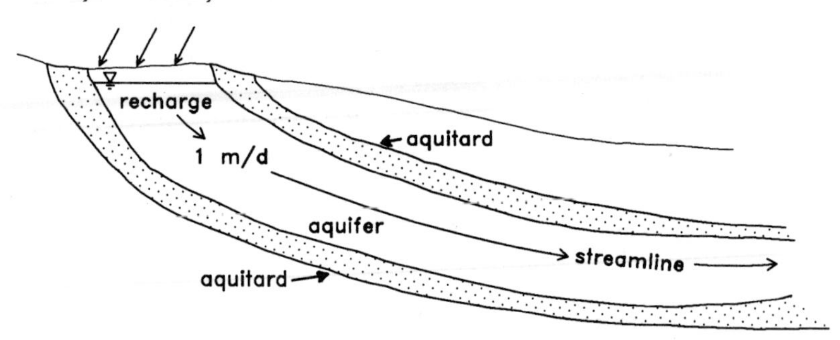

a. Draw boundaries along the streamline in the preceding diagram, clearly showing distance from the recharge area at which each of the preceding electron acceptors is expected to disappear (neglect the possible role of iron for now).

b. Estimate pϵ along the flow line, assuming pϵ is determined by the dominant electron acceptor/reduced electron acceptor pair at each point.

c. Many domestic wells tap the aquifer; some of the users complain of iron in their tap water. What is the closest these wells are likely to be located to the recharge area?

22. A new municipal well is installed in an alluvial aquifer having transmissivity of 0.1 m^2/sec, f_{oc} of 0.5%, porosity of 0.25, and thickness of 20 m. The well is put into operation at a pumping rate of 0.5 mgd (million gallons per day). Neglect regional flow.

Unfortunately, a pool of gasoline lies on the water table beneath the site of an abandoned gas station 150 m south of the well. Although the health department has pronounced the well "safe," it may not remain so. When will dissolved gasoline reach the municipal water system? (Model the gasoline, a complex mix of hydrocarbons, as though it were benzene.)

23. Explosion and fire have damaged the motor pool at an army base. As a renowned safety expert, you are retained to be part of the investigative team. You learn that 10 years ago, an underground storage tank released over 2000 gal of gasoline to the subsurface environment. You decide to see if an accumulation of gasoline vapor could have led to the explosion, which evidently occurred in the compressor room (compressor controls might spark), a space 6 ft high with a dirt floor. Depth to the water table as indicated on soil logs averages 10 ft below the floor; the leaking tank was about 50 ft away upgradient. Treat the gasoline as if it were hexane, C_6H_{14}, which is flammable in air at concentrations between 1.2 and 7.4% by volume (or by mole fraction or by pressure). Molecular weight of hexane is 86, and vapor pressure is 152 mm Hg at 20°C.
    a. What is the maximum possible diffusive flux of hexane into the compressor room?
    b. If the ventilation rate of the room is extremely low (e.g., 0.1 air changes per day), could the flammable limit be reached?
24. A manufacturer wishes to install a 200 m^3/day water supply well in the sand and gravel of an alluvial river valley. Because the river traversing the valley is contaminated, the engineer in charge proposes locating the well far enough away from the river so that it does not draw in any river water. Regional flow is directly toward the river; the water table has a gradient of 0.0017, and aquifer transmissivity is 75 m^2/day. How far from the river must the well be located?
25. A packed chromatographic column is a piece of tubing through which a carrier fluid flows; it is packed with a porous medium that has a significant tendency to sorb chemicals of interest. A pulse of a mixture of chemicals is injected into one end of the column, and each chemical then moves at a velocity according to its distribution coefficient between the moving fluid and the porous medium. Note that this is identical to a sorbing solute being transported by groundwater.

    Let the column have the following characteristics:

Length	=	220 cm
Inside area	=	0.06 cm^2
Diameter of porous medium particles	=	0.02 mm
Porosity of porous medium	=	0.3
Bulk density of porous medium	=	1.8 g/cm^3
Flow rate of fluid	=	7 ml/min
K_d for chemical A	=	0
K_d for chemical B	=	2.8 ml/g

If 0.1 μg of each of chemicals A and B is instantaneously and simultaneously injected,

a. How long does it take the center of mass of chemical A to travel 200 cm along the column?
b. How long does it take the center of mass of chemical B to travel 200 cm along the column?
c. What is C_{max} and the spatial distribution of chemical A after its center of mass has traveled 2 m? Assume the only mixing process is mechanical dispersion.

26. For purposes of capturing a plume of contaminated groundwater, you are asked to design a well that will create a steady-state hydraulic gradient of 0.009 at a distance of 75 m. (The gradient will be superimposed on the preexisting gradient.) Assume a fully penetrating, fully screened well is to be installed.
    a. Propose the necessary pumping rate of the well, if the aquifer transmissivity is 15 m^2/day and aquifer thickness is 14 m.
    b. To not violate the assumption of a linear aquifer, one must avoid excessive drawdown relative to the aquifer thickness. In the present case, what drawdown do you expect if your well has a casing diameter of 25 cm (i.e., calculate drawdown for a radius of 1/8 m)? Assume a radius of influence of 40 m.
27. You are called to consult on a LUST problem; an estimated 2000 kg of gasoline was lost from a local service station before the leak was discovered. A layer of gasoline now floats at the water table 7 m below ground surface. A detailed examination of records suggests that the leak began 2 years ago.
    a. Describe what you would do first to recover the NAPL. What will limit how much you expect to recover?
    b. Some of the gasoline has dissolved in water and migrated downgradient. Specific discharge associated with natural groundwater flow under the station is 30 m/year. It is 25 m downgradient from the edge of the bolus of gasoline to the property line. Do you think the gasoline is contained on the property? Aquifer material is a silty fine sand with 0.5% organic carbon. Assume gasoline has an effective log K_{oc} of two (it is actually a mix of compounds, each having its own K_{oc}), and make other reasonable and necessary assumptions.
    c. Your proposed cleanup plan includes enhanced *in situ* biodegradation.
        i. What environmental factor is probably limiting the present biodegradation rate most?
        ii. You are contemplating the addition of water containing mineral nutrients and electron acceptors. You naturally want as many equivalents of oxidizing capacity as possible in each liter of water

that you inject. If you assume a temperature of 10°C, what is the capacity (in equivalents/liter) of

- a solution containing 250 g/liter of H_2O_2, which decomposes to O and H_2O?
- water equilibrated with air at one atmosphere?
- water equilibrated with 1 atm of O_2?
- water containing KNO_3 at its solubility limit (2.1 *M* at 10°C)?

28. In 1990, gasoline contamination is discovered during a routine site assessment at an urban site shown in the figure that follows. It is not known how long before 1990 this contamination reached the site, assuming that it originated off-site.
   a. Based on the following data, make a justified recommendation as to which gas station or stations are the likely culprits (include all assumptions): Station A was established in 1978; Station B in 1985; and Station C, in 1980.
      At the site, hydraulic conductivity is 100 ft/day, porosity is 0.3, gradient (in direction indicated) is 0.00225, and transverse dispersivity is 8 ft.
      Hints:
      i. Consider advection and dispersion in a vertically well-mixed aquifer.
      ii. Transverse dispersivity is horizontal dispersivity in the direction perpendicular to groundwater flow.
      iii. There is no water supply or other pumping well at the site.
      iv. You may neglect longitudinal dispersion (dispersion in the direction of groundwater flow).
   b. What are the inherent weaknesses in this analysis? What specific warnings or caveats would you convey to the lawyers? In what ways could this analysis be improved given more time and money?

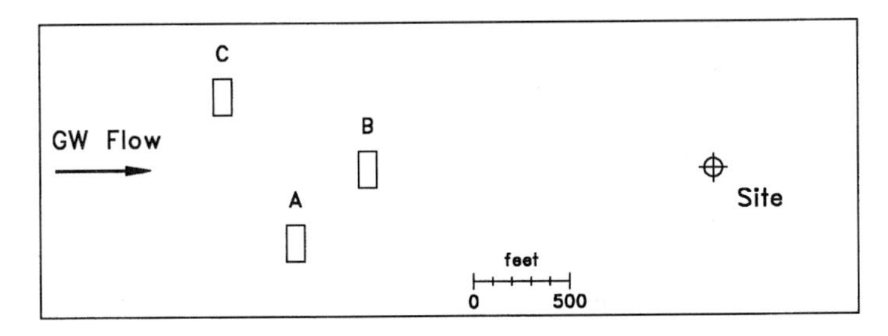

29. Assume log K_{oc} for a certain solvent is 2.1 ml/g. Estimate the speed with which a cloud of solvent, dissolved in water, moves along a saturated porous medium column 2 m in length, if head difference from end to end

is 0.01 m, porosity is 0.3, f_{oc} is 1.7%, and K is 2.5×10^{-3} cm/sec. The density of the porous medium grains is 2.6 g/cm^3.

30. A greenhouse owner purchases a 55-gal drum containing a solution of the sodium salt of metham, a soil fumigant made famous by the Sacramento River spill of 1991. Unskilled in the use of the forklift (the regular driver is home sick), the owner punctures the barrel, and 30 mol of sodium metham spills down a storm drain and into a creek. It is too late to contain the spill, but you decide to make the best of the situation and to do a little pesticide geochemistry. This uniform creek has a discharge of 1 m^3/min and an average velocity of 20 cm/sec at the time.
    a. A half-kilometer downstream, you set up a sampling station on a footbridge, and begin your measurements before the peak concentration of metham is reached. Given that the chemical was shipped as the sodium salt, each mole of the chemical enters the creek as a mole of conservative sodium ion accompanied by a mole of metham anion. The peak sodium concentration that you measure at the bridge is 8 mol/m^3.
        i. What is the average longitudinal dispersion coefficient for the creek?
        ii. How wide, in space, is the cloud of sodium as its center of mass passes your bridge? Use as your criterion the width that contains 68% of the total mass.
    b. By recalling that k_a for hydrolysis of metham is 300/($M \cdot$sec), and k_n' is 10^{-8}/sec (k_b is negligible), what peak concentration of metham anions do you expect to measure at the bridge? This is a soft water region where acidic deposition is a problem, so the pH of the creek is only 5.2.
    c. Instead of the peak concentration estimated in (b), you actually measure a peak metham concentration that is 10% lower. Can you credibly attribute this amount of difference to
        i. photolysis (indirect photolysis, typically at a rate of 0.0001/sec)?
        ii. error in your estimate of how much metham entered the creek?
        iii. variance in chemical analysis?

31. (Refer to Exercise 30.) On the way home, it occurs to you that you should warn stream abutters about the spill. The local Motel $5\frac{1}{2}$ draws water from a well installed 5 m from the creek in the 4 m thick sandy aquifer, whose porosity is 0.3. The well pumps an average of 5000 liters/day. The natural specific discharge in the aquifer is 5 cm/day toward the creek.
    a. Estimate how much of the well water comes from the creek, based on the simple flow net provided.
    b. You estimate that peak metham concentrations in the creek will be

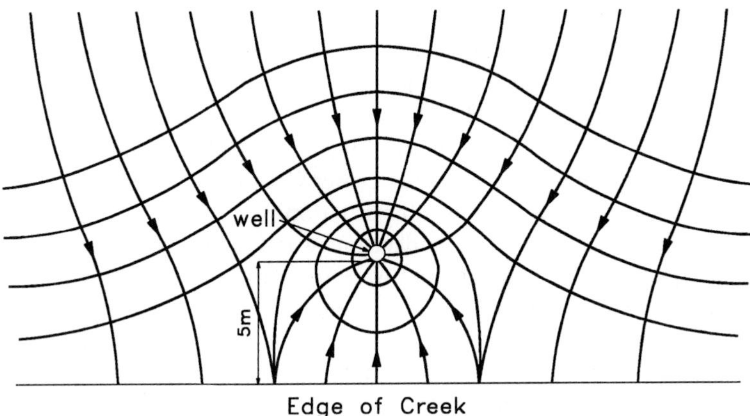

down to the levels deemed safe for drinking water by 8 P.M. about the time the metham in the creek should reach the vicinity of the motel. Further safety is provided by the fact that the water drawn from the creek will be mixed with clean groundwater. You tell the motel manager not to worry about metham in the motel water.

One of the guests is taken ill at 11 P.M. that evening. Five days later the guest's lawyer files suit, and you must defend your decision not to recommend shutting off the water. The plaintiff's lawyer is attempting to discredit your testimony, claiming that you erroneously used a steady-state tool (the flow net) to analyze a time-varying situation. Use sound hydraulic arguments to defend your opinion that the sick guest could not have ingested metham via the well.

32. It is desired to recover naphthalene from a plume of groundwater moving from a former coal tar disposal site. Based purely on hydraulic considerations, you have determined that all the naphthalene-contaminated water in the plume can be removed with an interception well (capture well) in a month. This, however, does not factor in retardation. How long would you expect it to take to recover all the naphthalene if the soil contains 0.5% organic matter? Make reasonable assumptions about other parameters.
33. A rectangular building foundation is to have dimensions of 18 × 24 m. The bottom of the footings will lie below the water table. To be able to work in dry conditions, it is desired to temporarily lower the water table by at least 2.5 m throughout the area proposed for excavation.
    a. If T is estimated to be 10^{-3} m^2/sec, what rate of pumping from a well in the center of the excavation area is necessary to accomplish this dewatering? Assume a radius of influence of 200 m.

b. The contractor wants to know how long it will take to dewater the soil. If she pumps the well at 20 liters/sec, how much will the water table be lowered at a corner of the proposed foundation 2 days after pumping begins? (Assume $S_y = 0.3$.)
c. Would the required total pumping rate (sum of four wells) be higher or lower if a dewatering well were installed at each corner, instead of the single larger one in the center (let the water table in the center of the excavation area be your criterion)?

34. (Refer to Exercise 33.) A decision was finally made to install a single dewatering well and to pump at 10 liters/sec. A leaking underground storage tank at a gas station 150 m directly downgradient poses the threat of supplying contaminated water to the dewatering well. The natural gradient is 0.006. If the well draws water contaminated with benzene, toluene, ethylbenzene, xylenes (BTEX), it will be illegal to discharge to a nearby stream without treatment.
    a. Can you rule out the possibility of drawing in water from beneath the gas station?
    b. How confident are you of your conclusions? (T is estimated to be 10^{-3} m^2/sec on the basis of a single pump test conducted at a site 0.5 km away.)
35. (Refer to Exercise 34.) Thanks to a strong recommendation from the contractor, your firm is chosen to design a remediation plan for the gas station. Fortunately, the contaminated water occupies a small region. Unfortunately, the only possible location for a capture well is 100 m from the contamination. Regional flow is at right angles to a line between the capture well location and the center of contamination. What pumping rate will you specify for the capture well to intercept all the contaminated water?
36. (Refer to Exercise 35.) You have begun operation of a pump-and-treat system for the gas station. According to a computer model for conservative solute transport, 1.5 travel times have now elapsed and the capture well should be pumping contaminated water. However, the laboratory reports that your capture well is still pumping essentially clean water, although the pH has dropped from 8 to 7.
    a. Discuss two possible reasons for this observation (other than model error and incorrect aquifer parameters). Be quantitative if possible. The aquifer contains 0.3% organic carbon.
    b. Explain how the pH could be dropping as an indirect result of gasoline contamination. If alkalinity (Alk) of the water now being pumped is 0.5 meq/liter, how much must C_T have risen to account for the observed pH decrease?

37. (Refer to Exercise 36.) To deal with residual saturation, you propose to stimulate biodegradation by supplying mineral nutrients and an oxidant (electron acceptor). You receive permission to make an experimental injection of sodium nitrate ($NaNO_3$). The hydrocarbon contaminants will be oxidized as the nitrate is reduced to N_2 (denitrification is expected to occur).
    a. Theoretically, how much $NaNO_3$ will be needed to oxidize the 25 kg of heptane (C_7H_{16}) that you guess remains in the aquifer?
    b. While denitrification is occurring, is the pϵ greater than 4, between zero and 4, or less than zero?
38. Water moves through a sand column 2 m long with a specific discharge of 15 cm/hr.
    a. Sketch concentration versus distance (only peak height, location, and standard deviation need be specified) for a pulse injection of a tracer 2 hr after injection. At this time, C_{max} is 75 ppm. Porosity is 0.3 and median sand grain diameter is 2 mm.
    b. A hydrophobic organic chemical is introduced instead of a tracer; K_{oc} is 25, organic carbon content of the aquifer is 0.6%, and bulk density is 2 g/cm^3. How far down the column is the maximum concentration, C_{max}, to be found after 2 hr?
    c. A nonsorbing but hydrolyzable chemical is injected at the same concentration as the tracer. The pH is 9, k_a is $7.8 \times 10^{-2}/(M \cdot \text{sec})$, k'_n is 6.6×10^{-6}/sec, and k_b is $14/(M \cdot \text{sec})$. What is the maximum concentration of this chemical in the column after 2 hr?
39. Analysis of vadose zone transport is more complex than analysis of saturated zone transport, in part because changes in soil water content have a strong influence on both hydraulic conductivity and pore pressure. How does a decrease in soil water content affect:
    a. hydraulic conductivity?
    b. piezometric head given constant elevation and pore gas pressure?
    c. soil gas content?
    d. Fickian transport coefficient for vapors in the soil?
    e. Retardation coefficient for a volatile solvent dissolved in pore water (for purposes of calculation, assume air spaces are neither continuous nor mobile; only dissolved solvent moves)?
40. Steady-state drawdown is 130 cm at a distance 12 m from a well whose radius of influence is judged to be approximately 400 m. The well is 120 cm in diameter, and is pumped at rate of 2 liters/sec.
    a. What is the transmissivity of the aquifer?
    b. The well is located 150 m directly upgradient from a junkyard where trichloroethane (TCA) was formerly disposed of into the ground. If

the natural gradient is 0.0004, is the well at risk of contamination from a plume of dissolved TCA emanating from a bolus of TCA pooled in the bedrock depression?

41. Imagine that you live in a remote mining colony in a polar region of Mars. It long has been suspected that liquid groundwater exists in this area, and your job is to develop a water supply for the colony. Permeameter tests, conducted on Earth, indicate that the soil has a porosity of 0.4, a bulk density of 2 g/cm^3, and a hydraulic conductivity of 10^{-4} cm/sec. Amazingly, also it contains 1% organic carbon. By using primitive augering equipment, you eventually succeed in drilling a hole 10 cm in diameter fully through a saturated zone 5 m thick. You estimate recharge from melting polar ice during the summer amounts to approximately 8 cm per Earth year.
    a. Estimate the steady-state well yield if drawdown 1 m from the well is not to exceed 0.3 m. (Assume g is the same as on Earth.)
    b. The groundwater has a pH of only 4.5, due to its high C_T (10^{-3} *M*) and low alkalinity. It is necessary to have C_T no greater than 10^{-4} *M* to minimize CO_2 buildup in the colony dwellings. You have available some membrane filters that will remove all ionic species from the water, but will not remove small uncharged molecules such as H_2CO_3. Thus, if you raise the pH of the water and then remove all ionic species, you may achieve the necessary low C_T. How many equivalents per liter of strong base must you add prior to putting the water through the filter? (*Hint:* Ignore carbonate ion.)

REFERENCES

Abramowitz, M. and Stegun, I. A. (1965). *Handbook of Mathematical Functions with Formulas, Graphs, and Mathematical Tables.* Dover, New York.

Bear, J. (1972). *Dynamics of Fluids in Porous Media.* American Elsevier, New York.

Bear, J. (1979). *Hydraulics of Groundwater.* McGraw-Hill, New York.

Beven, K. and Germann, P. F. (1982). "Macropores and Water Flow in Soils." *Water Resour. Res.* **18**(5), 1311–1325.

Birkeland, P. W. (1984). *Soils and Geomorphology.* Oxford University Press, New York.

Bohn, H. L., McNeal, B. L., and O'Connor, G. A. (1985). *Soil Chemistry,* 2nd ed. Wiley, New York.

Bouwer, E. J. and Wright, J. P. (1988). "Transformations of Trace Halogenated Aliphatics in Anoxic Biofilm Columns." *J. Contam. Hydrol.* **2**, 155–169.

Brady, N. C. (1990). *The Nature and Properties of Soils,* 10th ed. Macmillan, New York.

Buol, S. W., Hole, F. D., and McCracken, R. J. (1980). *Soil Genesis and Classification.* 2nd ed. Iowa State Univesity Press, Ames.

Carslaw, H. S. and Jaeger, J. C. (1959). *Conduction of Heat in Solids.* 2nd ed. Clarendon Press, Oxford.

Cooper, H. H., Jr. and Jacob, C. E. (1946). "A Generalized Graphical Method for Evaluating Formation Constants and Summarizing Well Field History." *Trans. Am. Geophys. Union* **27**, 526–534.

Darcy, H. (1856). *Les Fountaines Publiques de la Ville de Dijon.* Dalmont, Paris.

Davis, S. N. and de Weist, R. J. M. (1966). Hydrogeology. Wiley, New York.

Driscoll, F. G. (1986). *Groundwater and Wells.* 2nd ed. Johnson Filtration Systems Inc., St. Paul, MN.

Dzombak, D. A. and Morel, F. M. M. (1990). *Surface Complexation Modeling: Hydrous Ferric Oxide.* Wiley, New York.

Epstein, S. S., Brown, L. O., and Pope, C. (1982). *Hazardous Waste in America.* Sierra Club Books, San Francisco.

Freeze, R. A. and Cherry, J. A. (1979). *Groundwater.* Prentice-Hall, Englewood Cliffs, NJ.

Fried, J. J. and Gombarnous, M. A. (1971). "Dispersion in Porous Media." *In Advances in Hydroscience* (W. Chou and V. T. Chow, Eds.), Vol. 9, pp. 169–282. Academic Press, New York.

Fuller, W. H. (1978). Investigations of Landfill Leachate Pollutant Attenuation by Soils. EPA COO /1-78-158. US EPA, Cincinnati, OH.

Gelhar, L. W. and Axness, C. L. (1983). "Three Dimensional Stochastic Analysis of Macrodispersion in Aquifers." *Water Resour. Res.* **19**(1), 161–180.

Güven, O., Molz, F. J., and Melville, J. G. (1984). "An Analysis of Dispersion in a Stratified Aquifer." *Water Resour. Res.* **20**(10), 1337–1354.

Hantush, M. S. (1964). "Hydraulics of Wells." *In Advances in Hydroscience* (V. T. Chow, Ed.), Vol. 1, pp. 281–432, Academic Press, New York.

Heath, R. C. (1984). "Basic Ground-Water Hydrology," U.S.G.S. Water-Supply Paper 2220.

Hillel, D. (1980). *Fundamentals of Soil Physics.* Academic Press, New York.

Hunt, B. (1978). "Dispersive Sources in Uniform Ground-water Flow," *J. Hydraul. Div., ASCE,* **104**(HY1), Proc. Paper 13467, pp. 75–85.

Javandel, I. and Tsang, C. F. (1986). "Capture-Zone Type Curves—A Tool for Aquifer Cleanup." *Groundwater* **24**(5), 616–625.

Kipp, K. L. Jr. (1985). "Type Curve Analysis of Inertial Effects in the Response of a Well to a Slug Test." *Water Resour. Res.* **21**(9), 1397–1408.

Klotz, D., Seiler, K. P., Moser, H., and Neumaier, F. (1980). "Dispersivity and Velocity Relationship from Laboratory and Field Experiments." *J. Hydrol.* **45**, 169–184.

Kruseman, G. P. and de Ridder, N. A. (1983). *Analysis and Evaluation of Pumping Test Data.* International Institute of Land Reclamation and Improvement, Wageningen, The Netherlands.

Lembke, K. F. (1886 and 1887). "Groundwater Flow and the Theory of Water Collectors" (in Russian). *The Engineer, J. Minis. Commun.* no. 2, no. 17–19.

Linsley, R. K., Jr., Kohler, M. A., and Paulus, J. L. H. (1982). *Hydrology for Engineers.* 3rd ed. McGraw-Hill, New York.

Lyman, W. J., Reehl, W. F., and Rosenblatt, D. H. (1990). *Handbook of Chemical Property Estimation Methods.* 2nd printing. American Chemical Society, Washington, DC.

Lyman, W. J., Reidy, P. J., and Levy, B. (1992). *Mobility and Degradation of Organic Contaminants in Subsurface Environments.* C. K. Smoley, Chelsea, MI.

McWhorter, D. B. and Sunada, D. K. (1977). *Ground-water Hydrology and Hydraulics.* Water Resources Publications, Fort Collins, CO.

Millington, R. J. (1959). "Gas Diffusion in Porous Media." *Science* **130**, 100–102.

Millington, R. J. and Quirk, J. M. (1961). "Permeability of Porous Solids." *Trans. Faraday Soc.* **57**, 1200–1207.

Polubarinova-Kochina, P. Ya. (1952). *Theory of Groundwater Movement* (Trans. by J. M. R. De Wiest). Princeton University Press, Princeton, NJ.

Reed, J. E. (1980). *Type Curves for Selected Problems of Flow to Wells in Confined Aquifers, Techniques of Water-Resources Investigations of the U.S.G.S.*, Book 3, Chap. B3. U.S. Geological Survey, Arlington, VA.

Schwartz, F. W. (1977). "Macroscopic Dispersion in Porous Media: The Controlling Factors." *Water Resour Res.* **13**(4), 743–752.

Schwarzenbach, R. P., Gschwend, P. M., and Imboden, D. M. (1993). *Environmental Organic Chemistry*. Wiley, New York.

Schwille, F. (1988). *Dense Chlorinated Solvents in Porous and Fractured Media*. Lewis, Chelsea, MI.

Strack, O. D. L. (1989). *Groundwater Mechanics*. Prentice-Hall, Englewood Cliffs, NJ.

Theis, C. V. (1935). "The Relation between the Lowering of the Piezometric Surface and the Rate and Duration of Discharge of a Well Using Ground Water Storage," *Transactions of American Geophysical Union*, 16th Annual Meeting, Part 2, pp. 519–524.

United States Geological Survey (1984). *National Water Summary 1984*, Water-Supply Paper 2275, Golden, CO.

Walton, H. F. (1969). "Ion Exchange Equilibria." *In Ion Exchange Theory and Practice* (F. C. Nachod, Ed.), Academic Press, New York.

Wang, H. and Anderson, M. P. (1982). *Introduction to Groundwater Modeling*. Freeman, San Francisco.

Wilson, J. L. (1993). "Induced Infiltration in Aquifers with Ambient Flow." *Water Resour. Res.* 29(10), 3503–3512.

Wilson, J. L. and Conrad, S. H. (1984). "Is Physical Displacement of Residual Hydrocarbons a Realistic Possibility in Aquifer Restoration?" *Proceedings of Petroleum Hydrocarbons and Organic Chemicals in Groundwater Prevention, Detection, and Restoration*. pp. 274–298. Houston, TX.

Wilson, J. L. and Miller, P. J. (1978). "Two-dimensional Plume in Uniform Ground-water Flow." *J. Hydraul. Div., ASCE*, **104**(HY4), Proc. Paper 13665, pp. 503–514.

Wu, S. C. and Gschwend, P. M. (1986). "Sorption Kinetics of Hydrophobic Organic Compounds to Natural Sediments and Soils." *Environ. Sci. Technol.* **20**(7), 717–725.

CHAPTER 4

The Atmosphere

4.1 INTRODUCTION

4.1.1 Nature of the Atmosphere

Many of the physical characteristics of the atmosphere, such as wind, temperature, cloud cover, humidity, and precipitation, are easily perceived. Sometimes, chemicals in the atmosphere also can be observed, as in smoke plumes and smog, and their physical transport tracked downwind just as downstream transport of substances in a river can be measured. Other atmospheric processes are less apparent to the unaided observer, however, occurring either on the microscopic scale of a chemical reaction, or on a global scale, or at high altitudes. Such processes may be detected only by instrumentation on satellites or some high-altitude aircraft.

Physical Attributes of the Atmosphere

The advent of aviation provided both the means and a greater incentive to investigate the atmosphere. Physical phenomena, such as winds, turbulence, atmospheric electricity, and the behavior of water (especially when it forms clouds or ice), were of great practical concern to early aviators (and atmospheric scientists). Although recent interest tends to focus on atmospheric chemicals, the physics of fluid behavior influence chemical transport and fate in the atmosphere no less profoundly than they do in surface waters and porous media; to understand atmospheric chemistry, a knowledge of atmospheric physics is essential. For the price of an airplane ticket, the modern student of the atmosphere can visit firsthand 75% (by mass) of Earth's atmosphere and directly observe processes that earlier generations of scientists had to infer from theory or indirect observation.

The initial part of an airplane flight is in the *troposphere,* the lowest layer of the atmosphere (Fig. 4-1). The word "troposphere" is based on the Greek word "tropos," meaning change. Immediately after takeoff, the turbulent nature of tropospheric airflow manifests itself; the ride is often bumpy as the plane passes through large eddies, sometimes called *air pockets.* These eddies may be due to wind flowing past irregularities on Earth's surface or due to *convective instability* (Section 4.2). Such eddy or turbulent diffusion causes significant Fickian transport and dominates the vertical mixing of atmospheric chemicals. Chemicals also are transported "via" or "by" advection by the wind. [Recall that eddy diffusion and advection are dominant transport processes in flowing surface waters (Section 2.2.1) as well.]

Clouds occur frequently in the troposphere, and from an airborne vantage point it is evident that clouds are not randomly scattered in the vertical direction. Usually, clouds have bases at relatively well-defined altitudes, because rising air containing water vapor (moisture) systematically becomes cooler with height, eventually reaching its *dew point,* the temperature at which water vapor condenses (Section 4.2.2). Water in both liquid and solid phases in the clouds facilitates a host of chemical reactions.

Climbing through the gray murk of a deep cloud layer, the plane emerges above into brilliant sunlight, which appears all the brighter because of reflection of light (*visible albedo*) from the clouds below. Light is a potent factor in chemical reactions in the atmosphere, and it is evident that clouds significantly influence the visible light flux.

Clouds reflect visible light and other portions of the incoming solar radiation, as well as absorb some fraction of all the incoming solar radiation. *Albedo* usually refers to the percentage of reflection of all incident solar radiation wavelengths, not just the percentage of reflection of the incident visible portion. (Albedo also can be expressed as a ratio of reflected radiation to incident radiation.) Reflection and absorption of solar radiation by clouds reduce the amount of solar radiation received at Earth's surface. Some clouds may reflect as much as 90% of incoming solar radiation, while others may absorb as much as 40% of incoming solar radiation (Rosenberg *et al.,* 1983). Clouds also

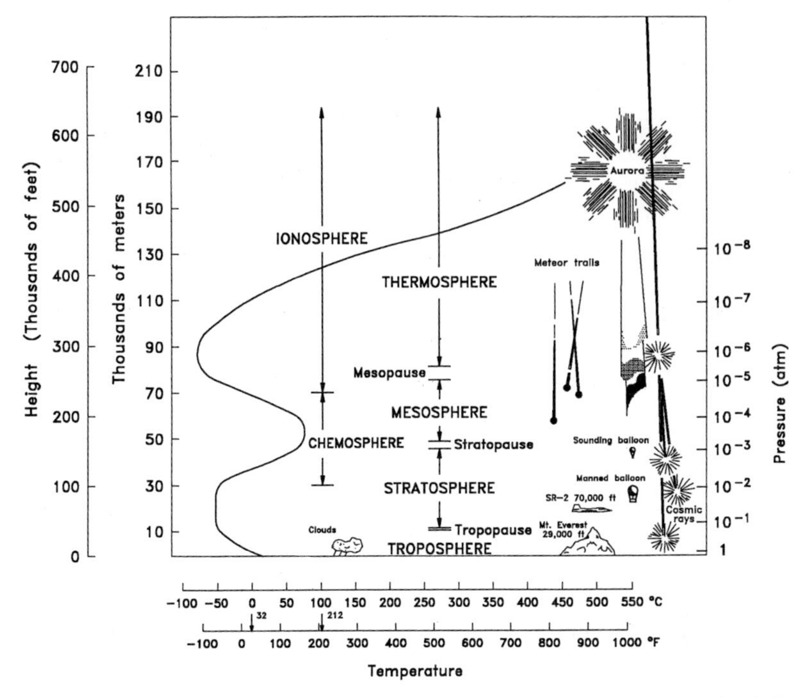

FIGURE 4-1 Vertical structure of the atmosphere. Weather phenomena are confined almost entirely to the troposphere, as are most air pollutants, which are removed by various processes before they can mix into the stratosphere. Certain long-lived pollutants, however, such as the chlorofluorocarbons (CFCs), do mix into the stratosphere, and other pollutants can be injected physically to stratospheric altitudes by processes such as volcanic eruptions or nuclear explosions. Note that more than one term may refer to a given layer of the atmosphere (adapted from *Introduction to Meteorology*, by F. W. Cole. Copyright © 1970, John Wiley & Sons, Inc. Reprinted by permission of John Wiley & Sons, Inc.).

radiate long-wave infrared energy (heat), both out toward space and toward Earth's surface.

As the plane continues to climb, the ride becomes smooth. Coffee served in open cups does not spill, providing testimony that in the *tropopause* [at approximately 10,000 m (33,000 ft)] and the still higher *stratosphere* (Fig. 4-1), the size and energy of atmospheric eddies decrease. Weather phenomena are confined almost entirely to the troposphere. Being at the edge of the stratosphere is comparable to being in the thermocline of a stratified lake (Section 2.2.2); turbulent diffusion is suppressed, and vertical Fickian transport is slowed. Chemicals released into the air near Earth's surface may mix throughout the troposphere in a few weeks, but take years to move into the stratosphere.

As the plane approaches the stratosphere, the pilot's outside air temperature gauge may read as low as − 55°C (− 67°F). The sky is clear blue because most of the water vapor has been condensed and removed as precipitation far below. Rarely, a massive thunderstorm in the distance may be observed; thunderstorms (along with volcanic eruptions) are one of the few natural agents capable of carrying materials by vertical advection (convection) into the stratosphere.

The passenger airplane reaches its maximum altitude capability in the stratosphere; special aircraft, high-altitude balloons, or rockets are needed to go higher. Stratospheric temperature is fairly uniform up to approximately 30,500 m (100,000 ft), at which height the temperature begins to increase with altitude. The temperature increase is due to the absorption of ultraviolet solar radiation, primarily by ozone molecules. This upper portion of the stratosphere is considered part of the *chemosphere* because much of the absorbed energy initiates chemical reactions. For example, a diffuse protective layer of ozone (O_3) is being produced by ultraviolet solar radiation interacting with oxygen, and some of the ozone is being destroyed by reaction with anthropogenic chlorofluorocarbons (CFCs). Important nuclear chemical reactions are occurring in the stratosphere as well. For example, extremely energetic cosmic rays are colliding with nitrogen molecules, producing radionuclides such as carbon-14 (^{14}C), which are used by scientists to date natural substances and cultural artifacts as far back in time as 10 to 20 millennia.

The stratosphere ends at the stratopause, at approximately 49,000 m (160,000 ft); above lies the *mesosphere,* within which temperature drops with increasing altitude until reaching the mesopause, at about 75,000 m (250,000 ft). Above the mesopause occurs the *thermosphere;* temperature again rises with altitude in this rarified layer of the atmosphere. The upper mesosphere and the beginning of the thermosphere also mark the end of the chemosphere and the beginning of the *ionosphere,* at approximately 70,000 m. The ionosphere is so named because absorption of solar radiation by atmospheric gases creates ionized layers, such as the E, F_1, and F_2 layers at approximately 120,000, 220,000, and 380,000 m, respectively; these ionized layers enable global radio communications to be carried out by refracting radio waves back to Earth (Fig. 4-2). Also in the ionosphere, solar particles (charged particles such as protons) are directed to the Earth's magnetic poles by Earth's magnetic field. These solar particles interact with the sparse air molecules above approximately 150,000 m to cause the spectacular displays of the aurora borealis in the Northern Hemisphere and the aurora australis in the Southern Hemisphere. Above the thermosphere, at approximately 500,000

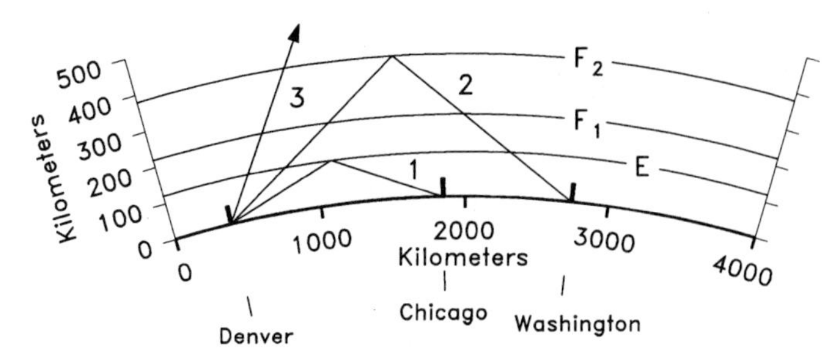

FIGURE 4-2 Refraction of radio waves in the ionosphere. Air molecules ionized by solar radiation are far enough apart to maintain their ionized state for a sufficient time period so that a significant ionization density occurs. The layers of ionized gas are responsible for most long-range radio transmission (except by satellite relay) on Earth. The degree of refraction is dependent on radio wave frequency as well as the angle at which the waves enter the ionosphere. In the figure, waves originating in Denver and following path 1 reach Chicago, whereas waves following path 2 skip Chicago, but are received in Washington. Radio waves following path 3 enter the ionosphere at an angle too close to vertical, and are not returned to Earth's surface (adapted from Howard W. Sams and Company, Inc., *Reference Data for Radio Engineers,* 6th ed., Englewood Cliffs, New Jersey, Prentice Hall, 1977).

to 600,000 m, is a region called the *exosphere,* in which air molecules can escape Earth's gravitational field into space.

The Standard Atmosphere

The boundaries between atmospheric layers are not rigidly fixed; for example, the boundary between the troposphere and the stratosphere varies from an average of about 7,500 m (25,000 ft) near the poles to 17,000 m (55,000 ft) near the equator, and fluctuates seasonally. A reference profile, the Standard Atmosphere, is defined by the International Civil Aviation Organization to represent typical atmospheric conditions at midlatitudes (Table 4-1). At sea level, the Standard Atmosphere exerts a pressure of 760 mm of mercury (1 atm) and has a temperature of 15°C (59°F). (Note that English units are still in widespread use in the meteorology and aviation communities in the United States.) Pressure decreases approximately exponentially with increasing altitude; at 5500 m (18,000 ft), pressure is half that at sea level.

The exponential relationship between pressure and altitude arises from the compressibility of air. [Recall that for water, which is essentially incompressible, hydrostatic pressure changes linearly with depth (Section 2.2.2).] The exponential relationship can be derived by referring to Fig. 4-3, which shows a sketch of a volume of air located directly above sea level. Pressure at the

TABLE 4-1 The Standard Atmosphere and Upper Atmosphere[a]

Altitude		Pressure	Temperature	
Feet	Meters	(atm)	°F	°C
0	0	1.000	59.0	15.0
2,000	610	0.943	51.9	11.0
4,000	1,219	0.888	44.7	7.0
6,000	1,829	0.836	37.6	3.1
8,000	2,438	0.786	30.5	−0.8
10,000	3,048	0.738	23.3	−5.0
15,000	4,572	0.564	5.5	−14.7
20,000	6,096	0.459	−12	−24.4
30,000	9,144	0.297	−48	−44.4
40,000	13,123	0.185	−67	−55.0
60,000	18,288	7.1×10^{-2}	−67	−55.0
80,000	24,384	2.7×10^{-2}	−67	−55.0
100,000	30,480	1.0×10^{-2}	−67	−55.0
140,000	42,672	2.0×10^{-3}	74	23.3
180,000	54,864	5.7×10^{-4}	170	76.7
220,000	67,056	1.7×10^{-4}	92	33.3
300,000	91,440	1.5×10^{-5}	27	−2.8
380,000	115,824	7.7×10^{-7}	188	86.7

[a] Adapted from Marks (1951).

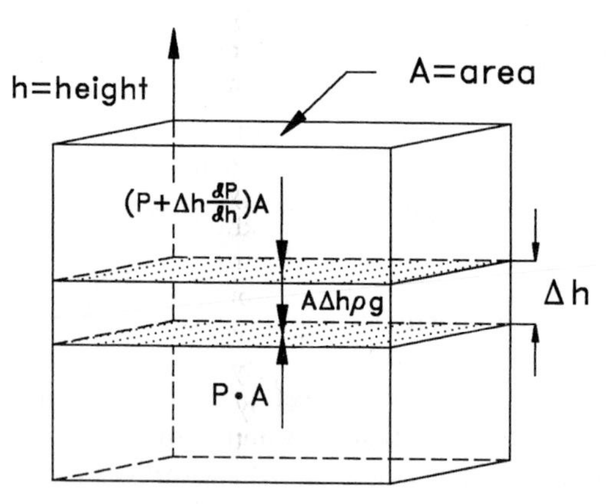

FIGURE 4-3 A schematic of a volume of air located directly above sea level. The schematic shows the balance of forces on a thin slice of air of thickness Δh. P is pressure, ρ is air density, and g is acceleration due to gravity.

bottom of the volume, P_0, is approximately 1 atm, but can vary somewhat in response to meteorological conditions. For simplicity, neglect vertical temperature variations, and assume a constant temperature of 288 K (15°C).

The balance of forces on a slice of air of thickness Δh is made up of pressure and gravity forces,

$$P \cdot A - (P + (dP/dh) \cdot \Delta h) \cdot A - A \cdot \Delta h \cdot \rho \cdot g = 0, \qquad [4\text{-}1]$$

where P is pressure [M/LT²], A is area [L²], h is height [L], ρ is air density [M/L³], and g is acceleration due to gravity [L/T²]. The air density (ρ) is equal to the number of moles of air per unit volume (n/V) multiplied by the average molecular weight (MW) of air (28.96 g/mol). Substituting this relationship into the ideal gas law, Eq. [1-29], gives:

$$\rho = n/V \cdot \text{MW} = P/(R \cdot T) \cdot \text{MW}. \qquad [4\text{-}2]$$

By combining Eqs. [4-1] and [4-2] and by simplifying,

$$\frac{dP}{P} = \frac{-\text{MW} \cdot g}{R \cdot T}\, dh. \qquad [4\text{-}3]$$

Integrating gives:

$$\ln P = -\left(\frac{\text{MW} \cdot g}{R \cdot T}\right) h + \text{constant}. \qquad [4\text{-}4]$$

Given that pressure equals P_0 when h equals 0 (at sea level), the constant must equal $\ln P_0$. Therefore Eq. [4-4] can be written as

$$P = P_0 e^{-(\text{MW} \cdot g/(R \cdot T)) \cdot h}. \qquad [4\text{-}5]$$

The approximate exponential decay rate for pressure with height is

$$\frac{\text{MW} \cdot g}{R \cdot T} = \frac{(28.96\ \text{g/mol}) \cdot (981\ \text{cm/sec}^2)}{(8.31 \times 10^7\ \text{erg/(mol} \cdot \text{K)}) \cdot (288\ \text{K})} = 1.2 \times 10^{-6}/\text{cm}. \qquad [4\text{-}6]$$

In actuality, temperature cannot be considered constant if the preceding calculation is performed for large altitude differences. In the Standard Atmosphere, temperature decreases at the rate of approximately 6.5°C per 1000 m (3.5°F per 1000 ft) up to an altitude of about 11,300 m (37,000 ft), the lower bound of the stratosphere. At that height, the temperature becomes constant at −55°C (−67°F). The constant temperature of the stratosphere greatly inhibits vertical mixing, thus leading to its name (think of stratification). Vertical temperature changes are discussed further in Section 4.2.

EXAMPLE 4-1

A student pilot takes off from an airport at sea level, and flies to an airport that is 1000 ft above sea level. (a) As the pilot enters the traffic pattern 1000 ft above ground level at the destination airport, what would an air pressure gauge aboard the plane read? Assume an air temperature of 15°C and an atmospheric pressure of 1 atm at sea level. (b) The plane altimeter is a gauge that senses pressure and translates the pressure value to an altitude readout. Suppose that during the flight, the atmospheric pressure at sea level in the region increases to 1.02 atm. If the pilot does not correct for this, at what altitude will he or she enter the traffic pattern?

(a) The air pressure 2000 ft above sea level can be estimated by using Eqs. [4-5] and [4-6]. First, convert the altitude of 2000 ft to 610 m by using Table A-2 in the Appendix. Then combine Eqs. [4-5] and [4-6]:

$$P = 1 \text{ atm} \cdot e^{-(1.2\times10^{-6}/\text{cm})\cdot 61{,}000\,\text{cm}} = 0.93 \text{ atm}.$$

(For comparison, from Table 4-1, the pressure at 610 m in a Standard Atmosphere is 0.94 atm. This small difference is partly attributable to the assumption of a constant temperature of 15°C with no vertical temperature variations. The temperature in a Standard Atmosphere at 610 m is only 11°C.)

(b) The altimeter, if uncorrected for the change in air pressure at sea level, will read 2000 ft at the altitude where the pressure equals 0.93 atm, calculated in (a). However, due to the atmospheric pressure rise, the pilot will be flying at too high an elevation. This elevation can be estimated by again using Eqs. [4-5] and [4-6]:

$$0.93 \text{ atm} = 1.02 \text{ atm} \cdot e^{-(1.2\times10^{-6}/\text{cm})\cdot(h)}$$

$$-\ln\left(\frac{0.93 \text{ atm}}{1.02 \text{ atm}}\right) = (1.2 \times 10^{-6}/\text{cm}) \cdot (h)$$

$$h = 7.8 \times 10^{4} \text{ cm or } 2550 \text{ ft (1550 ft above ground level).}$$

Chemical Composition of the Atmosphere

Air is primarily composed of two chemicals: nitrogen, N_2, constituting approximately 78% of the molecules found in air and oxygen, O_2, constituting approximately 21% of the molecules. Such molar percentages are often called *mixing ratios*, but also can be thought of as volume percentages or as partial pressure percentages, because the pressure–volume product of a gas is very nearly a function of the number of molecules present in it (recall the ideal gas law in Section 1.8.1). As discussed in Chapter 2, oxygen is not only metabolically critical for most forms of life but also is an important oxidant in abiotic reactions. By comparison, nitrogen is much less chemically reactive, although small amounts of it, incorporated into organic forms, are essential to organisms. Most of the nitrogen required by organisms is *fixed* (converted from inorganic N_2 into a usable organic form) by specific bacteria and blue-green algae. Some N_2, however, is fixed in the atmosphere by a series of abiotic reactions initiated by electric discharges in lightning. These reactions fix N_2 as nitric acid, HNO_3.

The third main constituent of air is argon, a *noble* (chemically inert) *gas* with a mixing ratio of approximately 0.9%. Other gases with mixing ratios less than 1% are frequently called *trace gases.* These trace gases include, in order of decreasing concentration, four other noble gases at fairly constant mixing ratios: neon, helium, krypton, and xenon. Other trace gases are more variable in their concentrations because of their reactivity (i.e., they are not inert): carbon dioxide (CO_2), methane (CH_4), hydrogen (H_2), nitrous oxide (N_2O), and ozone (O_3). Several other chemicals are present in the atmosphere with highly variable mixing ratios. These include water, carbon monoxide (CO), nitrogen dioxide (NO_2), ammonia (NH_3), sulfur dioxide (SO_2), and hydrogen sulfide (H_2S). The mixing ratios for all these atmospheric gases are provided in Table 4-2.

As shown in Table 4-2, the concentration of CO_2 in the atmosphere is approximately 356 ppm by volume (therefore, the current mixing ratio of CO_2 is approximately 0.0356%). Despite the relatively low abundance of CO_2, CO_2 is the source of carbon from which organic material is fashioned during photosynthesis. CO_2 is also a *radiatively active gas.* Radiatively active gases absorb long-wave infrared radiation emitted by Earth's surface, thereby trapping heat. Currently, the atmospheric mixing ratios are increasing for CO_2 and other radiatively active trace gases, such as CH_4, N_2O, and chlorofluorocarbons (CFCs). Therefore, these radiatively active gases are having an increased warming effect on the planet by enhancing the *greenhouse effect,* which is discussed in Section 4.7.

Table 4-2 also shows that water vapor is present in the atmosphere at mixing ratios ranging from a fraction of a percent to approximately 4%. (Recall that *vapor* refers to a gaseous chemical that also has a liquid phase at ordinary temperatures and pressures.) Atmospheric water is central to the global hydrologic cycle, and its phase transformations among vapor, liquid, and solid affect pollutant mixing and air movement. Water vapor also is a radiatively active gas that, globally, absorbs more long-wave infrared radiation than does CO_2. Water vapor concentration in a given parcel of air often is expressed by *relative humidity* (RH), which is the ratio of the actual partial pressure of water vapor in the atmosphere to the partial pressure of water vapor that would exist if the air were in equilibrium with liquid water at the

TABLE 4-2 Components of Atmospheric Air

Gas	Formula	Mixing ratio (percentage by volume)	Concentration [ppb by volume, ppb(v)]
		Major gases[a]	
Nitrogen	N_2	78.084	
Oxygen	O_2	20.946	
Argon	Ar	0.934	
		Inert trace gases[a]	
Neon	Ne		18,180
Helium	He		5240
Krypton	Kr		1140
Xenon	Xe		87
		Reactive trace gases[b]	
Carbon dioxide	CO_2		358,000
Methane	CH_4		1720
Hydrogen	H_2		500[a]
Nitrous oxide	N_2O		312
Ozone	O_3		10–50[c]
Water	H_2O	0.004–4[c]	
Carbon monoxide	CO		50–150[d]
Nitrogen dioxide	NO_2		0.1–5[c]
Ammonia	NH_3		0.1–10[c]
Sulfur dioxide	SO_2		0.03–30[c]
Hydrogen sulfide	H_2S		<0.006–0.6[c]

[a] Weast (1990).
[b] IPCC (1996), unless otherwise noted.
[c] Mészáros (1981).
[d] Prather *et al.* (1996).

same temperature (Table 4-3). For a parcel of air at a given temperature, as RH increases, the dew point also increases (i.e., as the concentration of water vapor increases, the air parcel does not have to be cooled as much for condensation to occur). In the limiting case of 100% RH, the dew point equals the temperature of the air parcel.

4.1.2 Sources of Pollutant Chemicals to the Atmosphere

Human activities contribute large fluxes of gases and particles to the atmosphere. Many of these fluxes are associated with combustion; such fluxes date back to prehistory, when humans first built fires. Historical records of air pollution from fossil fuel combustion appear in the late 13th century, when

TABLE 4-3 Vapor Pressure of Water[a]

Temperature (°C)	V.P. (mm Hg)	Temperature (°C)	V.P. (mm Hg)
−5	3.16	16	13.63
−4	3.41	17	14.53
−3	3.67	18	15.48
−2	3.96	19	16.48
−1	4.26	20	17.54
0	4.58	21	18.65
1	4.93	22	19.83
2	5.29	23	21.07
3	5.68	24	22.38
4	6.10	25	23.76
5	6.54	26	25.21
6	7.01	27	26.74
7	7.51	28	28.35
8	8.04	29	30.04
9	8.61	30	31.82
10	9.21	31	33.70
11	9.84	32	35.66
12	10.52	33	37.73
13	11.23	34	39.90
14	11.99	35	42.18
15	12.79	36	44.56

[a]Weast (1990).

King Edward I banned coal burning in kilns in London and Southwark due to the detrimental impact of coal combustion on air quality (Wilson and Spengler, 1996). Although King Edward I ordered that kilns be heated by brushwood or charcoal, the combustion of virtually any fuel produces a range of pollutants: CO_2, pollutants derived from impurities and additives (e.g., sulfur dioxide from sulfur impurities), oxides of nitrogen, and products of incomplete combustion (PICs) such as carbon monoxide, soot particles, and hydrocarbons. PIC production is greatly influenced by the conditions of combustion (e.g., the air to fuel ratio). Besides combustion, sources of pollutant gases and particles include evaporation of volatile chemicals; vapor emissions from metallurgical operations; and mechanical activities, such as grinding, sandblasting, and chipping, which produce coarse particles.

Gases and Vapors

Foremost among the many gases emitted by human activities is carbon dioxide. At natural levels, CO_2 is not an air pollutant, but an atmospheric component essential to the functioning of ecosystems. Anthropogenic emissions of CO_2 primarily originate from fossil fuel combustion in industries and vehicles and from the effects of *deforestation*. Deforestation refers to the large-scale removal of trees in a region by cutting or burning, and is followed by the oxidation of soil organic material. Both the oxidation of soil organic material and the burning of organic carbon release large amounts of CO_2 to the atmosphere. Furthermore, the destruction of trees removes their photosynthetic ability to serve as a sink for atmosphere CO_2. CO_2 fluxes to the atmosphere from deforestation and fossil fuel combustion are now a significant fraction of the natural production of CO_2 and may be threatening the stability of Earth's climate.

In many industrial areas, emissions of gaseous oxides of sulfur (SO_x), especially sulfur dioxide (SO_2), also rival natural sulfur gas emissions from volcanoes, wetlands, and oceans. SO_x are produced from the oxidation of sulfur in fuels, especially coals and residual oils, and are responsible in large part for acid rain (Section 4.6.3). In fuels, sulfur typically occurs either in organic compounds (organic S) or as pyrite (FeS_2). SO_x also are formed from the refining of the ores of the many metals that occur in the form of metal sulfides [e.g., copper (Cu), lead (Pb), and nickel (Ni)].

Oxides of nitrogen (NO_x) occur naturally as a result of bacterial activity, wildfires, and lightning, but their atmospheric loading is greatly enhanced by industrial combustion processes. Both organic nitrogen impurities in fuel and nitrogen from the air (N_2) can be oxidized during combustion to form nitric oxide (NO), nitrous oxide (N_2O), and nitrogen dioxide (NO_2). Oxidation of atmospheric N_2 is most pronounced during high-temperature combustion. Because high temperatures are desirable from the standpoint of power plant thermal efficiency, there is a tradeoff between higher efficiency and lower NO_x emissions in many combustion systems. Emissions of NO_x not only contribute to acid deposition (Section 4.6.3), but also contribute to ozone depletion and the greenhouse effect (Section 4.7).

Carbon monoxide (CO), a toxic gas, is produced during combustion, both in wildfires and in fuel-burning devices; CO also can be produced and consumed by bacterial activity. The presence of CO may indirectly increase the atmospheric mixing ratios of other gases by competing for oxidant species (such as the hydroxyl radical, OH·), thereby decreasing the oxidation rates of the other gases. This competition for oxidant species is believed to be one reason for the current increase in atmospheric methane, whose major atmospheric sink is reaction with the hydroxyl radical.

Methane (CH_4), a key greenhouse gas, has many natural sources, including wetlands and termites. Methane is also released by the petroleum industry, landfills, cattle, rice cultivation, and wastewater treatment plants. (Wastewater treatment plants often burn off CH_4 at the top of a tall stack in a process termed *flaring*.) Collectively, these sources are significant to the global budget (see Table 4-17 later). Many other *nonmethane hydrocarbons* (chemicals composed of carbon and hydrogen) are emitted in large amounts by industrial processes and as by-products of combustion. Hydrocarbon fuels and solvents escape by spillage, or are deliberately released into the atmosphere by usage or disposal; for example, the release of solvent from drying paint represents a purposeful use of the atmosphere as the primary sink for an anthropogenic chemical. Natural releases of certain hydrocarbon molecules, such as terpenes $(C_{10}H_{16})_n$ from pine trees and other vegetation, and ethylene from ripening fruits, also occur. Anthropogenic releases of hydrocarbon fuels and solvents, however, usually overwhelm natural sources in industrialized areas.

Halogenated solvents are an example of a class of chemicals that with a few exceptions (such as methyl chloride, CH_3Cl, produced by some marine algae) are not naturally present in the atmosphere. Currently, such solvents are released in large amounts (Table 4-4). Many are degraded in the lower atmosphere (Section 4.6). Some, however, such as the CFCs, pose a unique set of problems because their low reactivity allows them to escape destruction in the troposphere; therefore, they can enter the stratosphere, where they take part in a catalytic cycle that destroys beneficial high-altitude ozone (Section 4.6.4).

Particulate Matter

Particulate matter (PM) refers to discrete particles that exist in either the solid or the liquid phase in the atmosphere. These particles range in size from a few nanometers to 100 μm in diameter (Fig. 4-4). Note that diameter is not strictly defined for irregularly shaped particles. Typically, particle diameter is assigned on the basis of filtration or aerodynamic separation methods.

Aerosols refer to particles generally less than 50 μm in diameter, and often are classified according to their source as either primary or secondary. Primary aerosols are particles directly emitted into the atmosphere, often from fossil fuel or wood combustion processes or from other high temperature processes, such as are found at smelters and steel mills. Secondary aerosols are produced by the conversion of gases, such as SO_2, NO_x, and NH_3, to particles [e.g., the formation of ammonium nitrate (NH_4NO_3) from ammonia (NH_3) and nitrogen oxides (NO_x)]. Many natural processes contribute aerosols to the atmosphere; these include volcanic activity, which produces both primary and secondary aerosols; wildfires; blowing of soil dust and desert sand; release of spores, pollen, and volatile organic compounds (VOCs) by vegetation; injection of particles of sea salt into the atmosphere due to wave action; and emission of gases from ecosystems (e.g., reduced sulfur from marshes).

TABLE 4-4 U.S. National Air Emissions as Reported by Facilities to the Toxics Release Inventory[a]

Chemical	1996 Air emissions (million lb)
Methanol	206
Ammonia	155
Toluene	125
Xylenes	83
Carbon disulfide	73
n-Hexane	72
Chlorine	66
Hydrochloric acid	65
Methyl ethyl ketone	59
Dichloromethane	53
Styrene	42
Trichloroethylene	21
Benzene	8
Manganese	7
Zinc	2
CFC-12	1
CFC-11	0.7
Lead	0.6
Arsenic	0.04
Mercury	0.01
Cadmium	0.005

[a] Facilities that employ 10 or more persons, and annually manufacture or process at least 25,000 lb or otherwise use at least 10,000 lb of any of 606 chemicals listed in the Toxics Release Inventory, must estimate and report their emissions to air (as well as other environmental media). US EPA (1998).

Table 4-5 shows the major sources of atmospheric particulate matter, for different aerosols and aerosol precursors.

Ultrafine aerosols are commonly defined as particles with diameters less than 0.1 μm; *fine particles* have diameters of approximately 0.1 to 2 μm. Fine particles may be formed by coagulation of ultrafine particles or through gas to particle conversion processes. In industrialized areas, fine particles tend to be mainly composed of SO_4^{2-}, NO_3^-, NH_4^+ (all from gas to particle conversion processes), elemental carbon, organic carbon, and trace metals. Particles formed from gases are typically less than 1 μm in diameter and cause reduced

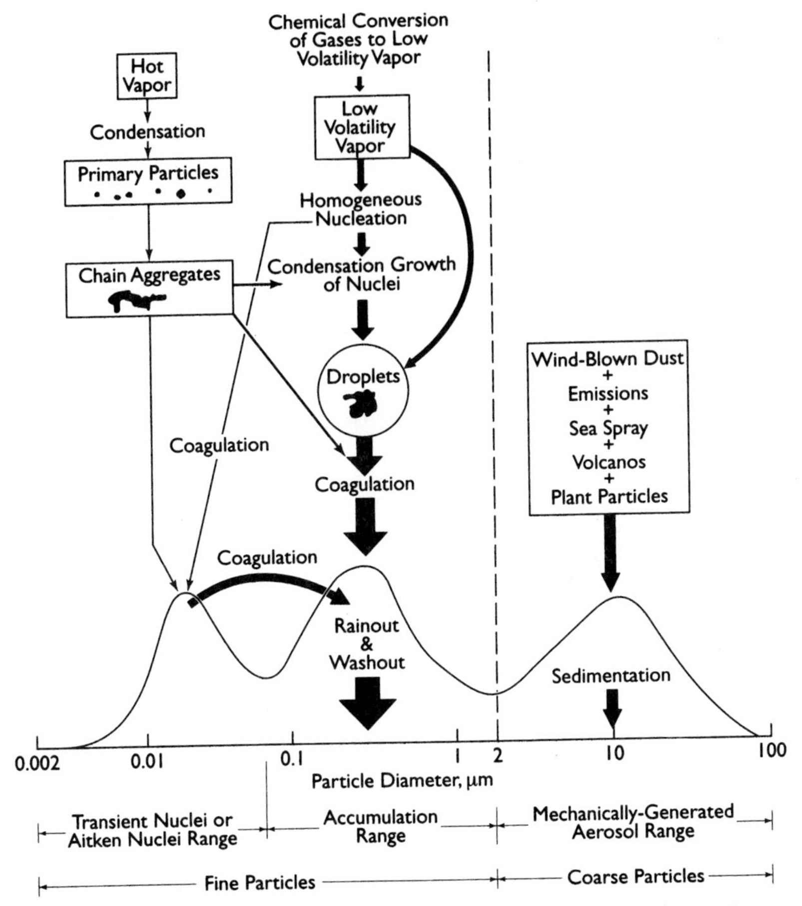

FIGURE 4-4 Origins and size distribution of particles in the atmosphere. Of particular importance are fine particles, which have a peak abundance in the range of 0.1 to 2 μm in diameter. These particles generally have their origin in chemical processes, such as the conversion of gases (e.g., SO_x, NO_x, and NH_3) to particles or the aggregation of ultrafine particles (less than 0.1 μm in diameter), which are formed by the condensation of hot vapors from combustion. A peak also occurs in the vicinity of 10 μm due to the presence of coarse particles, which are typically produced by physical processes (Wilson and Spengler, 1996).

visibility in the atmosphere (Friedlander, 1977). Fine particles, commonly referred to as $PM_{2.5}$, can enter the lungs easily, and are a public health concern. Fine particles have long residence times, on the order of days to weeks, because they settle out of the atmosphere extremely slowly by gravity, but they are too large to coagulate further. Figure 2-11 shows estimated

TABLE 4-5 Sources of Particulate Matter to the Atmosphere for the United States in 1990[a]

Sector	VOCs[b] (1000 ton/year)	NO_x (1000 ton/year)	SO_2 (1000 ton/year)	PM_{10}[b] (1000 ton/year)	$PM_{2.5}$ (1000 ton/year)	Secondary organic aerosols (1000 ton/year)
Electric utility sources						
Boilers fired by coal	27	6,700	15,000	270	99	0.5
Boilers fired by gas/oil/other	7.8	680	610	11	5.9	0.2
Gas turbine and internal combustion	1.9	57	31	4.1	3.7	0
Industrial point sources						
Fuel combustion	140	2,000	2,700	240	170	2.1
Chemical and allied product manufacturing	1,100	280	440	62	43	5.0
Metals processing	72	81	660	140	96	0.2
Petroleum and related industries	240	100	430	29	20	0.6
Solvent utilization	1,100	2.5	0.8	2.1	1.8	16
Storage and transport/waste disposal and recycling/other industrial processes/misc.	830	330	420	460	260	11
Area sources						
Fuel combustion—residential wood	660	46	6.3	480	480	29
Fuel combustion—industrial/other	41	1,800	1,000	64	35	0.5
Industrial processes[c]	980	24	3.1	36	24	2.2
Solvent utilization	4,700	0	0	0	0	45
Storage and transport	1,200	0	0	0	0	14
Waste disposal and recycling	2,200	60	15	220	190	1.3
Agricultural production—crops	79	10	0.2	7,000	1,500	0.1
Agricultural production—livestock	76	0	0	400	60	0
Other fugitive dust[d]	0	0	0	27,000	5,000	0
Prescribed burning/miscellaneous	180	120	4.7	460	380	0.1
Nonroad vehicles[e]	2,100	2,800	240	340	290	23
On-road motor vehicles	6,800	7,400	570	350	290	48
Natural biogenic sources[f]	26,000	0	0	0	0	3,300
Other natural sources[g]	250	89	1.3	5,400	990	0.2
Total[h]	49,000	23,000	22,000	43,000	9,900	3,500

[a] US EPA (1997). All values rounded to two significant figures.
[b] VOCs refers to volatile organic compounds. PM_{10} refers to particulate matter with a diameter of 10 μm or less, and therefore includes $PM_{2.5}$.
[c] Chemical and allied product manufacturing, petroleum and related industries, and other industrial processes.
[d] From paved roads, unpaved roads, and construction.
[e] Includes aircraft, marine vessels, and railroads.
[f] Includes pollen, mold spores, leaf waxes, fragments from plants, and plant emissions of gaseous species such as terpenes. Terpenes, which are photochemically reactive, form secondary organic aerosols through reactions with ozone or hydroxyl radicals; US EPA (1996).
[g] Includes sea salt, natural forest fires, and windblown dust from soil undisturbed by human activity; US EPA (1996).
[h] Emission estimates may not sum to total due to independent rounding.

settling velocities for particles of various diameters in air. Fine particles can travel hundreds to thousands of kilometers from their sources (Wilson and Spengler, 1996).

Coarse particles are commonly defined as particles with diameters greater than 2.5 μm. Coarse particles often contain metal oxides of silicon (Si), aluminum (Al), magnesium (Mg), titanium (Ti), and iron (Fe), as well as calcium carbonate ($CaCO_3$) and salt (NaCl). Anthropogenic sources of coarse particulate matter include coal and oil fly ash from combustion, and fugitive dust emissions from mechanical crushing and grinding activities such as occur in agriculture, metal refining, construction, and cement manufacturing. Coarse particles have atmospheric residence times on the order of only minutes to hours, because they tend to settle out of the atmosphere fairly rapidly by gravity. Therefore, they rarely travel more than a few tens of kilometers from their sources.

4.2 ATMOSPHERIC STABILITY

In any discussion of air movement in the atmosphere, the issue of atmospheric stability must be considered. The role of stability in lakes was discussed in Section 2.2.2; recall that in a lake, the occurrence of higher temperatures near the lake surface greatly inhibits vertical mixing. Similarly, the vertical temperature profile of the atmosphere can either enhance or suppress vertical mixing of air and the chemicals contained in it. In a lake, the essentially incompressible water column is in a condition of neutral stability when water is at a *constant temperature* throughout; in contrast, in the atmosphere, due to the compressibility of air, neutral stability occurs when the *actual* vertical temperature gradient (the *actual lapse rate*) is equal to the *adiabatic lapse rate*. The adiabatic lapse rate is the rate at which the temperature of an air parcel changes in response to the compression or expansion associated with elevation change, under the assumption that the process is *adiabatic*; that is, no heat exchange occurs between the given air parcel and its surroundings. (This is the same phenomenon that is responsible for the warmth of a tire pump after it has been used to pump up a tire; the pump barrel becomes warmest at the air outlet where the highest compression occurs. Friction, which also causes heat, is more evenly distributed along the barrel.)

4.2.1 The Dry Adiabatic Lapse Rate

The adiabatic lapse rate for a *dry atmosphere*, which may contain water vapor but which has no liquid moisture present in the form of fog, droplets, or

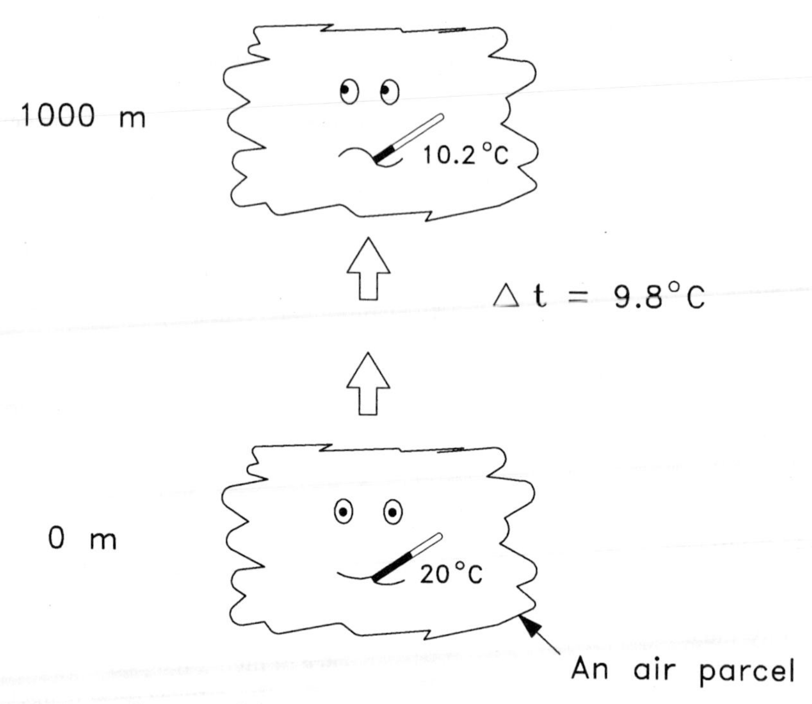

FIGURE 4-5 Illustration of the adiabatic lapse rate. As this air parcel is raised in altitude by 1000 m, the air pressure decreases and the parcel expands and cools by 9.8°C (5.4°F for an altitude increase of 1000 ft). Assuming no heat is lost or gained by the parcel (i.e., the process is *adiabatic*), its temperature will increase to its original value on being lowered to its original height.

clouds, is approximately 9.8°C/1000 m (5.4°F/1000 ft). A hypothetical parcel of air placed in an insulated but expandable container and raised in height by 1000 m would become 9.8°C cooler than its initial temperature (Fig. 4-5). This value, borne out by experiment, is predictable on the basis of the atmospheric pressure profile and thermodynamic principles.

In Section 4.1.1, Eqs. [4-1] and [4-2] were used to estimate the relationship between air pressure and altitude, assuming temperature to be constant with height. When combined with a third equation, Eqs. [4-1] and [4-2] also can be used to calculate the dry adiabatic lapse rate. The third equation, presented as the following Eq. [4-7], is based on an adiabatic process for air that rises and expands due to a decrease in pressure. By definition for an adiabatic process, heat flow into the rising air is assumed to be zero. Therefore, conservation of energy requires that the mechanical (pressure–volume) work performed by the air as it expands equal the decrease in the internal energy of the air. The change in the internal energy of the air is given by its heat capacity, C_v (energy per unit mass per degree), multiplied by the change in temperature,

$$P \cdot \frac{\partial}{\partial h}\left(\frac{1}{\rho}\right) = -C_v \cdot \frac{\partial T}{\partial h}, \qquad [4\text{-}7]$$

where P is pressure [M/LT^2], h is height [L], ρ is density [M/L^3], C_v is heat capacity [L^2/T$^2 \cdot$ K], and T is absolute temperature [K]. (Note that $1/\rho$ is volume per unit mass.)

Equation [4-2], which is derived from the ideal gas law, can be rearranged as

$$T = \frac{\text{MW}}{R} \cdot \frac{P}{\rho}, \qquad [4\text{-}8]$$

where MW is the molecular weight and R is the gas constant. (Note that V/n equals MW/ρ.) By differentiating Eq. [4-8]:

$$\frac{\partial T}{\partial h} = \frac{\text{MW}}{R} \cdot \left(P \cdot \frac{\partial}{\partial h}\left(\frac{1}{\rho}\right) + \frac{1}{\rho} \cdot \frac{\partial P}{\partial h}\right). \qquad [4\text{-}9]$$

From Eqs. [4-2] and [4-3]:

$$\frac{\partial P}{\partial h} = -\rho \cdot g. \qquad [4\text{-}10]$$

Substituting Eqs. [4-7] and [4-10] into Eq. [4-9] gives:

$$\frac{\partial T}{\partial h} = \frac{\text{MW}}{R} \cdot \left(-C_v \cdot \frac{\partial T}{\partial h} + \frac{1}{\rho} \cdot (-\rho \cdot g)\right). \qquad [4\text{-}11]$$

Eq. [4-11] can be rearranged:

$$\frac{\partial T}{\partial h} = \frac{-g \cdot \text{MW}}{R + C_v \cdot \text{MW}}. \qquad [4\text{-}12]$$

The quantity ($C_v \cdot$ MW) is the heat capacity expressed in units of energy *per mole* per degree, and equals 5/2 $\cdot$ R for diatomic gases. Thus,

$$\frac{\partial T}{\partial h} = \frac{-g \cdot \text{MW}}{R + 5/2 \cdot R} = \frac{(981\ \text{cm/sec}^2) \cdot (28.96\ \text{g/mol})}{(7/2) \cdot (8.31 \times 10^7\ \text{erg/mol} \cdot \text{K})} = 9.8\ \text{K/km}.$$

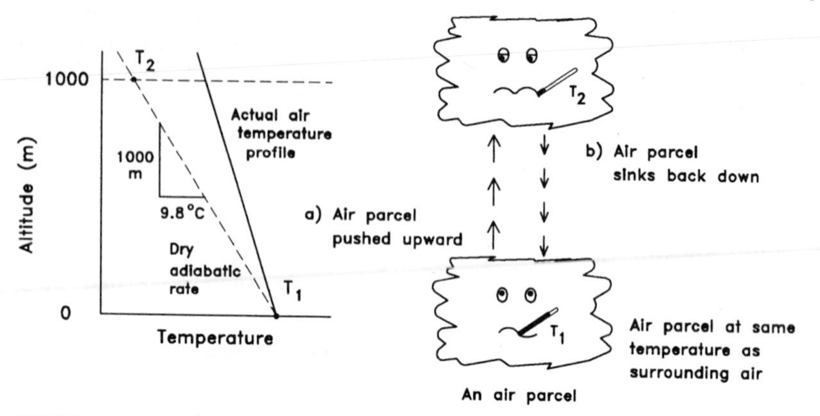

FIGURE 4-6a A *stable* atmosphere. The actual, measured temperature profile of the atmosphere decreases at a rate *less* than the adiabatic rate. Thus, when this air parcel is pushed upward, perhaps by an eddy in a turbulent atmosphere, it finds itself in warmer, and hence less dense, surroundings; being cooler and denser it sinks back down. This stable situation tends to suppress vertical mixing.

It is important not to confuse this dry adiabatic lapse rate with the rate of change in temperature with height in a Standard Atmosphere. The latter represents average conditions in Earth's atmosphere, where heating, mixing, and wet adiabatic processes also are occurring.

Actual temperature profiles are measured routinely by *sondes,* which are devices that measure temperature, pressure, and often water vapor content of the air. Sondes are either carried by weather balloons or dropped from airplanes and transmit their data to ground stations as they travel vertically through the atmosphere. If the actual measured lapse rate of the atmosphere is less than 9.8°C/1000 m, a parcel of air initially at a temperature equal to that of the surrounding air becomes warmer than the surrounding air on being pushed downward (Fig. 4-6a). By being warmer and therefore less dense than its surroundings, it experiences an upward force tending to restore it to its original height. (Forces that arise from density differences in a fluid are called *buoyancy* forces.) Likewise, if the parcel of air is pushed upward, its temperature increases with height more slowly than that of the surrounding air, so that it experiences a downward buoyant force, again tending to restore it to its original height. The net result is that vertical air movements, such as those associated with eddies, are suppressed by buoyant forces; the atmosphere is *stable,* and vertical transport of chemicals is suppressed.

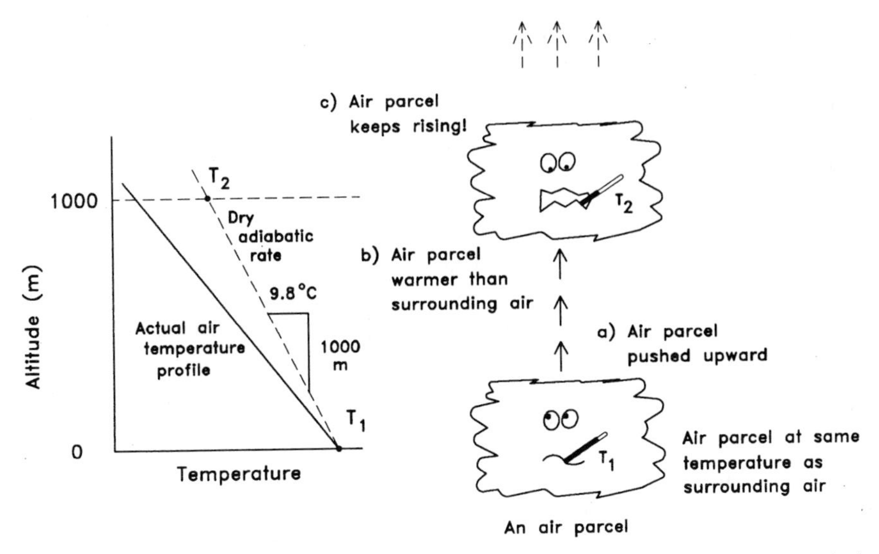

FIGURE 4-6b An *unstable* atmosphere. The same air parcel pictured in Fig. 4-6a is again pushed upward. This time, the temperature of the surrounding atmosphere decreases more rapidly with altitude than the adiabatic rate that the air parcel follows. As it rises, the air parcel finds itself in air that is increasingly cooler and denser than itself and consequently it is pushed up further by buoyancy. Vertical mixing is enhanced by unstable conditions such as these; the mixing, however, tends to reduce the magnitude of the instability, unless an external influence, such as heating by a warm surface, maintains the unstable temperature gradient.

Conversely, if the actual measured lapse rate is greater than 9.8°C/1000 m, a parcel of air displaced upward from its initial height becomes warmer than its surroundings and therefore tends to rise (Fig. 4-6b). If pushed downward, the parcel becomes colder than its surroundings and therefore tends to keep sinking. In this case, buoyant forces amplify any initial upward or downward movement of the air parcel, thus creating an *unstable* atmosphere.

It might be inferred from the preceding discussion that, based on its average lapse rate, the Standard Atmosphere is stable with little vertical mixing; yet, on average, the troposphere is reasonably well mixed. Tropospheric mixing occurs in part because of atmospheric variability; although widespread areas of stable air exist, there are also areas where, even in a "dry" atmosphere, the actual lapse rate exceeds the adiabatic lapse rate. As described in the following section, moisture also contributes to tropospheric mixing. In clouds, the adiabatic lapse rate can be as small as 3.6°C/1000 m (2°F/1000 ft). When this occurs, an actual lapse rate of, for example, 7.2°C/1000 m (4°F/1000 ft), which would correspond to a stable condition in a dry atmosphere, represents an unstable condition.

4.2.2 The Wet Adiabatic Lapse Rate

Energy in the form of heat is required to convert liquid water into water vapor; 589 calories are absorbed per gram of water evaporated at 15°C. The evaporation of water thus removes large amounts of heat, a process that is important to transpiring plant leaves, sweating people, and steam boilers. In the atmosphere, this heat energy is released upon condensation of water vapor into liquid water during the formation of clouds. When a parcel of air whose RH is 100% (i.e., it is at the dew point) is moved upward, adiabatic cooling causes its temperature to drop, at which time its RH exceeds 100% and water condenses. The heat released by condensation mitigates the adiabatic cooling effect, resulting in the parcel of air being warmer than it would be if it experienced only the adiabatic cooling. The rate at which adiabatic cooling occurs with increasing altitude for "wet" air (air containing clouds or other visible forms of moisture) is called the *wet adiabatic lapse rate*. The wet adiabatic lapse rate lies in the range of 3.6 to 5.5°C/1000 m (2 to 3°F/1000 ft), depending on temperature and pressure. Note that any actual lapse rate less than 9.8°C/1000 m in dry air results in a stable atmosphere, whereas a "wet" atmosphere must have an actual lapse rate of less than 3.6 to 5.5°C/1000 m to be stable.

Wet adiabatic lapse rates can be determined from Fig. 4-7, which is a *skew T-log P diagram* (or *adiabatic chart*). On this chart, *dry adiabats* are lines having a nearly constant slope of 9.8°C/1000 m (5.4°F/1000 ft). The *wet adiabats* are curved and have slopes that not only vary with the temperature at which the adiabat originates but also change along the length of the adiabats. Note that the wet adiabats tend to approach the slope of the dry adiabats at low temperatures, where the absolute amount of moisture in saturated air is small (see Table 4-3).

Frequently, both kinds of adiabats must be employed to follow the behavior of a parcel of air. A parcel of dry air may rise by its own buoyancy, or may be pushed up by *orographic* lifting, which occurs when winds meet mountains and are deflected upward. The temperature of the air parcel follows the dry adiabat until water vapor condensation is incipient; with further cooling, condensation can occur. If it is assumed that *supercooling* (a nonequilibrium situation in which air cools below the dew point without condensation) does not occur, the parcel of air then moves upward following the corresponding wet adiabat. *Conditional stability* refers to conditions under which dry,

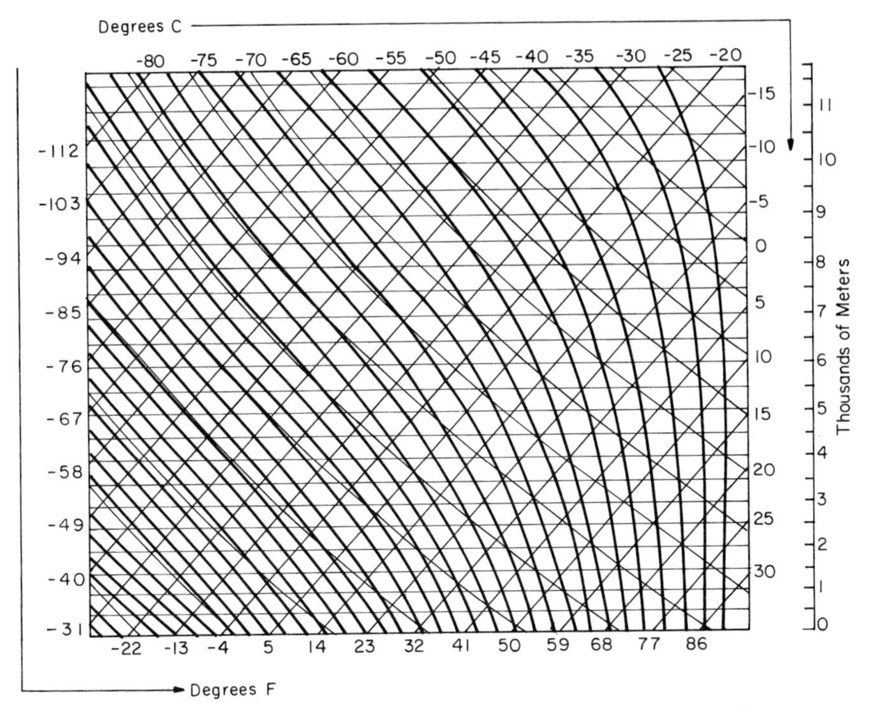

FIGURE 4-7 The skew T-log P diagram, or adiabatic chart. On this chart all lines of temperature versus altitude for dry conditions (i.e., no condensation of water vapor) are nearly straight lines with a slope corresponding to the dry adiabatic lapse rate of 9.8°C/1000 m (5.4°F/1000 ft). These are called *dry adiabats* and slope upward to the left. *Wet adiabats* have a variable slope and appear as curved lines. Horizontal lines denote altitude; lines of constant temperature slope upward to the right.

stable air is pushed upward to its dew point and consequently becomes unstable.

The formation of cumulus clouds (Fig. 4-8) is a situation in which rising air parcels initially follow dry adiabats, but subsequently follow wet adiabats after the dew point is reached. Solar heating of the ground warms the adjacent air, and "bubbles" of this warmed air rise due to buoyancy, their temperature decreasing with altitude according to the dry adiabatic lapse rate. At the altitude where condensation begins, a cloud forms, and the stability of the air decreases as the air begins to follow a wet adiabat. Instability can lead to extensive vertical development of the cloud, strong upward air currents within the cloud, and possibly production of a thunderstorm.

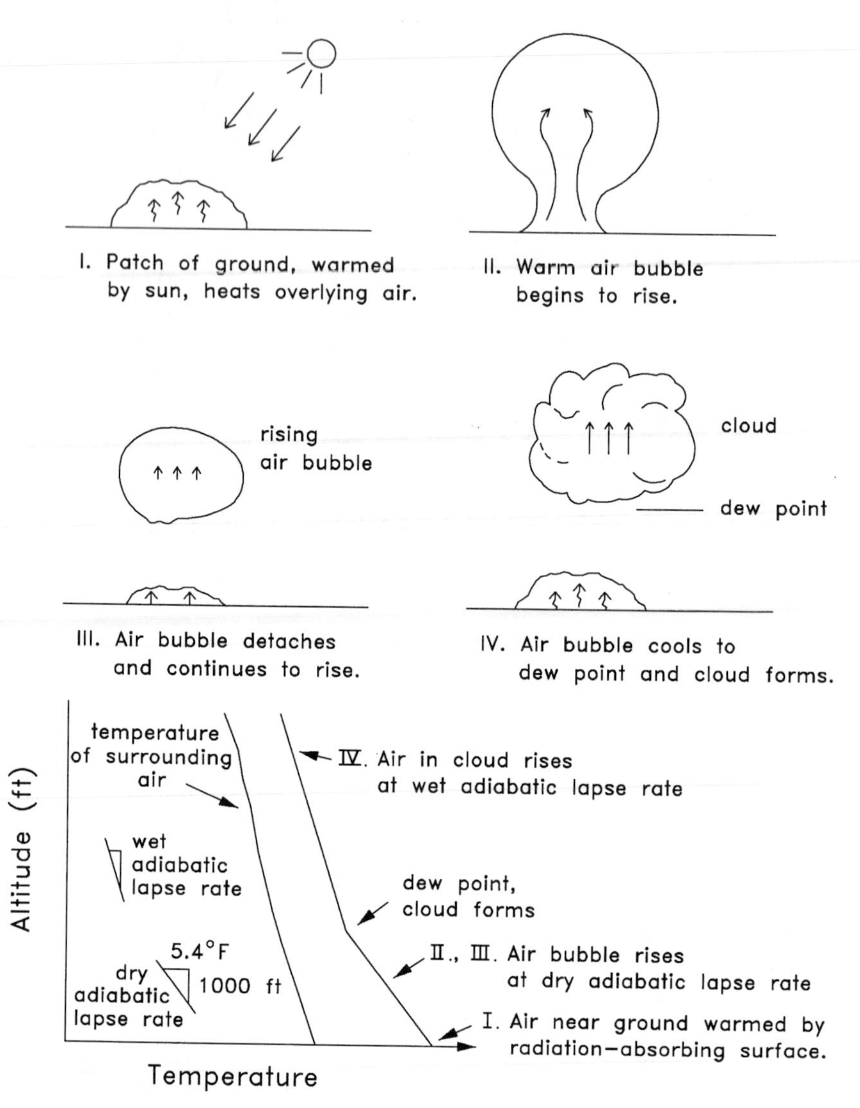

FIGURE 4-8 Cumulus cloud formation. Steps I to III illustrate the warming of near-surface air as the ground is heated by the Sun, and the subsequent rising of the heated bubble of air (the "thermal" in the parlance of glider pilots and soaring birds). In step IV, the rising bubble of air reaches its dew point, and a cloud begins to form. Thereafter, the rising bubble of air cools at a *wet adiabatic lapse rate,* which is less than the dry adiabatic lapse rate, and accentuates the tendency of the air bubble, now a cloud, to rise farther.

EXAMPLE 4-2

On a summer afternoon, air heated over a plowed field 1000 m above sea level has a temperature of 25°C and RH of 80%. At what altitude does the rising air begin to form a cloud? What is the air temperature 2000 m above the cloud base (inside the cloud)? What is the average wet adiabatic lapse rate in this portion of the cloud?

As shown in Table 4-3, at 25°C the vapor pressure of water is 23.8 mm Hg. Therefore, in air at 25°C with 80% RH, the partial pressure of water vapor is

$$80\% \times 23.8 \text{ mm Hg} = 19 \text{ mm Hg}.$$

Condensation will begin when the air is cooled to the temperature at which the vapor pressure of water is 19 mm Hg (from Table 4-3, approximately 21°C).

Use Fig. 4-7, beginning at 25°C and 1000 m, and follow the slope of the dry adiabats until the air parcel reaches 21°C, which occurs at an altitude of approximately 1500 m. (Strictly, it must cool a bit more, because the expansion of the rising air causes the partial pressure of water to decrease somewhat.) Thereafter, the air follows a wet adiabat; at 3500 m (2000 m above the cloud base) the temperature is approximately 12°C. The average wet adiabatic lapse rate in the bottom 2000 m of cloud is therefore 9°C/2000 m, or 4.5°C/1000 m.

4.2.3 Mixing Height

Many common meteorological processes influence the actual lapse rate and thus the stability of the atmosphere. Even the daily cycle of daytime heating by the sun and nighttime cooling produces periods of instability and stability in the atmosphere. As previously mentioned, during the day solar radiation warms the ground, which in turn warms air near the ground surface. Such air tends to rise by buoyant forces until it reaches a height where its temperature, and hence its density, are equal to those of the surrounding air. This height is called the *mixing height;* between this height and the ground lies a layer of air (often called the *boundary layer*) within which atmospheric mixing is aided by buoyancy. Mixing height can be inferred from simple graphic construction of the applicable adiabatic lapse rate on the atmospheric temperature profile (Fig. 4-9).

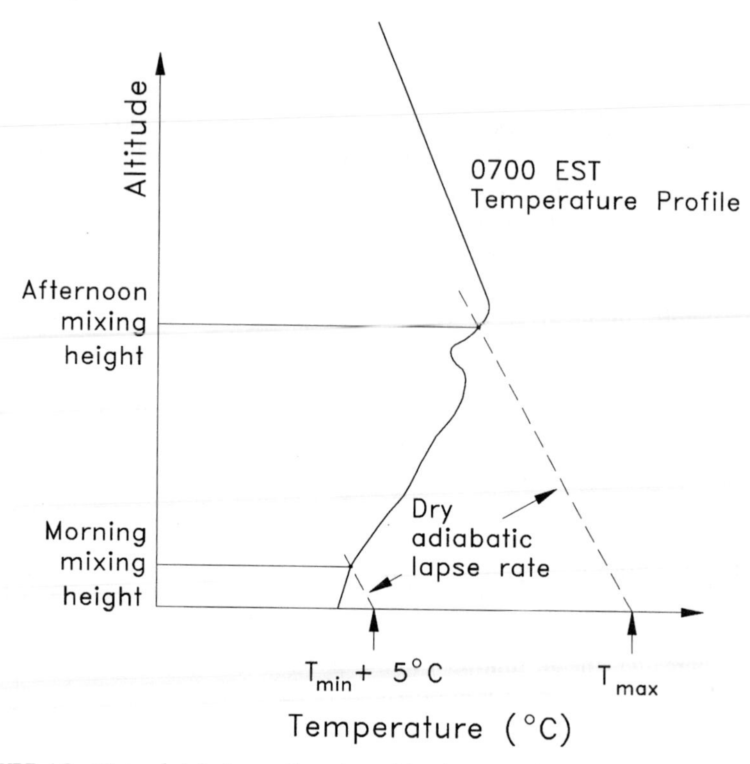

FIGURE 4-9 Mixing height is usually estimated by drawing a line originating at the surface temperature and having a slope equal to the applicable adiabatic lapse rate; the intersection of the line with the actual temperature profile approximates the mixing height. In this figure, which shows a set of approximations commonly used in urban air quality work, only one measured temperature profile from a rural area is available; it is assumed that the upper portions of this 1200 GMT (700 EST) profile are applicable for both the A.M. and P.M. mixing height estimates. An estimated correction for urban surface temperature is made by offsetting the rural surface temperature by 5°C in the graphic construction, giving a somewhat higher A.M. mixing height than if the directly measured surface temperature were used. The afternoon maximum surface temperature (T_{max}) in the urban area is used to estimate the maximum daytime mixing height.

At night, cooling of the land surface lowers the temperature of the air close to the ground and decreases the actual lapse rate, in many cases creating an *inversion,* which is a layer of atmosphere in which the actual lapse rate is negative; air becomes warmer with increased altitude. Nighttime inversion development is favored by clear nighttime skies that radiate little long-wave radiation downward to offset the loss of heat through long-wave radiation upward from the ground surface. This represents an exceptionally stable situation. Nighttime inversions trap pollutant chemicals in a relatively thin layer of atmosphere and lead to lowered air quality near the ground surface. As examples of the role of atmospheric stability in influencing pollutant dispersal, diagrams showing stability effects on the plume of emissions from a smokestack are presented in Fig. 4-10.

EXAMPLE 4-3

What is the mixing height associated with the temperature profile of the following figure at 2:00 P.M.? At 8:00 A.M.? Assume the air is "dry," as defined in Section 4.2.1.

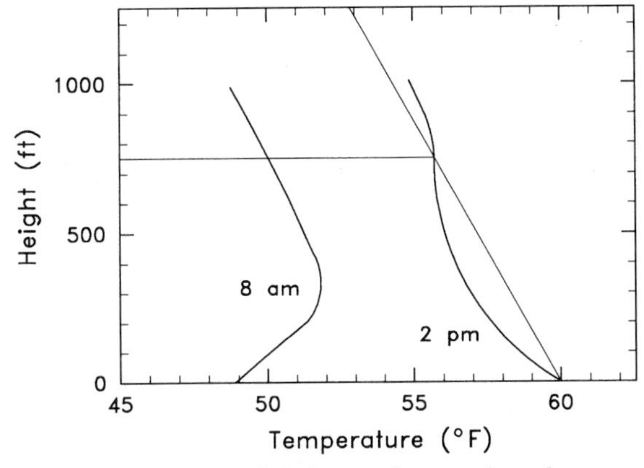

To determine the mixing height, draw a line with a slope equal to the dry adiabatic lapse rate beginning at the temperature and elevation of the ground surface. The height where the line intersects the 2 P.M. temperature profile is the mixing height, at approximately 750 ft. Pollutants released at the surface in the afternoon will not mix readily beyond this height. The mixing height at 8:00 A.M. is zero; inversion conditions prevail.

4.3 CIRCULATION OF THE ATMOSPHERE

The movement of air in the atmosphere is directly responsible for both the transport of chemicals and the production of weather. Circulation in the atmosphere occurs at many different scales, from the microscale (tens of meters) to the global scale. The smallest scale processes can often be observed

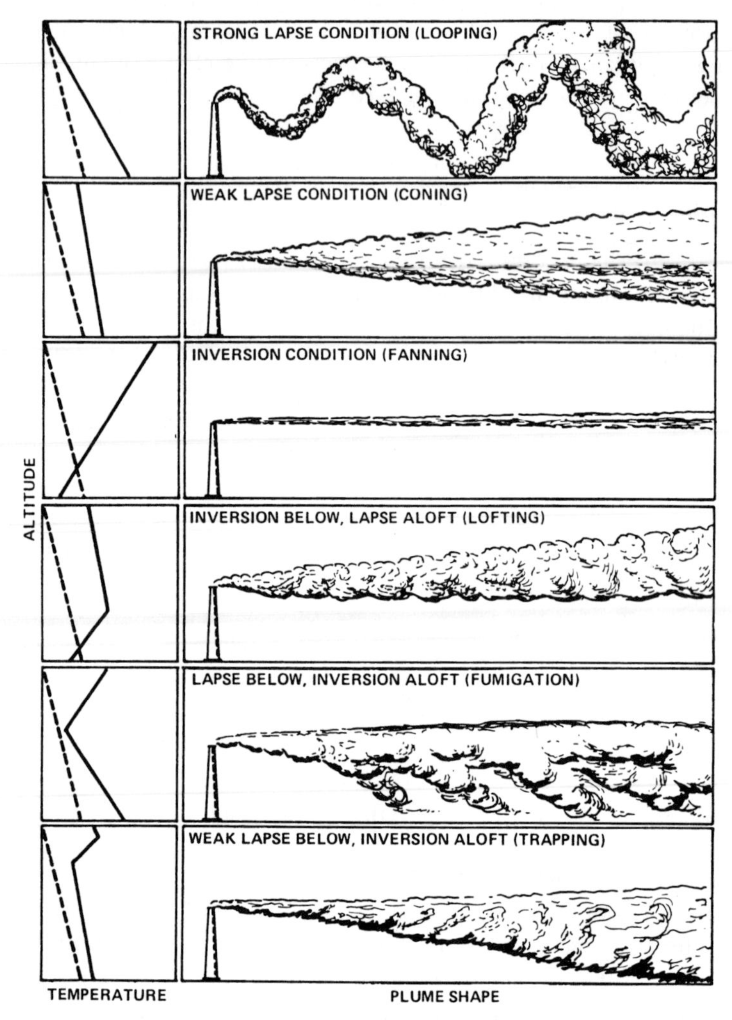

FIGURE 4-10 Emission of pollutants from a smokestack, a typical continuous source, under a variety of meteorological conditions. The dry adiabatic lapse rate is represented as a dashed line and the actual measured lapse rate as a solid line in the left panels. Vertical mixing is strongest when the adiabatic lapse rate is less than the actual measured lapse rate and the atmosphere is unstable (top). Weak lapse is a term used to express the existence of a stable atmosphere, which results in less vigorous vertical mixing. An inversion, in the third panel from the top and in part of the last three panels, results in a very stable atmospheric layer in which relatively little vertical mixing occurs (Boubel *et al.*, 1994).

directly. Common examples include the formation of "dust devils" or "leaf devils," and the advection of campfire and barbecue smoke. (It is not known why air currents are almost always in the direction that blows barbecue smoke toward people.) The advent of satellite and high-altitude aircraft imagery, which has enabled humans to see entire weather systems at once, has helped create an appreciation for atmospheric processes occurring at the larger scales. For example, Fig. 1-4 not only illustrates both advective and turbulent transport, but also shows clearly that variation of wind direction may occur with altitude (note the curving paths of smoke plumes as they rise). Satellite imagery shown on television weather broadcasts reveals the organized structures of major storms on the scale of thousands of kilometers.

4.3.1 Global Air Circulation

On the very longest scale, air circulation over the planet carries heat from the warm equatorial regions to the colder polar regions. Outside the atmosphere, the *solar constant,* or amount of power delivered by the Sun to a unit area oriented perpendicular to the Sun's rays, is approximately 1400 W/m^2 (based on Earth's mean distance from the Sun). Approximately half of this power is delivered directly to Earth's surface (see Fig. 2-30). This is an enormous rate of energy flow; averaged over the globe, it dwarfs the power output of all engines, generators, power plants, and furnaces combined. Solar heating of both the atmosphere and the oceans is greatest at low latitudes near the equator. The warmed fluids (both air and water) tend to rise due to buoyancy, and cooler fluid coming from the direction of the poles moves toward the equator to replace the warm, rising air or water. In exchange, the warmed fluid moves poleward. Planetary-scale convection cells thus are created in each hemisphere, in both the atmosphere and the oceans, and serve to transport heat toward the poles. If Earth did not rotate, one might expect relatively simple convective flow of fluids to occur, as shown in Fig. 4-11.

Because Earth does rotate, however, another set of forces also acts on the fluids; these *geostrophic* forces are the consequence of the fact that Earth's surface forms an accelerating reference frame. Relative to Earth's surface, all fluid motions are deflected perpendicular to their velocity by the *Coriolis force;* in the Northern Hemisphere, fluid motions are deflected to the right, and in the Southern Hemisphere, to the left. A parcel of fluid in motion in the Northern Hemisphere, under the influence of only the Coriolis force, experiences acceleration equal to

$$A = 2\,\omega(\sin\phi)v, \qquad [4\text{-}13]$$

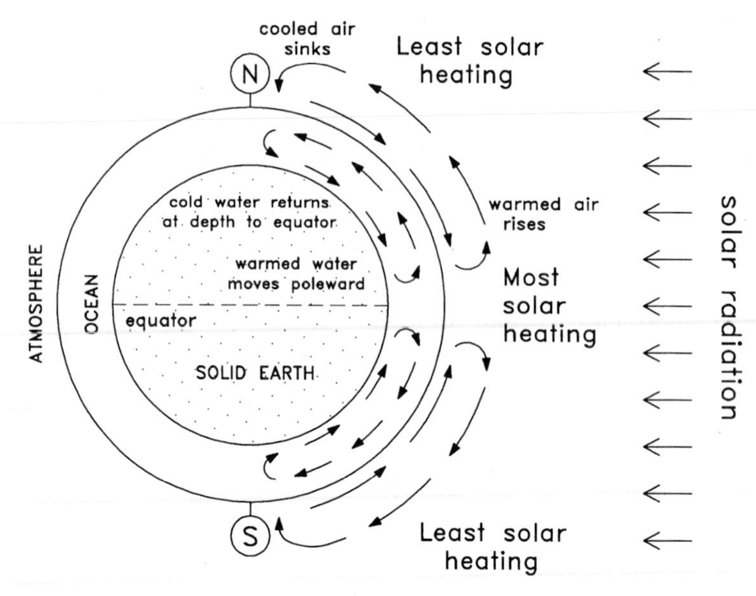

FIGURE 4-11 Global-scale tropospheric circulation as it would be if Earth did not rotate. Heat is transported from the equatorial area to the cold polar regions by both atmospheric and oceanic currents in each hemisphere.

where A is the magnitude of acceleration [L/T^2], ω is the angular speed of Earth's rotation (2π radians/24 hr) [T^{-1}], ϕ is the latitude of the parcel of fluid, and v is the velocity of the fluid relative to Earth's surface [L/T].

The Coriolis force becomes particularly important when fluid motion occurs at large scales, as in vast lakes such as Lake Michigan (Fig. 2-6 b), or in atmospheric circulation (Fig. 4-12). Figure 4-12 explains the origin of the Coriolis force in terms of the radial and tangential components of the velocity of a fluid parcel in a rotating mass of fluid.

The Coriolis force, in conjunction with solar heating, creates a more complex global circulation pattern than that shown in Fig. 4-11. Three major latitudinal bands of surface winds result from these combined forces. In the Northern Hemisphere, the *trade winds* lie between the equator and approximately 30° latitude and are generally from the northeast (Fig. 4-13). North of the trade wind latitudes, between approximately 30° and 60° latitude, poleward-moving surface winds are deflected to the right by the Coriolis force, giving rise to the *westerlies*. Finally, from approximately 60° poleward, a third global-scale convective flow moves southward near Earth's surface and returns

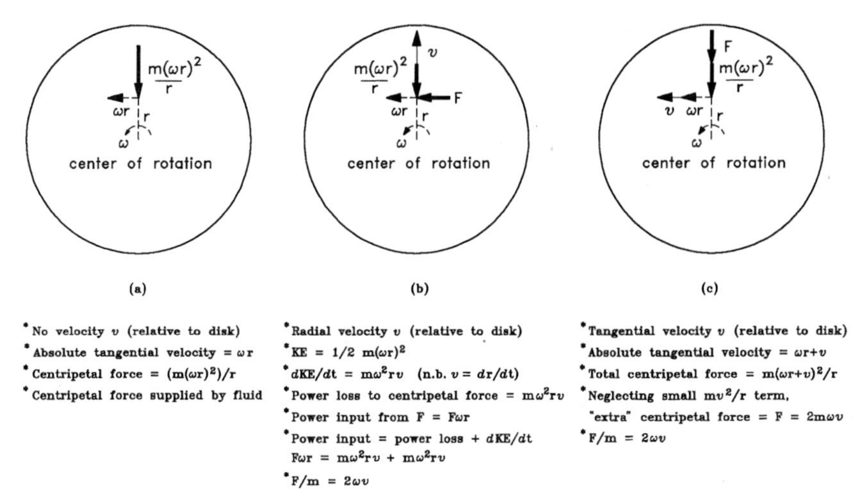

FIGURE 4-12 Origin of the Coriolis force. Consider a small parcel of water of mass m within a body of fluid rotating on a disk. In (a), the parcel is not moving with respect to the disk, although its tangential velocity ωr is evident to a fixed observer, and a centripetal force of magnitude $m(\omega r)^2/r$ is being exerted by the surrounding fluid to keep the parcel on a circular path. The Coriolis force is the *additional* force that would have to be applied to the parcel to give it a uniform velocity v with respect to the rotating disk. Radial velocity and tangential velocity are considered separately in (b) and (c); any velocity can be expressed as a sum of these two components. Light lines represent velocity; heavy lines represent forces.

In the case of *radial* velocity (b), the magnitude of the required additional force can be calculated from the rate at which energy must be added to (or removed from) the parcel. (Recall that energy per time is power.) A fixed observer can see that the velocity of the parcel, and hence its kinetic energy, must increase as it moves radially outward and its tangential velocity increases.

In the case of *tangential* velocity relative to the disk (c), extra centripetal force (or less force, if the parcel is moving opposite to the direction of rotation of the disk) is required to keep the parcel traveling at a constant radius.

In each case, if the additional force is not applied to keep the parcel traveling in a straight path relative to the disk, the parcel will experience an acceleration of magnitude F/m with respect to the disk; this is the Coriolis acceleration. For this direction of disk rotation, the Coriolis acceleration is always to the *right* of the direction the parcel travels relative to the disk, and is always of magnitude $2\omega v$.

to the poles at higher altitude; the Coriolis force deflects these surface winds so that they come from the east (the *polar easterlies*), as seen by an earth-bound observer.

Prior to the 20th century, when trade across the Atlantic Ocean depended on sailing ships, ships leaving Europe often would sail south to pick up the

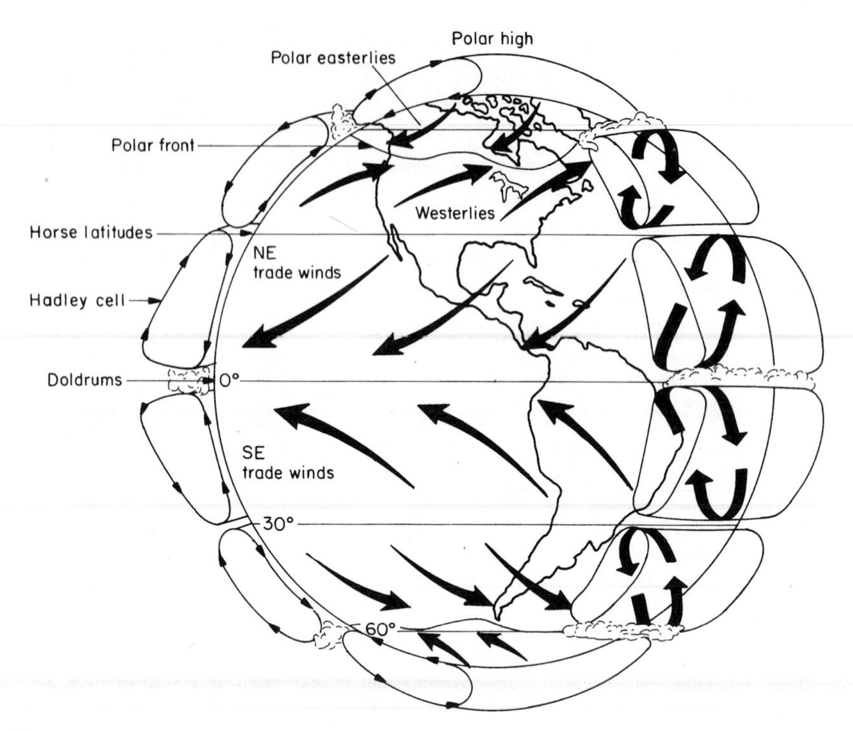

FIGURE 4-13 A simplified diagram of Earth's global-scale tropospheric circulation, including the Coriolis force. Note the three major bands of winds, as well as the cloud-producing effects of rising air near the equator and the polar fronts, and the desertifying effects of descending air at the poles and in the vicinity of 30° north and south latitude (adapted from Frederick K. Lutgens/ Edward J. Tarbuck, *The Atmosphere: An Introduction to Meteorology*, 5th ed., © 1992, p. 170. Reprinted by permission of Prentice Hall, Englewood Cliffs, New Jersey.)

trade winds for the western crossing, and then sail north to pick up the westerlies for return to Europe (see Fig. 4-13). Following prevailing winds along lines of constant latitude was well suited to the large sailing ships of the time, with their square rigs optimized for downwind sailing. (It also was suited to early navigational tools that could accurately measure latitude, but could not determine longitude, due to the unavailability of sufficiently accurate timepieces (Sobel, 1995). Note that mariners used the conversion of 1 min latitude to 1 nautical mi. A *knot* is 1 nautical mph, or approximately 1.15 mph.)

These three cells of convective circulation occur within the troposphere, contributing to relatively rapid and complete mixing of this layer of the atmosphere. The tropical cells, located north and south of the equator, often are called *Hadley cells.* The circulation patterns immediately north and south of the Hadley cells are often called *Ferrel cells.* Mixing occurs within the northern and southern hemispheres faster than between hemispheres, because there is no systematic large-scale, north–south advection of air in the vicinity of the equator, as shown in Fig. 4-13.

Many features of atmospheric chemical transport can be inferred by inspection of the global-scale atmospheric circulation pattern. The west-to-east movement of industrial chemicals, such as acid rain precursors, is a familiar example in the midlatitudes of the Northern Hemisphere. Of course, these belts of wind, such as the midlatitude westerlies, are subject to modification on a smaller scale by a variety of local and regional conditions. For example, a *wind rose* from Chicago (Fig. 4-14) shows that despite the dominant southwesterly direction of the wind, wind from other directions also occurs. To explain such departures from the average global circulation pattern, atmospheric processes on smaller scales must be considered.

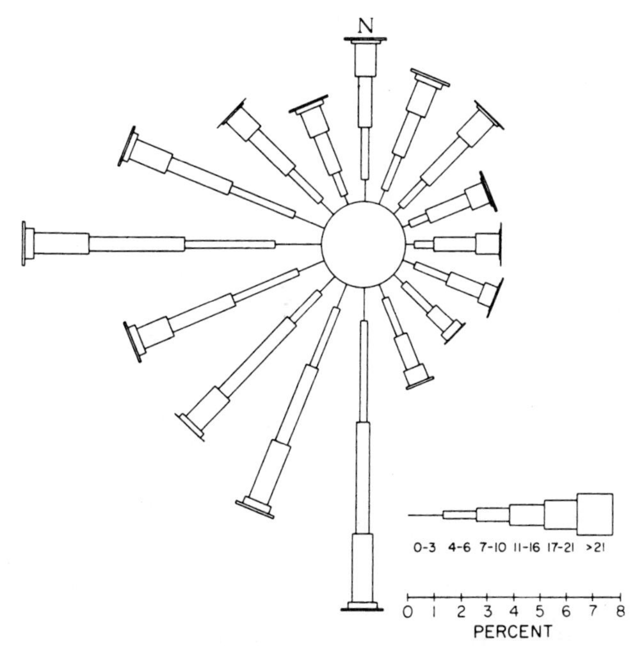

FIGURE 4-14 A wind rose, which graphically portrays the statistical pattern of wind velocities (in knots) at O'Hare Airport in Chicago from 1965 to 1969. Note the large fraction of time when winds are from the west to the south, as would be expected in the region of westerlies. At times, however, local and regional effects (e.g., storm systems) override the air movement attributable to global circulation (Boubel *et al.*, 1994).

4.3.2 Synoptic-Scale Air Circulation

Air Masses

At the *synoptic scale,* which corresponds to roughly a thousand kilometers, the formation and migration of tropospheric *air masses* are responsible for many of the day-to-day changes in weather and air quality. Air masses form when air remains over a large, fairly homogeneous region of Earth's surface, such as a desert or tropical ocean, for a sufficient period of time to acquire distinctive characteristics, especially temperature and moisture content, from the surface. Terrestrial and oceanic regions give rise to *continental* and *maritime* air masses, respectively. A continental air mass usually has a lower moisture content, and often a more extreme temperature (hot or cold, depending on its latitude of origin), than a maritime air mass; a maritime air mass generally is closer to moisture saturation than a continental air mass. Polar air masses are relatively colder, while tropical air masses are relatively warmer, than other air masses that they encounter.

The migration of an air mass results in transport of its associated temperature, moisture, and chemical contents. Along boundaries between air masses of differing temperature and moisture content, called *fronts,* the warmer, less

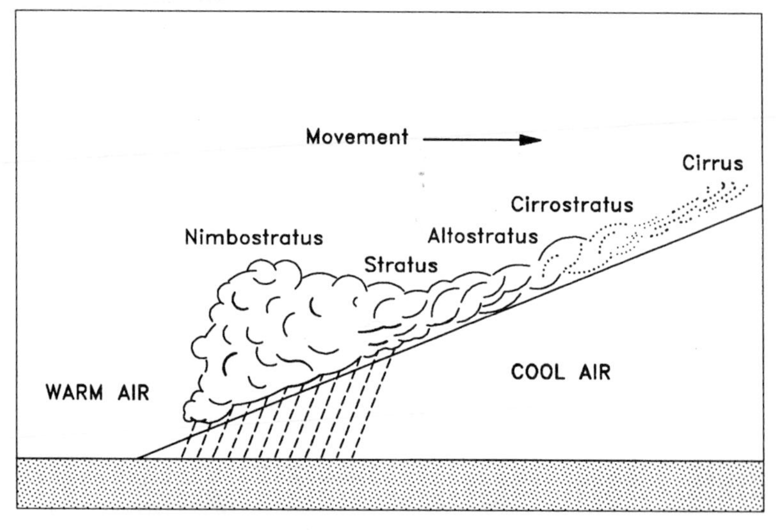

FIGURE 4-15a A warm front occurs when a warm air mass overtakes an adjoining cold air mass and rises over it. Adiabatic cooling tends to produce condensation, and ultimately precipitation, along the front. In the situation shown, the warm air mass is unconditionally stable. If the warm air mass is conditionally stable, instability is produced when clouds form along the front, and vertical cloud development (often including thunderstorms) occurs.

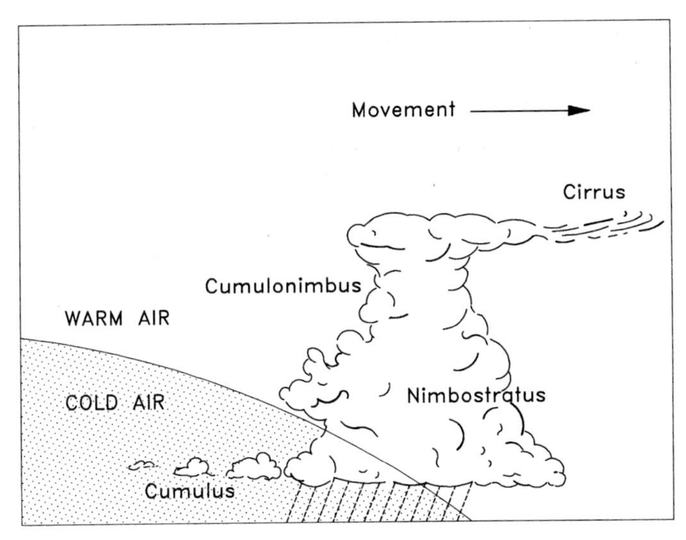

FIGURE 4-15b A cold front occurs when a cold air mass overtakes a warm air mass, which it displaces upward. Again, adiabatic cooling of the rising air tends to produce precipitation. Cold fronts are usually faster moving than are warm fronts; they are also narrower, with the slope of the interface between air masses typically about 80 : 1.

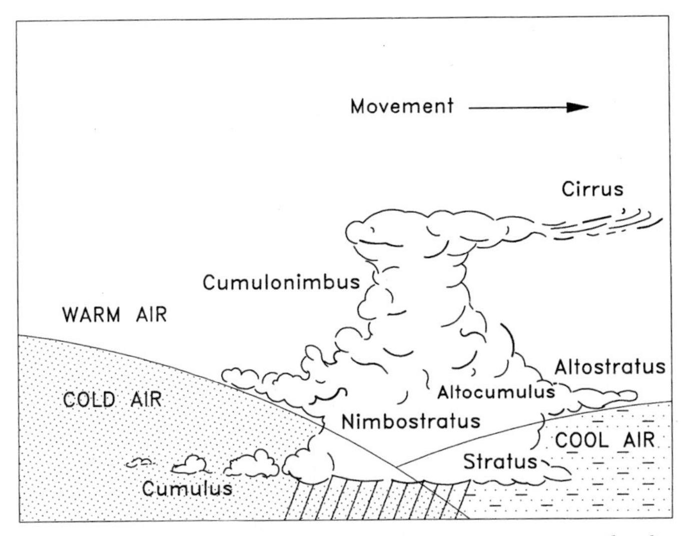

FIGURE 4-15c An occluded front occurs when a warm air mass rises under the combined influence of two colder air masses, causing extensive cloud development and precipitation.

dense air mass tends to rise over the cooler, denser air; because of adiabatic cooling, this frequently results in precipitation. When a warm air mass overtakes a cold air mass, the front is called a *warm front* (Fig. 4-15a); when a cold air mass overtakes a warm air mass, a *cold front* results (Fig. 4-15b). A cold front is usually narrower and faster moving than a warm front, but both can produce clouds and precipitation. An *occluded front* results when a cold front catches up to a more slowly moving warm front, and warmer air becomes trapped between two colder air masses (Fig. 4-15c).

Cloud patterns associated with fronts frequently can be seen from the perspective of a satellite, but a person on the ground, who cannot observe a front in its entirety, must infer its presence from the characteristic progressions of cloud types and weather that occur.

Cloud types often are classified based on altitude. *High clouds* have their bases above 7 km (23,000 ft) and include the wispy mare's tail clouds known as *cirrus;* the *cirrocumulus,* known as "mackerel sky"; and the layers of *cirrostratus. Middle clouds* have altitudes between 2 and 7 km (6500 to 23,000 ft), and are either the rounded *altocumulus* or the layered *altostratus. Low clouds* have bases from near Earth's surface to about 2 km (6500 ft), and include *stratocumulus, stratus,* and *nimbostratus.* Nimbostratus clouds usually bring rain or snow. Clouds with vertical development extend from about 2 to 7 km or more, and include *cumulonimbus* (thunderhead clouds) and *cumulus.*

A passing cold front is heralded by clouds, a drop in temperature, and precipitation; cooler air and clear skies occur behind the cold front (see Fig. 4-15b). The slower moving warm front is characterized by a more gradual lowering of cloud heights, followed by rain or snow (Fig. 4-15a). As air masses pass, so do their burdens of airborne chemicals. The clouds and precipitation formed along the front act as sinks for certain atmospheric chemicals because of rainout and washout processes. These processes, which remove particles, gases, and dissolved chemicals from the atmosphere and deposit them on Earth's surface, are discussed in Section 4.5.

Cyclones and Anticyclones

While some aspects of weather can be explained in terms of air mass movements, other synoptic-scale features of weather systems are associated with *cyclones* and *anticyclones.* Cyclones and anticyclones are large eddies (hundreds of kilometers across) in the atmosphere. Within their boundaries they influence and often control the direction and speed of wind and dominate the weather on a regional scale. They also influence the advective transport of air pollutants and subsequent air quality. On a global scale, cyclones and anticyclones contribute to tropospheric mixing.

Contrary to lay terminology and certain regional usages, a cyclone does not necessarily connote a violent windstorm. In its most general usage, a cyclone is an eddy whose direction of rotation is the same as that of the hemisphere within which it lies (i.e., counterclockwise as viewed from above in the Northern Hemisphere, clockwise in the Southern Hemisphere).

The characteristics of air circulation in a cyclone can be explained in terms of four forces: air pressure gradient, the Coriolis force, friction, and centrifugal force. Two of these forces, the Coriolis force and centrifugal force, are so-called virtual forces that exist because of the acceleration of the reference system. These two forces are treated differently in a nonaccelerating, or *inertial,* reference system. (To an observer in an inertial reference system, the stars appear motionless.) In the following discussion the chosen reference frame is an air parcel that is part of a cyclone or an anticyclone; this reference frame has acceleration due to both the rotation of Earth (giving rise to the Coriolis force) and the circular movement of air within the cyclone or the anticyclone (giving rise to centrifugal force). Within this reference system the sum of all forces is zero. (Note that if an inertial reference frame is chosen, the sum of all forces acting on a parcel moving at velocity v with radius R must have an acceleration of magnitude v^2/R, directed toward the center of the curved path.)

Thus, within the reference frame of a parcel of air circulating about the center of a cyclone, the air parcel experiences the forces shown in Fig. 4-16. The Coriolis force acts radially outward, in a direction perpendicular to, and to the right of, the velocity of the reference system. The centrifugal force also acts radially outward. If the air parcel is located higher than a few hundred meters above Earth's surface, friction may be neglected in an approximate analysis. When friction is neglected, the sum of the Coriolis force and the centrifugal force is balanced by an air pressure force that acts in an inward direction due to the air pressure being lower in the core of the cyclone than outside the cyclone.

In many cyclones, the Coriolis force is larger than the centrifugal force. However, because the centrifugal force is proportional to v^2/R while the Coriolis force is proportional to v, the centrifugal force increases more rapidly than the Coriolis force as a cyclone becomes small and/or its velocity increases. Thus, in a *hurricane* (a violent tropical cyclone over the Atlantic Ocean), the centrifugal force may substantially exceed the Coriolis force. In a *tornado* (a localized, violent midlatitude storm usually originating over land), the Coriolis force may be negligible compared with the centrifugal force because a tornado is smaller and has even higher wind speeds than a hurricane. As testimony to the dominance of centrifugal force under such conditions, *clockwise* tornadoes are sometimes observed in storm systems in the Northern Hemisphere.

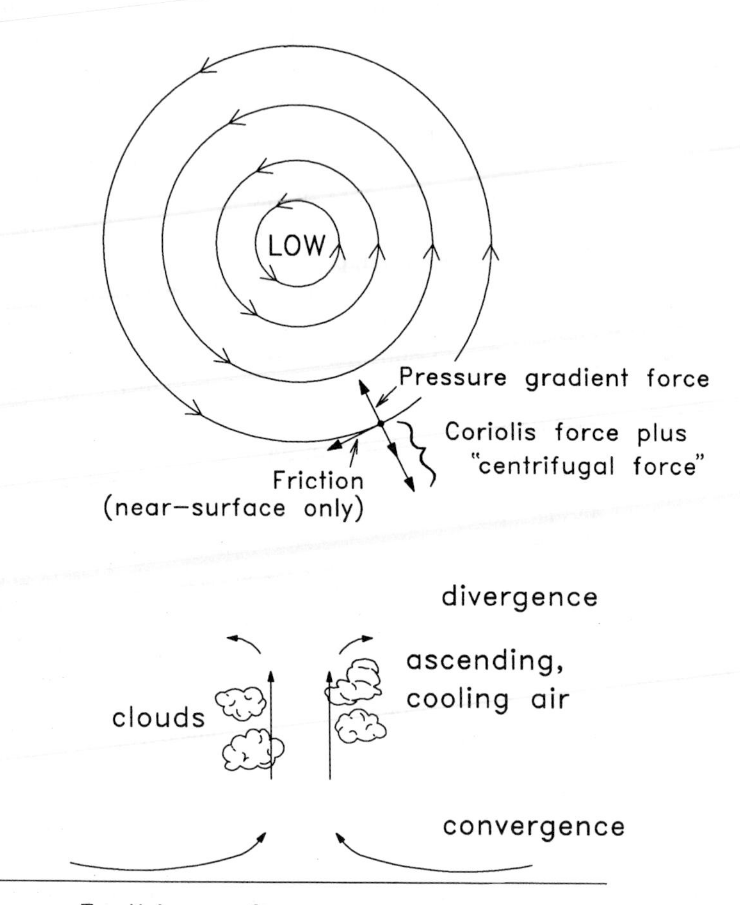

FIGURE 4-16 Forces acting on air in a cyclone. Wind is counterclockwise as viewed from above in the Northern Hemisphere; therefore, both the Coriolis force and centrifugal force are outward. Their sum is balanced by the pressure gradient force. At low altitude, where friction with the ground causes the air velocity to decrease, the Coriolis force and centrifugal force are correspondingly weakened, while the pressure gradient force is not; this results in low-level air being drawn into the core, where it then must ascend to satisfy mass balance constraints. Condensation and precipitation are a likely result in the center of the cyclone.

The low pressure at the center of a cyclone results in large cyclones often being called *lows* or *low pressure areas.* Near the ground surface, frictional forces retard the movement of air, decreasing its velocity and therefore lessening the Coriolis force, Eq. [4-13]. Near-surface air slowed by friction is drawn by the air pressure gradient toward the center of the cyclone. This *convergence* of surface air must, by mass balance considerations, be accompanied by an upward flow of air through the core of the cyclone.

As air is lifted through the center of a cyclone, it often reaches its dew point, forming clouds and creating precipitation. If sufficiently warm and moisture-laden air is drawn into a cyclone, the heat energy released by condensation can augment the kinetic energy of the cyclone, causing it to intensify. This process occurs in hurricanes over warm, tropical seas.

The balance of forces on a parcel of air moving in an anticyclone is shown in Fig. 4-17. The reference system is the accelerating reference system of the air parcel itself. In this case, the Coriolis force acts radially inward due to the clockwise movement of the air. This inward force is balanced by the outward centrifugal force and the outward pressure gradient. Because the anticyclone has higher pressure at its core than outside, it is commonly called a *high* or *high-pressure area.* Friction causes the air velocity near Earth's surface to be lower than the velocity aloft; this in turn decreases the Coriolis force, allowing near-surface air to flow outward from the high-pressure center of the anticyclone.

Conservation of mass requires air to descend within the core of the anticyclone to replace that lost by outflow near Earth's surface. Adiabatic warming occurs as the air descends; precipitation in an anticyclone is therefore suppressed, and high-pressure areas are generally associated with fair weather. Anticyclones may, however, be associated with episodes of lowered air quality, in part because the downward movement of air can create atmospheric inversions, causing pollutants to build up to elevated concentration levels in urban areas. This air pollutant buildup is further aggravated by the light winds of high-pressure areas that lead to low rates of pollutant removal (ventilation). Although winds in a low-pressure area can reach very high velocities, because arbitrarily large pressure gradients can be balanced by arbitrarily large centrifugal forces associated with high wind speeds, winds in a high-pressure area are constrained to be light. In an anticyclone, a finite Coriolis force is balanced by both the pressure gradient and the centrifugal force; increasingly higher pressures cannot be balanced by increasingly higher wind speeds, because the outward centrifugal force also would increase rapidly (proportional to v^2) at higher wind speeds.

Three of the most acutely health-threatening episodes of air pollution in the 20th century were associated with high-pressure areas. These notorious

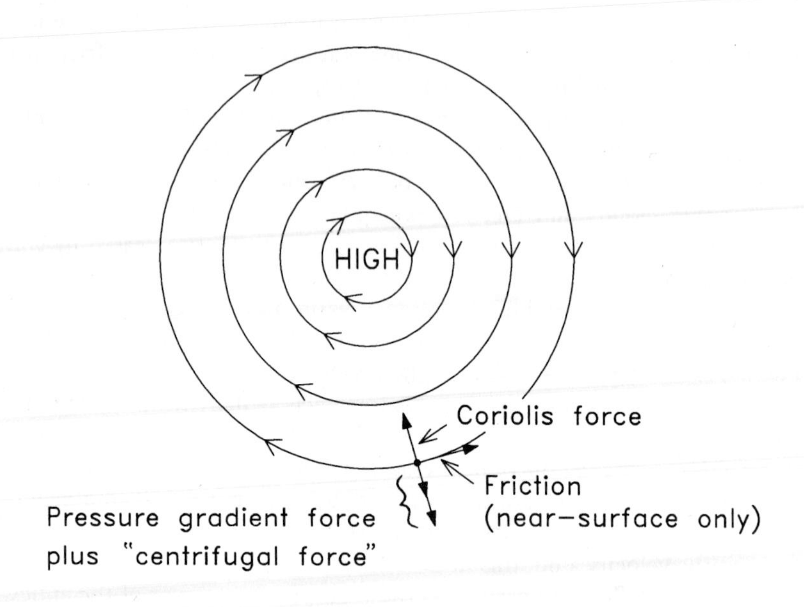

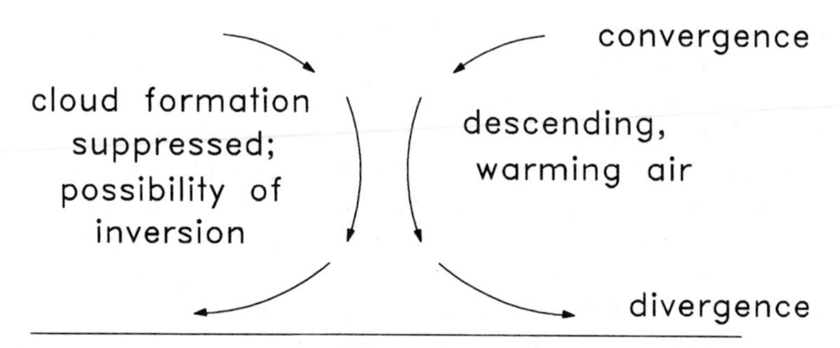

FIGURE 4-17 Forces acting on air in an anticyclone. In the Northern Hemisphere, the Coriolis force acts toward the center of the anticyclone, balanced by the outward centrifugal force and the outward pressure gradient force. Near ground level, Coriolis force and centrifugal force are both weaker, because the air velocity is slowed by friction with the ground; air flows outward near ground level under the influence of the pressure gradient force and therefore descends within the core of the anticyclone. Descending air is warmed adiabatically, inhibiting precipitation and sometimes creating an inversion.

episodes occurred in the Meuse Valley, Belgium, December 1 to 5, 1930; in Donora, Pennsylvania, October 25 to 31, 1948; and in London, United Kingdom, December 5 to 9, 1952. In these cases, strong atmospheric inversions associated with anticyclones helped trap pollutants such as SO_2, resulting in increased air pollutant concentrations. These elevated concentrations caused elevated rates of illness and mortality (Wilson and Spengler, 1996).

Relationships between Frontal Masses and Cyclones

Frequently, weather systems involve a close interaction among fronts, cyclones, and anticyclones. The weather system shown in Fig. 4-16 has a large cyclone (low-pressure area) located at 180° west. A warm front, indicated by a line with semicircles pointing in the direction of movement, extends eastward from the cyclone, while a cold front, indicated by a line with triangles pointing in the direction of movement, extends southward. An occluded front extends a short distance northward. Note that clouds and precipitation are associated with this system, especially to the north of the low-pressure area and along the fronts.

The spatial relationships among fronts and cyclones, such as seen in Fig. 4-18, are frequently more than just coincidental; they are often causal. Mid-

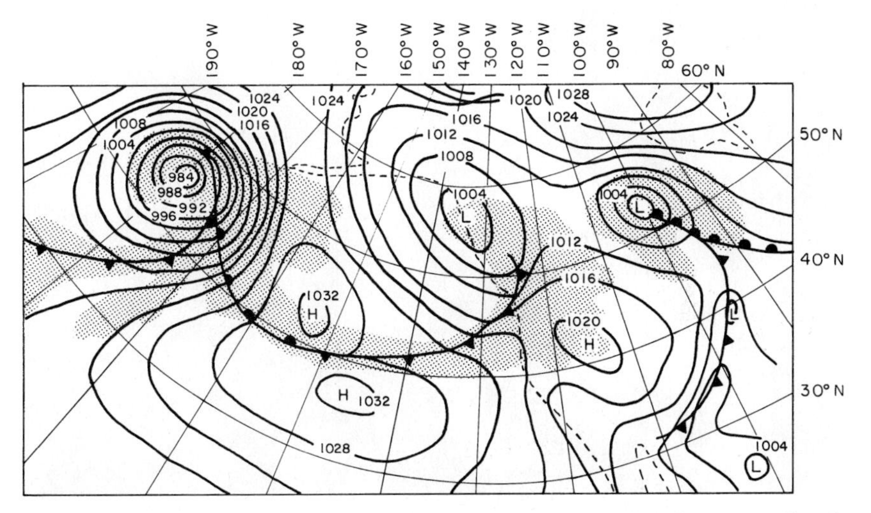

FIGURE 4-18 Tracing of a satellite photo showing cloud formation along fronts, as well as the relationship between low-pressure areas (cyclones) and fronts. Note the correspondence between the cyclones and fronts in this figure and the idealized patterns shown later in Fig. 4-20 for the genesis of a cyclone. Pressures are in millibars; dotted lines show the west coast of North America and Hudson Bay (adapted from FAA, 1965).

latitude cyclones can be formed by *shear* (motion parallel to the front) between the two air masses constituting a front. It is a general principle that shear in fluids tends to produce eddies; familiar small-scale examples include vortices behind the oars of a row boat, or behind a spoon while stirring a cup of coffee. In rivers, the turbulence that gives rise to mixing is generated in part by shear between the more slowly moving water near the river bottom and the faster moving water above it. In the ocean, temperature differences seen by satellite reveal eddies (*warm* and *cold rings*) created by shear between

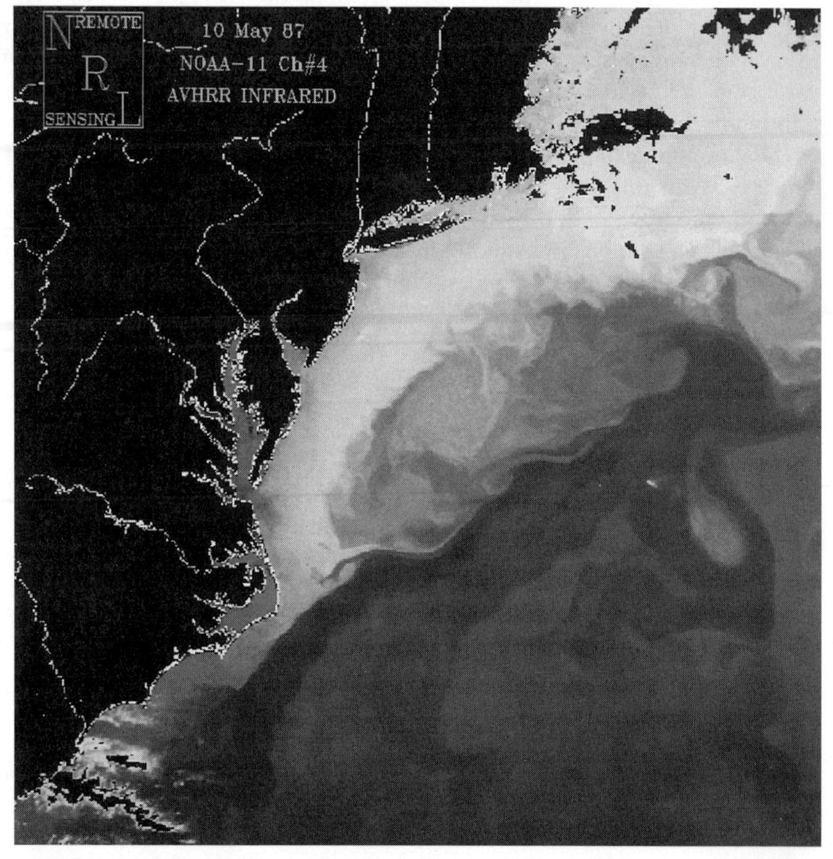

FIGURE 4-19 May 10, 1987 imagery of eddies along the boundary between the warmer (darker gray) Gulf Stream and colder (lighter gray) North Atlantic water. This large-scale eddy generation is similar to the generation of cyclones along the polar front, where shear between air masses occurs (courtesy J. Hawkins and D. May, Naval Remote Sensing Laboratory).

the warm northward-moving Gulf Stream and adjoining cooler waters of the North Atlantic (Fig. 4-19). Although less readily visible, the same process of eddy production by shear occurs in the atmosphere. In the Northern Hemisphere, a regular progression of low-pressure systems is spawned along the interface between the midlatitude westerlies and the colder polar air masses with their prevailing easterly winds. Steps in the genesis of a cyclone along this *polar front* are shown in Fig. 4-20.

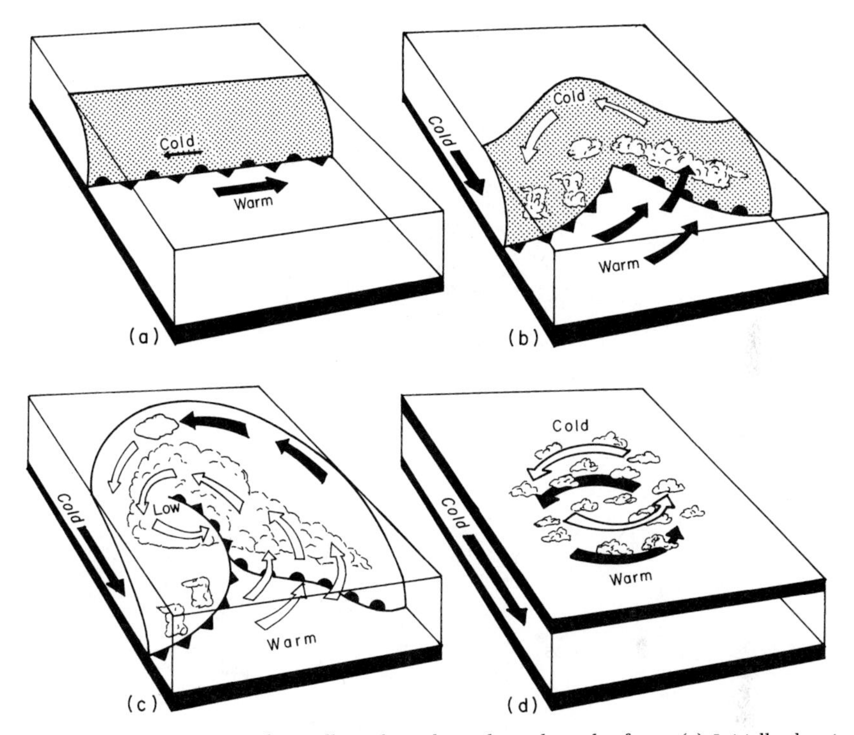

FIGURE 4-20 Formation of a midlatitude cyclone along the polar front. (a) Initially the air masses form a stationary front. Although the front is not moving, the prevailing winds cause shear along the front. (b) The cyclone (low-pressure area) begins as the warm midlatitude air at one location moves poleward over the polar air, creating a warm front; elsewhere the polar air moves southward, creating a cold front. The beginnings of cyclonic air circulation are evident. (c) In a mature system, part of the faster moving cold front may catch up with the warm front, creating an occluded front. Note the similarity of this arrangement of fronts to the weather system of Fig. 4-18. (d) In this dissipated stage, mixing between the air masses has caused the fronts to lose their identities. Moisture has been removed from the system as precipitation. (Adapted from Frederick K. Lutgens/Edward J. Tarbuck, *The Atmosphere: An Introduction to Meteorology*, 5th ed., © 1992, p. 217. Reprinted by permission of Prentice Hall, Englewood Cliffs, New Jersey.)

4.3.3 The Synoptic Weather Map

Because of the great importance of weather to many endeavors, including agriculture, ship navigation, surface travel, and aviation, extensive weather reporting and forecasting have been implemented over large areas of the globe. Over the oceans, where surface-based meteorological data are sparse, satellite imagery has enhanced coverage of weather phenomena. Synoptic-

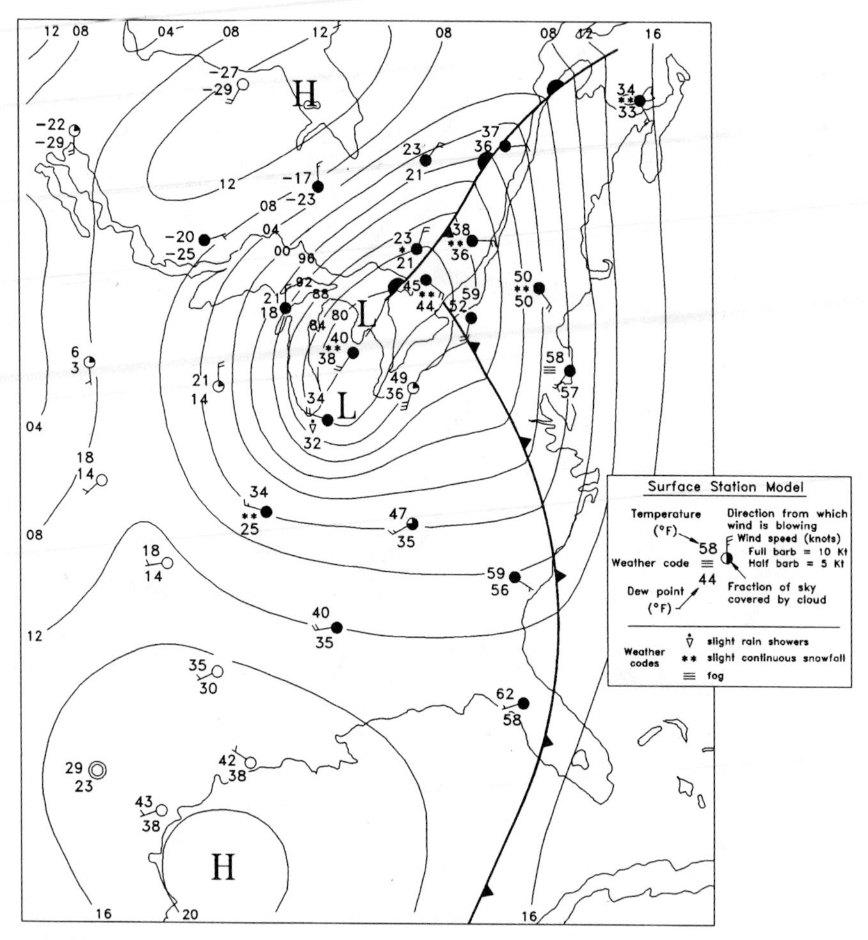

FIGURE 4-21 A surface weather chart for 1200 GMT (700 EST), March 5, 1964, showing information available at each observation station. The chart also shows high- and low-pressure areas, fronts, and isobars constructed by interpolation between stations. (900 or 1000 millibars need to be added as appropriate to values; see text. Adapted from FAA, 1965).

scale weather data are commonly summarized in *surface weather charts*, such as that shown in Fig. 4-21.

Cloud cover, surface wind direction and speed, temperature, dew point, and precipitation are indicated by symbols at each weather reporting station. Interpretation of the weather in a region is facilitated by the identification of high-pressure and low-pressure areas, the delineation of frontal systems, and the construction of *isobars*. Isobars are lines of constant atmospheric pressure that are interpolated between observation points, and are so named because pressure often is presented using the unit of the bar, which is equivalent to 10^6 dyn/cm^2 or 0.9869 atm. On some maps, such as shown in Fig. 4-21, pressure is given in millibars; as a shorthand notation, only the last two digits are presented. The user must add either 900 or 1000 millibars to the value shown, whichever brings the value closest to 1 bar.

Figure 4-21 shows a major low-pressure system over the Great Lakes and the eastern coast of the United States, with a cold front located to the southeast and an occluded front to the northeast. In this late winter storm system, surface winds are counterclockwise and blowing somewhat toward the center of the low, as would be expected from Fig. 4-16. By convention, the tails of the wind "arrows" extend in the direction from which the wind is blowing, and the barbs on the arrow tails indicate wind speed, which is generally greater for stations closer to the center of the low. The various precipitation symbols, and the filled-in or partially filled-in station circles that reflect the fraction of sky covered by clouds, show that precipitation and clouds are associated with this system, especially near the fronts. Southward, behind the cold front, clear skies and cooler temperatures prevail; RH also is lower, as shown by the larger differences between temperature and dew point.

The data presented in weather maps are relevant to chemical transport both at the local scale, where local weather governs atmospheric mixing, and at the regional scale, where synoptic-scale circulation governs the long-range transport of chemicals across state and national boundaries.

4.3.4 Local Effects

At scales smaller than the synoptic scale, topography and differences in the surface cover (e.g., forest, agricultural fields, open water, or urbanized land) influence local winds, precipitation, and temperature. One common example of a local effect due to surface cover is the *sea breeze*, which occurs because water bodies warm and cool more slowly than the land does. During the day in coastal areas, air over the land warms and rises more rapidly and is replaced by cooler air originating from over the water. The reverse may happen at night, as the land cools to a temperature lower than that of the water body,

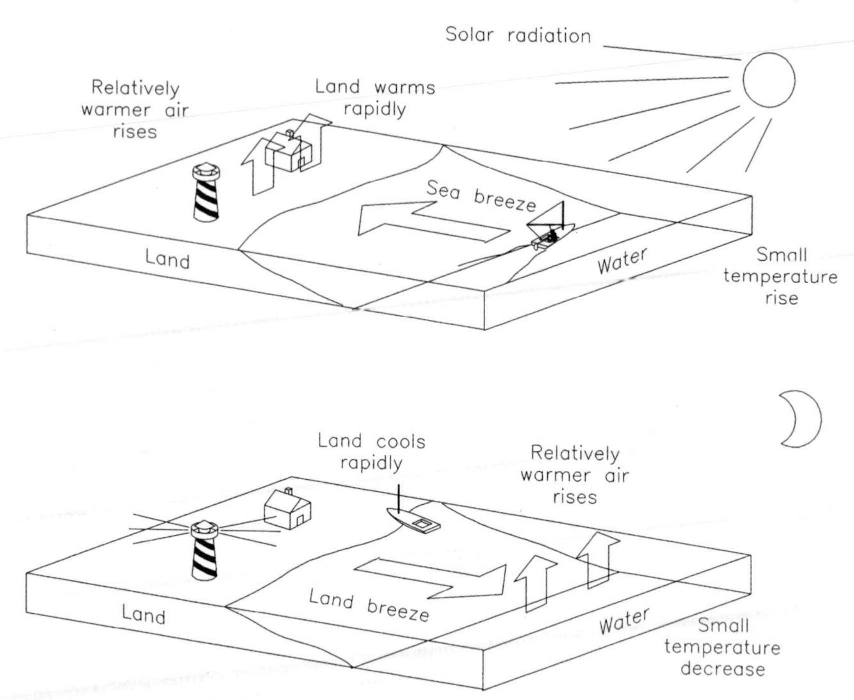

FIGURE 4-22a Sea and land breezes. The sea breeze is a local effect, caused by the faster heating of a land surface than a large water body nearby. The water body temperature rises more slowly than that of the land due to a higher heat capacity and the downward mixing of heat in the water. Sea breezes typically begin in midmorning as air rising over the land is replaced by cooler air from the lake or ocean. At night, when the land becomes cooler than the water, a land breeze may be established.

and air cooled by the land flows toward the water (Fig. 4-22a). Because of the great heat capacity of a large water body, coastal areas also are likely to experience more moderate temperature fluctuations throughout the year than are inland areas. In addition, areas located downwind of large water bodies may receive more precipitation than surrounding areas (the *lake effect*), because the water bodies supply moisture to the air more rapidly than does the land surface.

Topography also influences local winds, precipitation, and temperature. For example, *valley winds* can form as air along a hillside is warmed during the day and thus rises, drawing air up the valley, or as air cools at night and thus flows downhill toward the valley bottom (Fig. 4-22b). In another topographic effect, the windward side of a mountain range often receives extra

FIGURE 4-22b Valley winds. Like the sea breeze, a valley wind is created by uneven solar heating of Earth's surface. In response to heating by the Sun, air rises along the sides of the valley, drawing air up the valley. Under nighttime conditions, cold air may drain down the valley in response to rapid cooling along the hillsides.

precipitation due to orographic uplifting and associated adiabatic cooling (partially along a wet adiabat) of humid air. The leeward side of such a range tends to be drier and warmer as air, which already has lost moisture on the windward side, descends and warms adiabatically (along a dry adiabat). This *rain shadow* effect, shown in Fig. 4-22c, is prominent along the coastal mountain ranges of the northwestern United States.

Another local effect, in this case due to surface cover, is the urban heat island. Rapid heating of urban pavements and buildings occurs during the daytime because of the high absorbance of constructed surfaces and the absence of cooling from evapotranspiration. At night, rural areas cool more effectively than urban areas because of the relatively unobstructed exposure of the land surface, in contrast to the impediment to heat radiation presented by tall, closely spaced buildings. As a result, daily minimum and maximum

FIGURE 4-22c The rain shadow effect. Moist air, shown here as coming off the ocean, is forced to rise by the presence of a mountain range (orographic lifting). As it rises, it cools according to the dry adiabatic lapse rate, until the dew point is reached. Precipitation then can occur as the air continues to rise along the windward side of the mountain range, its further cooling corresponding to a wet adiabatic lapse rate. The air loses much of its original moisture to precipitation, and as it descends on the leeward side of the mountains, it warms according to the dry adiabatic lapse rate. Thus, when the air has descended on the leeward side to its original altitude (sea level in this example), it is drier and warmer than it was on the windward side. The result is a drier, warmer climate and a dearth of precipitation on the leeward side.

temperatures in urban areas are often higher than those found in outlying areas.

For further information on meteorological processes, the reader is referred to Houghton (1985), Oke (1978), Miller and Thompson (1975), Lutgens and Tarbuck (1992), and Hess (1959).

EXAMPLE 4-4

Humid, fog-laden air at a temperature of 65°F from over the Pacific Ocean is deflected up over a 4000-ft coastal mountain. On the inland side of the mountain, the air follows the land contours and descends to a valley 1000 ft above sea level. What is the temperature and humidity of the air in the valley?

Assume that the initial RH of the air is 100% because of the fog. Use Fig. 4-7, begin at 65°F and sea level (0 m), and follow the slope of the wet adiabat until the air parcel reaches 4000 ft (approximately 1200 m). At that altitude, the temperature is approximately 54°F. Because the air is still water saturated, it has a partial pressure of water vapor of approximately 11 mm Hg (see Table 4-3). Descending, the air warms along a dry adiabat to approximately 72°F at 1000 ft. The equilibrium vapor pressure of water at 72°F is approximately 20 mm Hg; thus, the new RH (neglecting any pressure correction) is approximately

$$11/20 \approx 55\%.$$

For increased accuracy, a correction for pressure change can be made. Note that the number of moles of water vapor per mole of air (the mixing ratio of water vapor) must be constant during the descent of the air. However, as the pressure of the air increases, the partial pressure of water vapor increases in the same proportion. From Table 4-1, the air pressure at 4000 ft is approximately 0.89 atm; at 1000 ft, it is approximately 0.97 atm. By correcting the partial pressure of water vapor for the compression during descent,

$$11 \text{ mm Hg} \cdot 0.97/0.89 \approx 12 \text{ mm Hg}.$$

The RH, therefore, is more accurately calculated as

$$12 \text{ mm Hg}/20 \text{ mm Hg} \approx 60\%.$$

4.4 TRANSPORT OF CHEMICALS IN THE ATMOSPHERE

A knowledge of both physical transport and chemical reaction rates is needed to predict concentrations of air pollutants as they are transported away from their sources. The spatial scales on which advective and Fickian transport occur in the atmosphere span an extraordinarily large range, from a few tens of meters in indoor air to thousands of kilometers globally. As in the case of chemical pollutants in surface waters and groundwater, it is important to compare chemical reaction rates with advective velocities and mixing rates. If chemical transformations are relatively slow compared with mixing rates, it may be appropriate to use simple box models in which air volumes are considered to be well mixed (i.e., no concentration gradients exist). When chemicals are transported significant distances in the amount of time it takes for the relevant chemical reactions to reach equilibrium, the most useful models are commonly based on solutions to the advection–dispersion–reaction equation (Section 1.5).

4.4.1 Indoor Air Pollution

To people who spend much of their time indoors, the quality of air in the indoor environment may be at least as important as the quality of outdoor air. The indoor environment lies at the small-scale end of the spectrum of spatial scales, and involves control volumes having well-defined boundaries and a comparatively well-defined geometry. A building is commonly modeled as a well-mixed box with input of outside air and an equal rate of exhaust of inside air (Fig. 4-23a). The measure of air exchange with the outside is called the *ventilation rate* and is expressed as the exhaust (or intake) rate divided by the building volume. If the exhaust rate is given in units of volume per hour, the ventilation rate is equal to the number of air changes per hour (ACH). The ventilation rate depends on building construction and operation, as well as on outside wind velocity and the difference between inside and outside

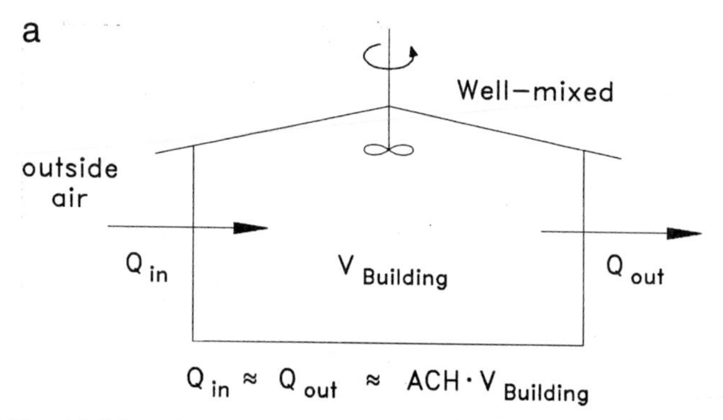

FIGURE 4-23a Modeling the steady-state concentration of a nonreactive air pollutant in a building. (a) The simplest model for air exchange in a building is a well-mixed volume, into which some outside air enters via leaks or through a ventilation system and out of which an equal flow of inside air exits, transporting airborne chemicals with it. Mixing is symbolized by the propeller; some buildings actually have such propellers inside them in the form of ceiling fans. (b) Indoor radon concentration as a function of the ventilation rate. Although radon-222 undergoes radioactive decay, its half-life, 3.8 days, is long compared with the ventilation rates in the graph. Therefore, radioactive decay is a relatively minor sink for indoor radon. The graph also can be applied for other indoor chemicals whose concentration is determined primarily by the balance between source strength and loss by ventilation (Fig. 4-23b from US EPA, 1986).

temperature. Typical values of ACH range from 0.25 or lower for a "tight" building to 1.0 or higher for a "leaky" building. Many new buildings are constructed to be "tight" for energy efficiency, while older frame-construction houses tend to be "leaky."

A mass balance expression for an indoor air pollutant may be written as

$$\text{Input} - \text{output} + \Sigma \text{ sources} - \Sigma \text{ sinks} = \text{rate of change of storage}$$

or

$$C_{outside} \cdot V_{building} \cdot \text{ACH} - C_{inside} \cdot V_{building} \cdot \text{ACH} + \Sigma \text{ sources} - \Sigma \text{ sinks} = \frac{d(C_{inside})}{dt} \cdot V_{building}, \qquad [4\text{-}14]$$

where $C_{outside}$ is the pollutant concentration in outside air [M/L^3], $V_{building}$ is the volume of the building [L^3], ACH is the number of air changes per hour [T^{-1}], C_{inside} is the pollutant concentration in indoor air [M/L^3], Σ sources is the sum of all sources of pollutant [M/T], Σ sinks is the sum of all sinks of

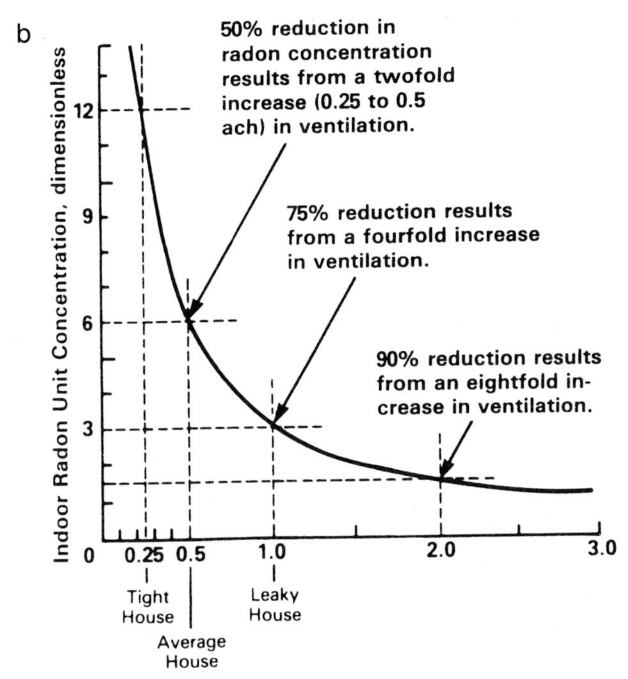

FIGURE 4-23b *(Continued)*

pollutant [M/T], and $d(C_{inside})/dt \cdot V_{building}$ is the rate at which the storage of pollutant mass in indoor air is changing.

Now consider a nonreactive pollutant having a constant source, no appreciable sink, and a negligible concentration in outside air. Such pollutants may be emitted by building materials and furniture components, such as adhesives and particleboard (Godish, 1989). To estimate the pollutant concentration inside the building at steady state [i.e., $d(C_{inside})/dt \cdot V_{building}$ equals zero], Eq. [4-14] can be simplified to

$$C_{inside} = \frac{\Sigma \text{ sources}}{V_{building} \cdot \text{ACH}}. \qquad [4\text{-}15]$$

It is of great practical significance that C_{inside} is directly controlled by the ventilation rate (e.g., see Fig. 4-23b).

EXAMPLE 4-5

A new office building has synthetic carpet laid in all its offices. Formaldehyde (HCHO) is released from the carpet at a rate of approximately 10 μg/hr in an office, and one worker in this 72 ft^2 office with an 8-ft ceiling complains of feeling nauseous from the chemical. Ambient air concentration of formaldehyde is 5 ppb. What ventilation rate is needed in this office to decrease the formaldehyde concentration to 0.1 ppb, assuming a density of air at room temperature of approximately 1.2 kg/m^3?

By assuming there is no significant concentration of formaldehyde in the outside air, and no sink for it in the office, Eq. [4-15] can be used:

$$\text{ACH} = \frac{\Sigma \text{ sources}}{V_{office} \cdot C_{inside}}.$$

First convert the office volume of 576 ft^3 to a metric equivalent of 16.3 m^3. Then:

$$\text{ACH} = \frac{10\ \mu\text{g formaldehyde/hr} \cdot 1\ \text{kg}/10^9\ \mu\text{g}}{16.3\ \text{m}^3 \cdot 0.1\ \text{kg formaldehyde}/10^9\ \text{kg air} \cdot 1.2\ \text{kg air}/1\ \text{m}^3\ \text{air}}.$$

$$\text{ACH} = 5/\text{hr}.$$

Of course, not all pollutants have steady sources, and many have sinks that are likely to be increasing functions of pollutant concentration. One example of a transient, or unsteady, case is the estimation of a pollutant concentration at a given time after the pollutant source begins emitting, as illustrated in Example 4-6.

EXAMPLE 4-6

A gas broiler is turned on in a closed kitchen whose volume is 35 m^3. The oven burner produces 12,000 Btu/hr by burning "bottled" gas (mostly propane, C_3H_8, yielding approximately 20,000 Btu/lb). If the ventilation rate is 1.3 ACH, estimate: (a) the steady-state CO_2 concentration in the kitchen, and (b) the CO_2 concentration in the kitchen $\frac{1}{2}$ hr after the broiler is turned on.

(a) By using Eq. [4-14] and by assuming that there are no sinks for CO_2 in the kitchen and that the outside CO_2 concentration is negligible, the mass conservation equation for CO_2 is

$$\frac{d(C_{\text{inside}})}{dt} \cdot V_{\text{building}} = -C_{\text{inside}} \cdot V_{\text{building}} \cdot \text{ACH} + \Sigma \text{ sources},$$

or

$$\frac{d(C_{\text{inside}})}{dt} = -C_{\text{inside}} \cdot \text{ACH} + \frac{\Sigma \text{ sources}}{V_{\text{building}}}.$$

The source strength is the CO_2 production rate and can be expressed as

$$\frac{12{,}000 \text{ Btu}}{\text{hr}} \cdot \frac{1 \text{ lb } C_3H_8}{20{,}000 \text{ Btu}} \cdot \frac{454 \text{ g}}{1 \text{ lb}} \cdot \frac{1 \text{ mol } C_3H_8}{44 \text{ g } C_3H_8} \cdot \frac{3 \text{ mol } CO_2}{1 \text{ mol } C_3H_8} \cdot \frac{44 \text{ g } CO_2}{1 \text{ mol } CO_2}$$

$$= 816 \text{ g/hr}.$$

The steady-state case occurs when $d(C_{\text{inside}})/dt$ equals zero:

$$C_{\text{inside}} = \frac{\Sigma \text{ sources}}{V_{\text{building}} \cdot \text{ACH}} = \frac{816 \text{ g/hr}}{35 \text{ m}^3 \cdot 1.3/\text{hr}} = 18 \text{ g/m}^3.$$

To express C_{inside} in terms of a mixing ratio, assume a room temperature of 20°C, at which temperature the average volume of 1 mol of an ideal gas is approximately 24 liter:

$$C_{\text{inside}} = \frac{18 \text{ g } CO_2}{\text{m}^3 \text{ air}} \cdot \frac{1 \text{ mol } CO_2}{44 \text{ g } CO_2} \cdot \frac{1 \text{ m}^3 \text{ air}}{1000 \text{ liter air}} \cdot \frac{24 \text{ liter air}}{1 \text{ mol air}}$$

$$= \frac{0.0098 \text{ mol } CO_2}{1 \text{ mol air}} = 9800 \text{ ppm(v)}.$$

This concentration is approximately twice the acceptable level of 5000 ppm(v) at 25°C. In practice, the CO emitted by the burner may be of even greater concern.

(b) In the transient, or unsteady, case, the mass conservation equation must be integrated with respect to time to obtain a solution. The mass conservation equation is of the following form:

$$\frac{dC}{dt} = -kC + r.$$

When integrated, this equation has the form,

$$C = \frac{r}{k} + \text{constant} \cdot e^{-kt},$$

where a constant arises from the integration. In this particular example, C corresponds to C_{inside}, r corresponds to the source term (Σ sources/V_{building}), and k corresponds to ACH. There is a constraint (boundary condition) that C_{inside} equals zero when time t equals zero; thus, the constant must equal $(-r/k)$. Therefore, for this example, the full solution is

$$C_{\text{inside}} = \frac{\Sigma \text{ sources}}{V_{\text{building}} \cdot \text{ACH}} - \frac{\Sigma \text{ sources}}{V_{\text{building}} \cdot \text{ACH}} \cdot e^{-\text{ACH} \cdot t},$$

or

$$C_{\text{inside}} = \frac{\Sigma \text{ sources}}{V_{\text{building}} \cdot \text{ACH}} \cdot (1 - e^{-\text{ACH} \cdot t}) = 18 \text{ g/m}^3 \cdot (1 - e^{-1.3 \cdot t}).$$

When $t = \frac{1}{2}$ hr,

$$C_{\text{inside}} = 18 \text{ g/m}^3 \cdot (1 - 0.52) \approx 8.6 \text{ g/m}^3 \approx 4700 \text{ ppm(v)}.$$

As time becomes large, the term $(1 - e^{-1.3 \cdot t})$ approaches one, and the preceding equation reduces to the steady-state equation, Eq. [4-15].

It is possible to measure building ventilation rates by using chemical tracers, or to assess building tightness by using a *blower door*, which blows air into a building with a fan of known pressure versus flow characteristics and measures the resulting pressure rise. In "tight," energy-efficient residential construction and in large modern buildings, ventilation is not left to the vagaries of miscellaneous cracks and joints; instead, controlled ventilation is provided mechanically. In such cases, engineering data may be available to indicate ventilation rates. Ventilation rates may be inadequate, however, due to incorrect design or operation (sometimes blowers are operated at lower ventilation rates to save energy costs). In such situations, the inability to open windows can result in the *sick building syndrome*; lack of sufficient fresh air causes people who work in such buildings to become ill when concentrations of chemicals released from carpets, furnishings, and cleaning compounds (Godish, 1989), or concentrations of CO_2 exhaled by people, increase to nuisance or toxic levels.

4.4.2 Local-Scale Outdoor Air Pollution

The smallest spatial scale at which outdoor air pollution is of concern corresponds to the air volume affected by pollutant chemical emissions from a single point source, such as a smokestack (Fig. 4-24). Chemicals are carried downwind by advection, while turbulent transport (typically modeled as Fickian transport) causes the chemical concentrations to become more diluted. Typically, smokestacks produce continuous pollutant emissions, instead of single pulses of pollutants; thus, steady-state analysis is often appropriate. At some distance downwind, the plume of chemical pollutants disperses sufficiently to reach the ground; the point at which this occurs, and the concentrations of the chemicals at this point and elsewhere, can be estimated from solutions to the advection–dispersion–reaction equation (Section 1.5), given a knowledge of the air (wind) velocity and the magnitude of Fickian transport.

The smokestack plume problem is entirely analogous to the groundwater plume problem, although the mechanism for mixing is turbulent diffusion instead of mechanical dispersion. The traditional steady-state Gaussian plume

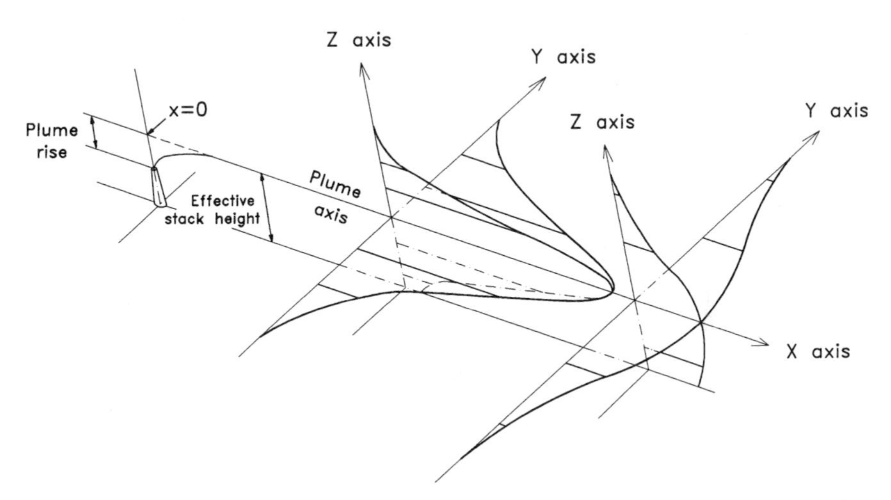

FIGURE 4-24 Cross sections of pollutant concentrations at two locations downwind of a smokestack. Note that physical height of the stack is typically less than the effective height of the stack, which takes plume rise into account. The total flux of pollutant is identical at each downwind location, although concentration decreases as the plume widens. The shape of each concentration versus distance plot is a normal, or Gaussian, curve; hence, this is often called a Gaussian plume (adapted from Boubel *et al.*, 1994).

model describing transport and mixing in air is presented in a different mathematical format than its counterpart in the groundwater literature; nonetheless, the two models are theoretically equivalent and interconvertible. Viegle and Head (1978) may be consulted for a derivation of the Gaussian plume model in air.

One classic Gaussian plume model for smokestack emissions is the Pasquill–Gifford model, which applies for steady emissions of a chemical over relatively level terrain. If no chemical sinks exist in the air (i.e., no reactions are degrading the chemical) and if there is an unlimited mixing height (i.e., no atmospheric inversion exists, and the plume can be mixed upward indefinitely), the Pasquill–Gifford model can be expressed in the form

$$C = \frac{Q}{u}\frac{g_1 g_2}{2\pi\sigma_y\sigma_z}, \qquad [4\text{-}16]$$

where C is the concentration of chemical in the air [M/L^3]; Q is the rate of chemical emission [M/T]; u is the wind speed [L/T] in the x direction, which is aligned with the wind direction; g_1 is the horizontal Gaussian distribution factor; g_2 is the vertical Gaussian distribution factor; σ_y is the standard deviation [L] of the horizontal distribution of the pollutant concentration (in the y direction); and σ_z is the standard deviation [L] of the vertical distribution of the pollutant concentration (in the z direction). [In a cross section of the pollutant plume (such as shown in Fig. 4-24) 68% of the pollutant mass or concentration in that slice lies within one standard deviation either side of the axis.] The horizontal and vertical Gaussian distribution factors are calculated as

$$g_1 = \exp(-0.5y^2/\sigma_y^2), \qquad [4\text{-}17a]$$

and

$$g_2 = \exp(-0.5 \cdot (z - H)^2/\sigma_z^2) + \exp(-0.5 \cdot (z + H)^2/\sigma_z^2), \qquad [4\text{-}17b]$$

where y is the distance along a horizontal axis perpendicular to the wind, z is the distance along a vertical axis having an origin at the ground surface, and H is the *effective stack height* [L] (defined next). Over relatively level terrain, the maximum impact of a smokestack usually occurs within 10 to 20 km.

It is important to recognize that the concentration C in Eq. [4-16] is a time-averaged value; at any instant, the actual concentration is likely to be higher or lower due to the eddies and billows that occur. H, the effective stack height, is usually higher than the physical stack height, because chemical pollutants are usually ejected with an upward momentum and, if the gases are less dense than the surrounding air (e.g., due to having a higher temperature), they also have positive buoyancy. Effective stack height is equal to physical stack height plus ΔH, the *plume rise*. One commonly used equation presented by Briggs (1972) to estimate ΔH for warm plumes at intermediate distances from the stack is

$$\Delta H = 1.6\, F_b^{1/3}\, x^{2/3}/u, \qquad [4\text{-}18]$$

where ΔH is the plume rise [L], F_b is the buoyancy flux parameter [L^4/T^3], x is the downwind distance [L], and u is the wind speed [L/T]. The buoyancy flux parameter is estimated as

$$F_b = g\,\frac{d^2V}{4}\left(\frac{T_s - T_a}{T_s}\right), \qquad [4\text{-}19]$$

where g is acceleration due to gravity [L/T^2], d is the stack diameter [L], V is the stack gas velocity [L/T], T_s is the absolute stack gas temperature, and T_a is the absolute ambient air temperature. Note that Eq. [4-18] is valid for any consistent set of units (e.g., meters and seconds).

The *final rise* is the plume rise occurring downwind at the distance where turbulent updrafts and downdrafts result in vertical motions of similar magnitude to those caused by the buoyancy forces. Other formulae for estimating ΔH under different conditions, and for estimating final rise, are given by Boubel *et al.* (1994).

EXAMPLE 4-7

A small industrial furnace emits 0.8 m³/sec of exhaust gases at a temperature of 80°C. Outside air temperature is 10°C, and wind speed is 2 m/sec. The stack is 20 m tall, with a diameter of 40 cm. What is the plume height at the edge of the property 30 m downwind?

First, estimate the stack gas velocity:

$$V = \frac{0.8\ \text{m}^3/\text{sec}}{\pi(0.2\ \text{m})^2} = 6.4\ \text{m/sec}.$$

Then use Eq. [4-19] to estimate the buoyancy flux parameter:

$$F_b = \frac{9.81\ \text{m}}{\text{sec}^2}\cdot\frac{(0.4\ \text{m})^2}{4}\cdot\frac{6.4\ \text{m}}{\text{sec}}\cdot\frac{353\ \text{K} - 283\ \text{K}}{353\ \text{K}} = \frac{0.50\ \text{m}^4}{\text{sec}^3}.$$

Finally, use Eq. [4-18] to estimate the plume rise:

$$\Delta H = (1.6)\left(\frac{0.50\ \text{m}^4}{\text{sec}^3}\right)^{1/3}(30\ \text{m})^{2/3}\left(\frac{1}{2\ \text{m/sec}}\right) = 6\ \text{m}.$$

Therefore, 30 m downwind, the plume height above the ground can be estimated as the stack height plus the plume rise:

$$20 \text{ m} + 6 \text{ m} = 26 \text{ m}.$$

When the Pasquill–Gifford plume model of Eq. [4-16] is compared with the equations given in Fig. 3-19, one immediately apparent difference is that dispersion is expressed in terms of σ_y and σ_z instead of Fickian dispersion coefficients (D_y, D_z), as are customarily used for surface water and groundwater. Recall that σ^2 equals 2 Dt for one-dimensional Fickian transport of a chemical in a fluid, and therefore σ increases with distance even though the turbulent diffusion coefficient D is constant. The turbulence that mixes air pollutants arises both from interactions of the wind with trees, buildings, and other surface obstructions and from heating of the ground surface by the Sun (see Fig. 4-8). Thus, σ_y and σ_z not only depend on distance from the source but also are a strong function of prevailing meteorological conditions, especially wind velocity and atmospheric stability. In theory, σ_y and σ_z could be obtained from chemical tracer experiments; in practice, it is more feasible to estimate σ_y and σ_z from physical measurements. A widely used approach for estimating σ_y and σ_z is to employ atmospheric stability categories, which are based only on average wind speed, cloudiness, and *insolation* (solar heat input). The commonly used Pasquill stability categories, shown in Table 4-6,

TABLE 4-6 Pasquill Stability Categories

	Insolation[a]			Night[b]	
Surface wind speed (m/sec)[c]	Strong[d]	Moderate[d]	Slight	Thinly overcast or ≥4/8 low cloud	≤3/8 cloud
<2	A	A–B	B	—	—
2–3	A–B	B	C	E	F
3–5	B	B–C	C	D	E
5–6	C	C–D	D	D	D
>6	C	D	D	D	D

[a] Strong insolation corresponds to sunny midday in midsummer in England; slight insolation to similar conditions in midwinter.
[b] Night refers to the period from 1 hr before sunset to 1 hr after sunrise.
[c] The neutral category D should also be used, regardless of wind speed, for overcast conditions during day or night and for any sky conditions during the hour preceding or following night as defined in footnote *b*.
[d] For A–B, interpolate between values in Fig. 4-25 for A and B, etc.
Source: Pasquill, 1974.

are based on standard meteorological data, such as are recorded at airports. Category A corresponds to periods of intense sunlight during which atmospheric mixing is augmented by instability due to solar heating of the ground surface. Categories B and C refer to successively less unstable conditions. Category D corresponds to an atmosphere of neutral stability, while categories E and F represent increasingly stable conditions associated with atmospheric inversion.

The effects of higher wind speed during periods of insolation at first may seem counterintuitive; higher wind speeds, which cause more atmospheric turbulence, shift the classification in the direction of *higher* stability categories, which one would intuitively expect to have less mixing. This shift occurs because the Pasquill stability categories are used to determine the width and height of a pollutant plume at a particular distance downwind, not to estimate a Fickian dispersion coefficient (D_y or D_z). Higher wind velocity can actually decrease the absolute amount of spreading a pollutant plume undergoes before reaching a fixed distance downwind if the effects of decreased travel time from source to downwind point override the effects of more intense mixing. Although there is more intense turbulence with higher wind speed, there is less time for mixing to occur, and therefore the latter effect may predominate.

As shown graphically in Fig. 4-25 (Gifford, 1976), σ_y and σ_z are correlated with the Pasquill stability categories through empirical relationships. The Pasquill stability categories are used to choose the appropriate curve that gives σ_y or σ_z as a function of x, the distance downwind of the source. As expected, at a given x the less stable categories correspond to more mixing. As stability categories shift from F to A, σ_y and σ_z increase by at least one order of magnitude. As also can be seen in Fig. 4-25, the less stable categories have slopes that change more rapidly with distance downwind. These increased slopes reflect the greater amount of mixing that occurs at larger scales. In terms of Fickian mixing, values of D_y and D_z become larger because as the plume expands, ever larger eddies contribute to mixing. Evidence of larger values of D at larger scales can be seen in Fig. 1-4; note that small smoke plumes spread more slowly with distance than do the large, composite plumes.

As an alternative to estimating σ_y and σ_z from a graph, σ_y and σ_z can be calculated from empirical equations. Seinfeld (1986) presents equations that are equivalent to the Pasquill–Gifford curves, as well as several other empirical models. An example of an empirical model presented as a set of formulae is that recommended by Briggs (Gifford, 1976), shown in Table 4-7.

There are limitations to estimating σ_y and σ_z using only information contained in atmospheric stability categories; further errors may result from applying these models to settings that differ in local topography or climatic conditions. Typically, these models apply to high plumes and downwind distances less than 10 km.

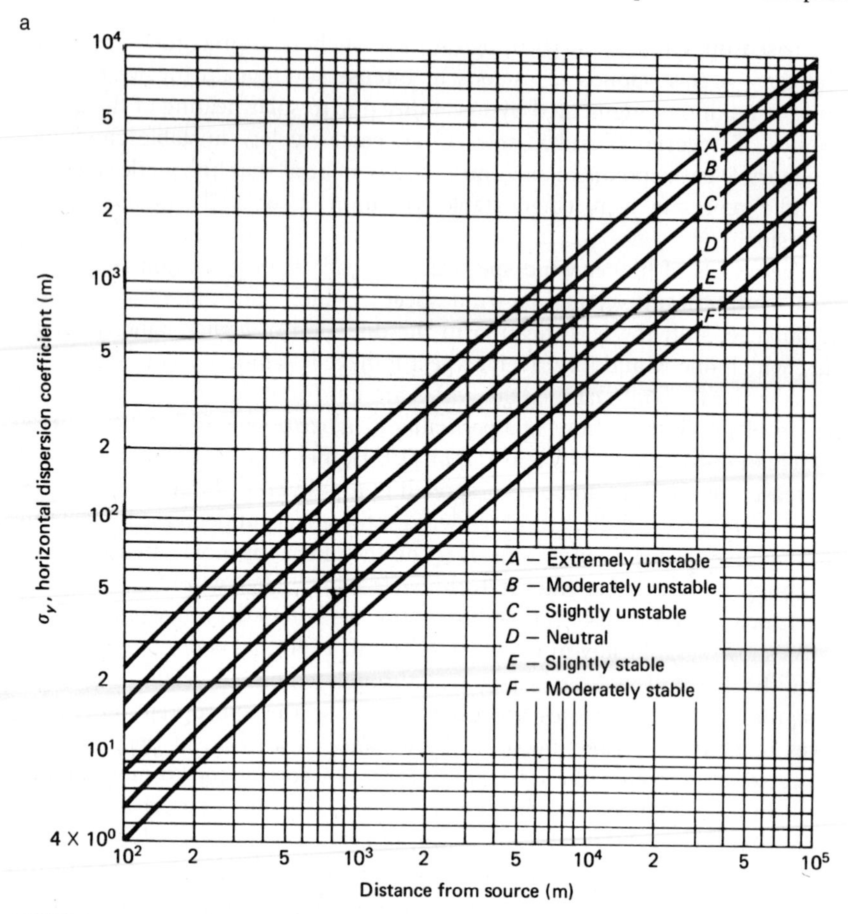

FIGURE 4-25a Standard deviations of mass distribution in a Gaussian plume, σ_y and σ_z, given as a function of both distance downwind from a point source and Pasquill stability categories. "Dispersion coefficient" as used in this figure means the standard deviation of the plume width or height [L] rather than a Fickian coefficient [L^2/T]. (From *Atmospheric Chemistry and Physics of Air Pollution,* by J. H. Seinfeld. Copyright © 1986, John Wiley & Sons, Inc. Reprinted by permission of John Wiley & Sons, Inc.)

EXAMPLE 4-8

A smokestack with an effective height of 25 m emits sulfur dioxide (SO_2) at a rate of 10 kg/hr. What is the contribution of this stack to ground-level SO_2 concentrations at a school yard 8 km downwind, if wind speed is 4.5 m/sec on a sunny midwinter day? Assume unlimited mixing height.

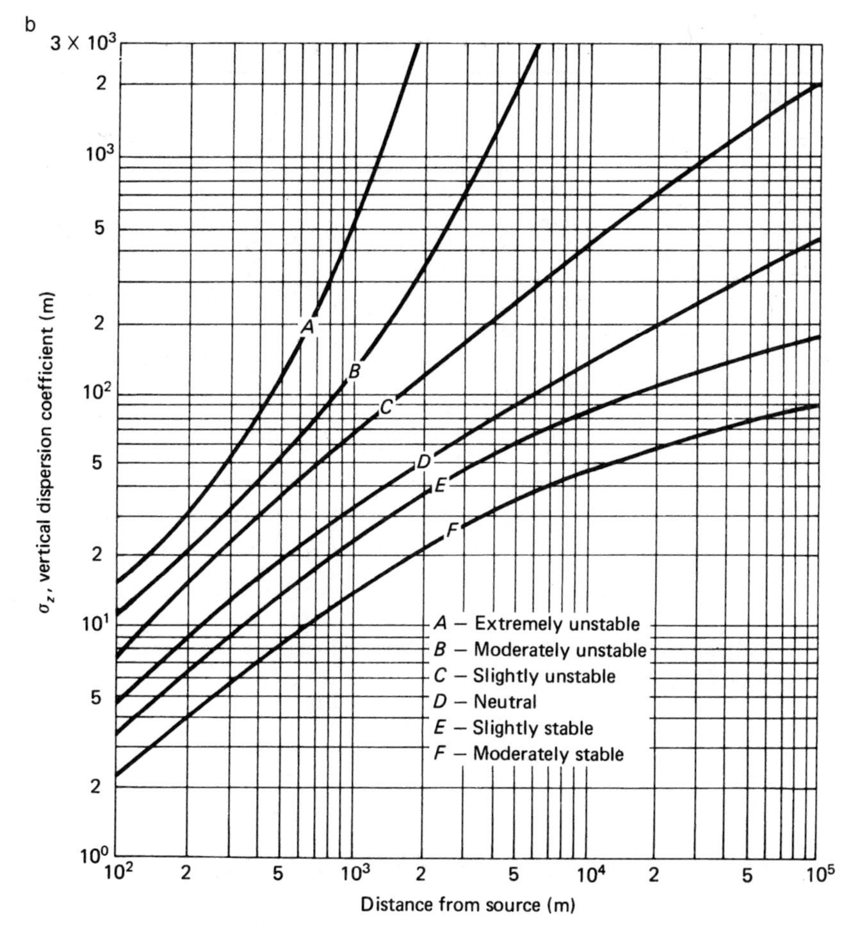

FIGURE 4-25b *(Continued)*

To calculate downwind concentrations, use Eq. [4-16]. From Table 4-6, a wind speed of 4.5 m/sec corresponds to Pasquill stability category C during the winter when insolation is low. By using Fig. 4-25, at a downwind distance of 8 km, σ_y is approximately 700 m and σ_z is approximately 400 m. Note that $y = 0$ directly downwind and $z = 0$ at ground level. First, estimate Q/u:

$$Q/u = (10 \text{ kg/hr})/(4.5 \text{ m/sec} \cdot 3600 \text{ sec/hr}) = 6.2 \times 10^{-4} \text{ kg/m}.$$

TABLE 4-7 Formulae for σ_y and σ_z Recommended by Briggs[a,b]

Pasquill stability category	σ_y (m)	σ_z (m)
	Open country conditions	
A	$0.22\ x/(1 + 0.0001\ x)^{0.5}$	$0.20\ x$
B	$0.16\ x/(1 + 0.0001\ x)^{0.5}$	$0.12\ x$
C	$0.11\ x/(1 + 0.0001\ x)^{0.5}$	$0.08\ x/(1 + 0.0002\ x)^{0.5}$
D	$0.08\ x/(1 + 0.0001\ x)^{0.5}$	$0.06\ x/(1 + 0.0015\ x)^{0.5}$
E	$0.06\ x/(1 + 0.0001\ x)^{0.5}$	$0.03\ x/(1 + 0.0003\ x)$
F	$0.04\ x/(1 + 0.0001\ x)^{0.5}$	$0.016\ x/(1 + 0.0003\ x)$
	Urban conditions	
A, B	$0.32\ x/(1 + 0.0004\ x)^{0.5}$	$0.24\ x \cdot (1 + 0.001\ x)^{0.5}$
C	$0.22\ x/(1 + 0.0004\ x)^{0.5}$	$0.20\ x$
D	$0.16\ x/(1 + 0.0004\ x)^{0.5}$	$0.14\ x/(1 + 0.0003\ x)^{0.5}$
E, F	$0.11\ x/(1 + 0.0004\ x)^{0.5}$	$0.08\ x/(1 + 0.0015\ x)^{0.5}$

[a] x is the downwind distance in meters; valid only for $100\ \text{m} < x < 10{,}000\ \text{m}$.
[b] Gifford (1976).

Then use Eqs. [4-17] to estimate the Gaussian distribution factors:

$$g_1 = \exp(-0.5 \cdot (0\ \text{m})^2/(700\ \text{m})^2) = 1.$$

$$g_2 = \exp(-0.5 \cdot (-25\ \text{m})^2/(400\ \text{m})^2) + \exp(-0.5 \cdot (25\ \text{m})^2/(400\ \text{m})^2) = 2.$$

Using Eq. [4-16] gives:

$$C_{\text{schoolyard}} = 6.2 \times 10^{-4}\ \text{kg/m} \cdot \frac{(1) \cdot (2)}{2\pi \cdot 700\ \text{m} \cdot 400\ \text{m}} = 7 \times 10^{-10}\ \text{kg/m}^3.$$

If the formulae in Table 4-7 are used instead of the graphs in Fig. 4-25 to estimate σ_y and σ_z, then for open country, or rural conditions:

$$\sigma_y = \frac{0.11 \cdot 8000}{(1 + 0.0001 \cdot 8000)^{0.5}} = 660\ \text{m},$$

and

$$\sigma_z = \frac{0.08 \cdot 8000}{(1 + 0.0002 \cdot 8000)^{0.5}} = 400\ \text{m}.$$

The concentration in the schoolyard then would be estimated as $8 \times 10^{-10}\ \text{kg/m}^3$. Note that this concentration is very close to that estimated using the graphs in Fig. 4-25.

If the conditions are urban, then

$$\sigma_y = \frac{0.22 \cdot 8000}{(1 + 0.0004 \cdot 8000)^{0.5}} = 860 \text{ m},$$

and

$$\sigma_z = 0.20 \cdot 8000 = 1600 \text{ m}.$$

The concentration in the schoolyard then would be estimated as 1×10^{-10} kg/m^3.

If mixing height, L, is low enough so that it restricts the ascension of a plume, it may be appropriate to assume that the plume becomes fully mixed vertically beneath L and spreads only horizontally. In this case, the concentration of a chemical in the air can be estimated as

$$C = \frac{Q}{u} \frac{g_1}{\sqrt{2\pi}\sigma_y} \cdot \frac{1}{L}. \qquad [4\text{-}20]$$

A more accurate method for estimating σ_y and σ_z takes into account direct measurements of atmospheric turbulence. Both σ_y and σ_z are increasing functions of the intensity of atmospheric turbulence, which can be estimated from continuous records of the components of wind velocity in the x, y, and z directions (see Fig. 4-26). Pasquill (1961) proposed the following measures

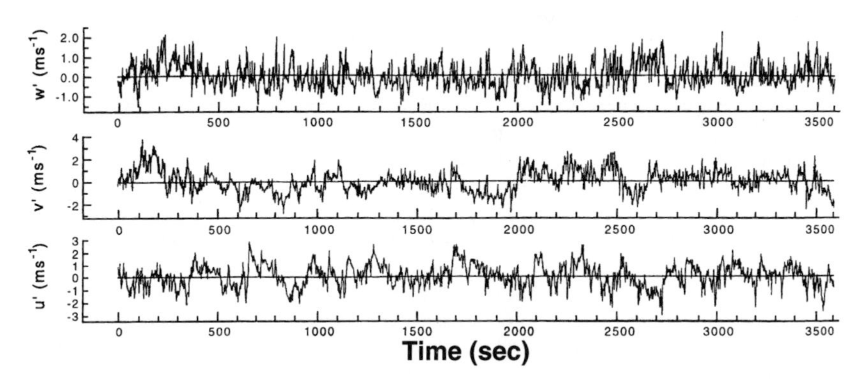

FIGURE 4-26 Measured time series of instantaneous wind velocities minus average wind velocities in the x, y, and z directions during moderately unstable conditions at a suburban site in Vancouver, British Columbia, Canada. (Arya, 1999.)

of turbulent intensity in the lateral and vertical directions,

$$i_y = \frac{\sigma_v}{u} \text{ and } i_z = \frac{\sigma_w}{u}, \tag{4-21}$$

where i_y is the turbulent intensity (unitless) in the y direction; i_z is the turbulent intensity (unitless) in the z direction; σ_v is the standard deviation [L/T] of the wind velocity in the y direction with respect to time; σ_w is the standard deviation [L/T] of the wind velocity in the z direction with respect to time; and u is the wind speed [L/T] in the x direction, which is aligned with the average wind direction. Then σ_y and σ_z are determined as

$$\sigma_y = i_y \cdot f_y \cdot x \text{ and } \sigma_z = i_z \cdot f_z \cdot x, \tag{4-22}$$

where f_y and f_z are semiempirical functions of distance and Pasquill stability category, for the lateral and vertical directions, respectively, as shown in Table 4-8 for rural and urban settings.

To summarize, Eq. [4-22] takes local measurements of turbulent intensity and extrapolates them, based on stability categories, to apply to downwind distances to estimate σ_y and σ_z. In contrast, the Pasquill–Gifford curves and similar models, such as Briggs' formulae in Table 4-7, rely on stability categories alone to estimate σ_y and σ_z. Estimates of σ_z also can be made from temperature and horizontal wind velocity data measured at several heights (e.g., from meteorological towers) (Gifford, 1976; Draxler, 1979).

TABLE 4-8 Formulae for f_y and f_z[a,b]

Pasquill stability category	f_y	f_z
	Rural conditions	
A, B	$1/(1 + 0.0001\,x)^{0.5}$	1
C	$1/(1 + 0.0001\,x)^{0.5}$	$1/(1 + 0.0002\,x)^{0.5}$
D	$1/(1 + 0.0001\,x)^{0.5}$	$1/(1 + 0.0015\,x)^{0.5}$
E, F	$1/(1 + 0.0001\,x)^{0.5}$	$1/(1 + 0.0003\,x)$
	Urban conditions	
A, B	$1/(1 + 0.0004\,x)^{0.5}$	$(1 + 0.001\,x)^{0.5}$
C	$1/(1 + 0.0004\,x)^{0.5}$	1
D	$1/(1 + 0.0004\,x)^{0.5}$	$1/(1 + 0.0003\,x)^{0.5}$
E, F	$1/(1 + 0.0004\,x)^{0.5}$	$1/(1 + 0.0015\,x)^{0.5}$

[a] x is the downwind distance in meters; valid only for x less than 10,000 m.
[b] Adapted from Hanna (1985).

EXAMPLE 4-9

The stack of a smelter in a rural area emits very fine particles containing metals such as nickel, copper, and lead. Wind speed is 5.5 m/sec, and insolation is slight. The emission rate of nickel is approximately 0.1 g/sec at an effective height of 120 m. Analysis of the data from a vertically oriented wind speed instrument, which produces data corresponding to the plot of w' versus time in Fig. 4-26, indicates that the standard deviation of vertical air velocity is 0.15 m/sec. The corresponding standard deviation of v' (perpendicular to the wind) is 0.25 m/sec. What is the nickel concentration at ground level directly downwind, 8 km from the stack?

From Table 4-6, the Pasquill stability category is D. Turbulent intensity in the vertical direction can be calculated from Eq. [4-21]:

$$i_z = \frac{\sigma_w}{u} = \frac{0.15 \text{ m/sec}}{5.5 \text{ m/sec}} = 0.027.$$

Likewise, Eq. [4-21] can be used to calculate the lateral turbulent intensity:

$$i_y = \frac{\sigma_v}{u} = \frac{0.25 \text{ m/sec}}{5.5 \text{ m/sec}} = 0.045.$$

To calculate σ_z and σ_y, use Eq. [4-22], together with estimates of f_z and f_y obtained using the appropriate formulae from Table 4-8. For rural conditions and Pasquill stability category D:

$$f_z = \frac{1}{(1 + 0.0015\, x)^{0.5}} = \frac{1}{(1 + 0.0015 \cdot 8000 \text{ m})^{0.5}} = 0.28.$$

From Eq. [4-22]:

$$\sigma_z = i_z \cdot f_z \cdot x = (0.027) \cdot (0.28) \cdot (8000 \text{ m}) = 61 \text{ m}.$$

For the lateral direction:

$$f_y = \frac{1}{(1 + 0.0001\, x)^{0.5}} = 0.75.$$

From Eq. [4-22]:

$$\sigma_y = i_y \cdot f_y \cdot x = (0.045) \cdot (0.75) \cdot (8000 \text{ m}) = 271 \text{ m}.$$

The nickel concentration at ground level can then be estimated by Eq. [4-16]:

$$C = \frac{0.1\ \text{g/sec}}{5.5\ \text{m/sec}} \cdot \frac{\exp(0) \cdot \left(\exp \dfrac{-0.5 \cdot (-120\ \text{m})^2}{(61\ \text{m})^2} + \exp \dfrac{-0.5 \cdot (120\ \text{m})^2}{(61\ \text{m})^2} \right)}{2\pi \cdot (271\ \text{m}) \cdot (61\ \text{m})}$$

$$C = 4.9 \times 10^{-8}\ \text{g/m}^3.$$

One commonly used suite of models that is based on Gaussian plume modeling is the Industrial Source Complex (ISC) Dispersion Models (US EPA, 1995). This suite includes both a short-term model (ISCST), which calculates the hourly air pollutant concentrations in an area surrounding a source, as well as a long-term model (ISCLT), which calculates the average air pollutant concentrations over a year or longer. ISCLT uses meteorological data summarized by frequency for 16 radial sectors (22.5° each); this data format is referred to as a stability array (STAR). Within each sector of STAR, joint frequencies of wind direction, wind speed, and atmospheric stability class are provided.

A specialized case of air pollutant modeling pertains to the transport and dispersion of very dense vapors that sink to the ground surface and are driven in large part by gravity forces. Such situations are uncommon, but may have catastrophic effects when they do occur; examples include release of very cold natural gas from liquified natural gas (LNG) carriers, massive venting of toxic vapors from chemical plants, or release of cold ammonia gas.

In the case of an instantaneous release of a mass of dense vapor, gravity forces cause the vapor mass to slump, or flatten out, at the same time spreading radially outward where it contacts the ground; the effect is not unlike that of pancake batter on a griddle. The horizontal velocity of a vapor mass near the ground due to slumping may exceed wind velocity, causing some of the vapor to move upwind. At the same time, vapor is mixing with air, and in some cases (e.g., LNG spills), the vapor temperature is increasing; both processes result in the density of the vapor cloud decreasing. Eventually the vapor becomes diluted enough with air that its transport is governed by the prevailing wind field and turbulence; until such time, specialized models that include gravity effects and dilution rates must be used to predict the behavior of the dense vapor (e.g., Kaiser, 1979).

4.4.3 Urban-Scale Air Pollution

At the scale of an urban area, covering tens to hundreds of square kilometers, air quality is likely to be affected by chemicals emitted from many locations.

Superposition of plumes, much as was done with perturbations to hydraulic head in Section 3.2.3, is a legitimate way to determine the chemical concentrations resulting from several sources, because in most cases the dispersal of one chemical has little effect on the behavior of other chemicals.

In practice, it is not often feasible to track individual plumes from hundreds of small sources in an urban area. An alternative approach that can yield reasonable city-wide averages is to add the total mass of emissions from all sources and assume that this mass fully mixes into a control volume of air over the city. The approach is similar to the well-mixed box model described for indoor air pollutants; the control volume has inflows and outflows of air, as well as chemical sources and sinks, and mass conservation is employed. The boundaries of the control volume, however, are less well defined than the walls of a building. Application of a box model is most appropriate when the mixing height is well defined and surface winds are light. Hills surrounding an urban area may help define the control volume further. The ventilation rate is less accurately quantifiable than in the case of a building; however, wind speed multiplied by area of the upwind side of the control volume yields an estimate of the air exchange rate. Steady-state conditions may or may not apply, depending on the rate at which meteorological conditions, source strengths, and photochemical reactions change. For further information on air pollution and atmospheric conditions, the reader is referred to Seinfeld (1975, 1986), Flagan and Seinfeld (1988), Dobbins (1979), Lyons and Scott (1990), Masters (1991), and Boubel *et al.* (1994).

EXAMPLE 4-10

During a summertime inversion, the mixing height over an urban area is 800 ft. Winds are calm. Estimate the rise in CO levels during the morning commute period, if 5000 vehicles all travel 1 km/km^2 of urban area. Assume autos travel 10 km/kg of fuel, and combustion products average 1% CO by weight. (Note that the measured concentration of CO in the exhaust gas would be less than 1% due to the presence of nitrogen in the combustion air.) Use octane (C_8H_{16}) to represent the complex hydrocarbon mixture in gasoline and assume the following reaction occurs:

$$C_8H_{16} + 12O_2 \longrightarrow 8CO_2 + 8H_2O + \text{small amount of CO.}$$

To solve this problem, first estimate the mass of CO produced per mass of fuel consumed. Given mass balance constraints, the mass of exhaust products and H_2O produced from combustion of 1 mol of octane must be 496 g (the mass of 1 mol C_8H_{16} and 12 mol O_2). Of this amount, 1%, or 5 g, is the mass of CO produced per 112 g of fuel (1 mol C_8H_{16}) consumed.

Second, estimate the total CO release per square kilometer:

$$5000 \text{ cars} \cdot \frac{1 \text{ km/km}^2}{\text{car}} \cdot \frac{1 \text{ kg fuel}}{10 \text{ km}} \cdot \frac{0.005 \text{ kg CO}}{0.11 \text{ kg fuel}} = 22 \text{ kg CO/km}^2.$$

Next, estimate the volume of air per square kilometer into which the mass of CO can mix:

$$1 \text{ km}^2 \cdot \frac{800 \text{ ft}}{1 \text{ km}^2} \cdot \frac{3.0 \times 10^{-4} \text{ km}}{1 \text{ ft}} = 0.24 \text{ km}^3/\text{km}^2.$$

Thus, the rise in CO levels can be estimated as

$$\Delta\text{CO concentration} = 22 \text{ kg}/0.24 \text{ km}^3 = 91 \text{ kg/km}^3.$$

Given that 1 m^3 of air has a mass of approximately 1.2 kg at 1 atm and 20°C, ΔCO can be expressed in parts per billion on a *mass* basis:

$$\Delta\text{CO} = \frac{91 \text{ kg}}{10^9 \text{ m}^3} \cdot \frac{1 \text{ m}^3}{1.2 \text{ kg}} = 76 \text{ ppb}.$$

4.4.4 Long-Range Transport Models

Some air pollutants are transported far beyond their points of release. For example, otherwise pristine areas have received acid precipitation originating from industrial smokestack emissions hundreds of miles away. Dust from the Sahara Desert in Africa has been detected in South America, and radioactive debris from the Chernobyl nuclear reactor meltdown has been deposited in countries throughout Europe.

In a few exceptional cases, the long-range transport of a discrete cloud of a chemical has been tracked. Figure 4-27, portraying data obtained by a spectrometer on the Nimbus 7 satellite, shows a time sequence of a cloud of sulfur dioxide (SO_2) emitted from the Cerro Hudson volcano in Chile (Doiron *et al.*, 1991). In this case, advection carried the cloud in an easterly direction around Earth, during which time the cloud became larger but more dilute as a result of dispersion and mass deposition onto Earth's surface.

More often in cases of long-range transport, the chemical source is continuous, and the effects of concern are chronic. Consequently, in the process of modeling long-range transport, averaging over long time intervals (often many months) not only is unavoidable, but also is often helpful. Several computer

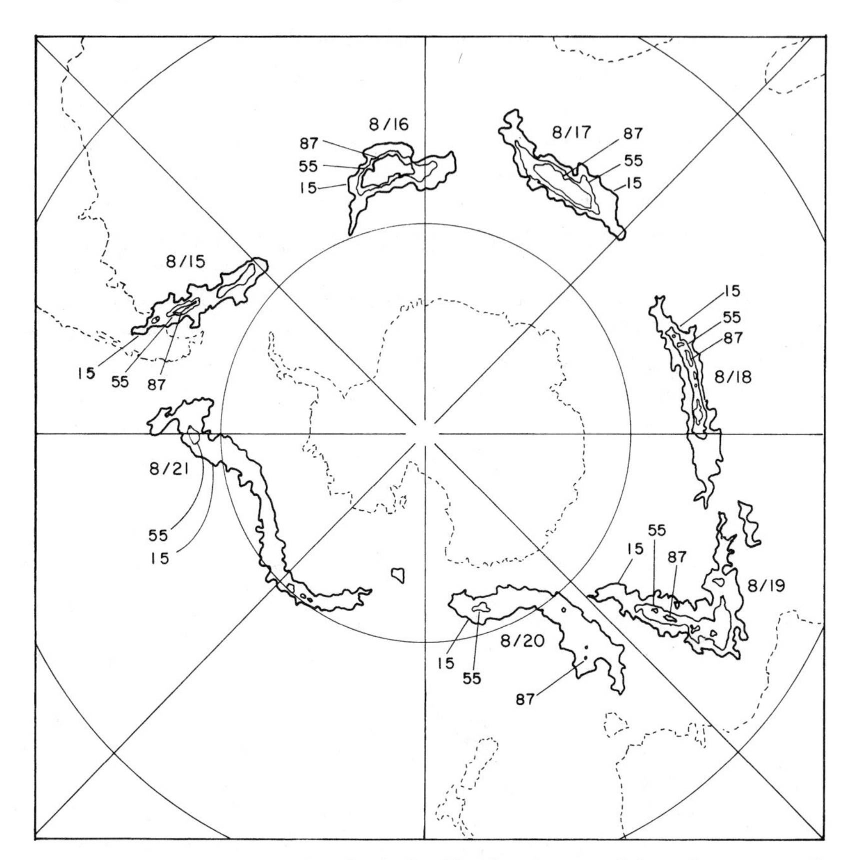

FIGURE 4-27 The transport of a cloud of sulfur dioxide emitted from the Cerro Hudson volcano in southern Chile on August 15, 1991. Images are taken from satellite data at 1-day intervals for 7 days. At this scale, the mixing observed cannot be modeled as purely Fickian (otherwise, mass would become more normally distributed about the center of mass); large-scale variability in the wind field clearly influences the shape of the cloud (adapted from Doiron *et al.*, 1991).

models have been constructed for the purpose of predicting chemical concentrations at locations hundreds of kilometers from the chemical source. Such models combine both advective transport and mixing, with averages taken over large time intervals and areas. Characteristically, such models rely on numerical solution techniques and require substantial computing capability. One example is the Regional Acid Deposition Model (RADM), which was constructed to represent the transport and fate of acidic chemicals over long distances in the United States (NAPAP, 1990). RADM is an Eulerian model, in which a framework of fixed coordinates is used to model fluid processes and estimate chemical concentrations at specific locations. Another Eulerian model is the Acid Deposition and Oxidant Model (ADOM), developed for Canada (Venkatram *et al.*, 1988). Both RADM and ADOM include three-dimensional transport, wet and dry deposition (discussed in Section 4.5), and gas-phase chemistry.

Other regional transport models, such as the Regional Lagrangian Model of Air Pollution (RELMAP; Eder *et al.*, 1986), use a different computational scheme than Eulerian models. In a Lagrangian model, the coordinate system moves with a parcel of air and mass balance of pollutant concentrations is computed on a parcel as it moves through space.

The predictions of such models often can be expressed graphically; an example is Fig. 4-28, which predicts the concentration of airborne sulfate

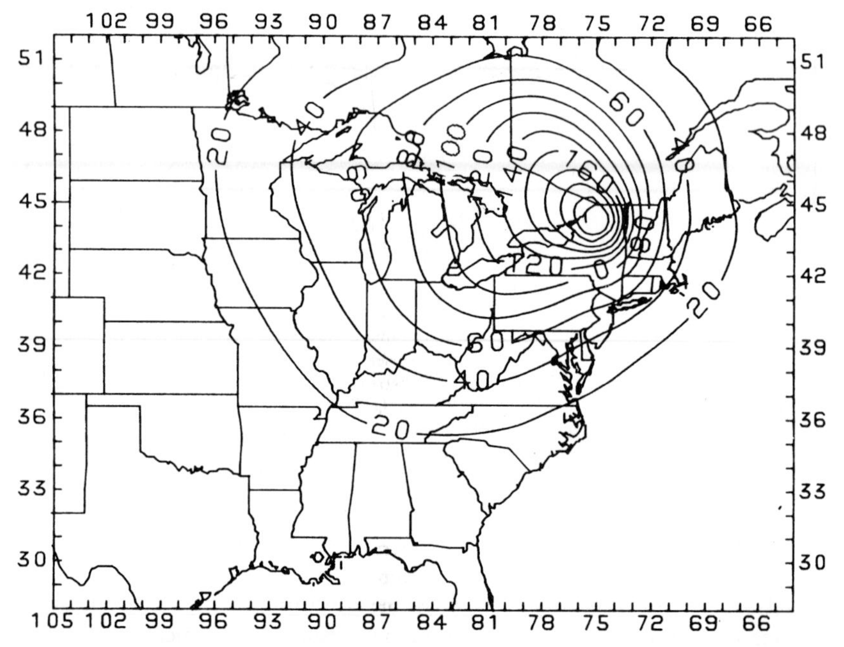

FIGURE 4-28 An example of output from ACID, a regional-scale model that simulates transport of SO_2 and sulfate, oxidation of SO_2 to sulfate, and sulfate deposition. Each contour represents the average airborne sulfate concentration in micrograms per cubic meter that would result in the Adirondacks region of New York per 10^{14} g of sulfur emitted annually anywhere along the contour line. For example, if a 10^{14} g/year source of SO_2 were sited in Tennessee, the resultant average addition to the airborne sulfate concentration in the Adirondacks would be 20 $\mu g/m^3$.

(SO_4^{2-}) in the Adirondacks region of New York State as the result of the siting of a 10^{14} g/year source of sulfur dioxide in the eastern United States or Canada. This graph is produced by a regional-scale model called ACID that simulates advective and turbulent transport of sulfur-containing chemicals in the atmosphere, as well as the deposition of sulfate from the atmosphere to the ground surface, as discussed further in Section 4.5.

4.4.5 Global-Scale Transport of Chemicals

The global presence of radioactive fallout from the nuclear bomb tests of the 1950s and 1960s clearly illustrated the capacity of the atmosphere to distribute chemicals around Earth. Global transport has been demonstrated for a large number of chemicals that do not rapidly degrade or settle out of the atmosphere. Examples are methane (CH_4), nitrous oxide (N_2O), and chlorofluorocarbons (CFCs), all of which have half-lives in the atmosphere ranging from years to decades.

Chemicals become well mixed throughout large portions of the troposphere in a matter of weeks. Exchange of atmospheric chemicals across the equator, however, is relatively slow compared with the mixing time within either hemisphere alone. Thus, for some purposes, the troposphere may be thought of as containing two boxes, the Northern and Southern Hemispheres. Within each box, chemicals mix in a few weeks; to mix between them, however, requires 1 to 2 years.

Transport of chemicals between the troposphere and the stratosphere is an even slower process. On average, nonreactive molecules reside in the troposphere for 20 to 30 years before entering the stratosphere. Therefore, only those few chemicals that are highly stable under tropospheric conditions enter the stratosphere in significant amounts. Although many details of the transport are still a subject of research, the major mechanism of troposphere-to-stratosphere transport appears to be upwelling across the tropopause in the tropics (Fig. 4-29). Although powerful convective storms can penetrate to the lower stratosphere in the tropics, they are evidently not the major mechanism of the upwelling; instead, air is pulled upward over the entire tropics by a wave-driven mechanism in the stratosphere that produces poleward pumping of air. This pumping is accompanied by large-scale subsidence of air in the polar regions, where the tropopause is lower and more variable over time. Much of the mixing of stratospheric air back into the troposphere appears to occur episodically in the midlatitudes, where horizontal transport across the tropopause can occur by several processes (see Fig. 4-29). An excellent review of the topic is given by Holton *et al.* (1995).

The average residence time of an air parcel in the upper stratosphere (above approximately 16 km) is approximately 2 years. The stratosphere should not be regarded as fully mixed, however. Upwelled layers of air, their exact chemical composition changing with season, have been shown to retain an identity for many months. Another example of stratospheric heterogeneity is the air in polar stratospheric vortices (see Section 4.6.4, Fig. 4-41 later); these vortices may have a trace gas chemistry that differs from that of stratospheric air in the tropics.

Whether a chemical in the stratosphere is predominantly degraded or is mixed back into the troposphere depends on whether its stratospheric resi-

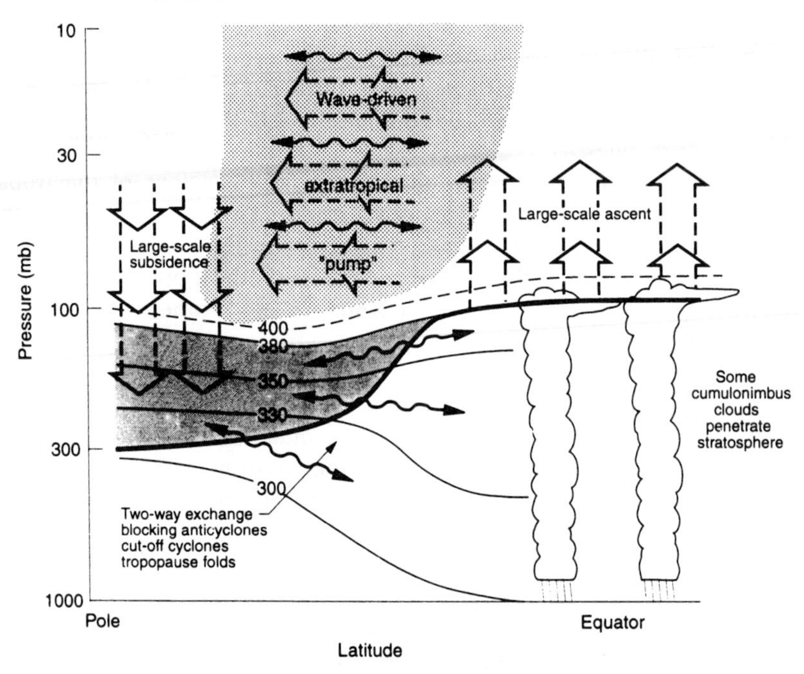

FIGURE 4-29 Schematic of troposphere–stratosphere air exchange. Although some very powerful cumulonimbus clouds (thunderstorms) may penetrate into the stratosphere in tropical regions, the major mechanism of air exchange is driven by the poleward pumping of air in the stratosphere. The result is a slow net upward flow of air from the troposphere into the stratosphere over the tropics, and a subsidence of air from the stratosphere into the troposphere over the poles. (J. R. Holton, P. H. Haynes, M. E. McIntyre, A. R. Douglass, R. B. Rood, and L. Pfister, *Reviews of Geophysics*, 33(4):403–439, 1995, copyright by the American Geophysical Union.)

dence time is longer than the average time required for its degradation. A few chemicals, such as some CFCs, may exchange between the troposphere and stratosphere several times before being fully degraded. Numerous light-driven reactions and physical removal processes occur in both the troposphere and the stratosphere, and models for atmospheric chemical behavior must include chemical reaction rates along with physical transport equations, described further in the next sections.

4.5 PHYSICAL REMOVAL OF CHEMICALS FROM THE ATMOSPHERE

A chemical may be removed from the atmosphere either by physical processes or by chemical transformation. If a chemical exists in the atmosphere in the form of sufficiently large particles or liquid droplets, physical deposition of the chemical to soil, vegetation, or water bodies may occur by gravitational settling. Smaller particles with negligible settling velocities also may be deposited to surfaces by impaction or diffusion. Chemicals in gaseous form may either sorb onto surfaces directly or sorb onto airborne particles that are subsequently removed.

Many of these processes are analogous to processes that deposit aquatic chemicals to the sediments of a lake or a river. Chemical transformation in the atmosphere typically results in the production of more oxidized species, and the complete mineralization of many organic chemicals is possible. In this section, physical processes that result in the removal of chemicals from the atmosphere without changes to their chemical structure are discussed. In Section 4.6, chemical transformation processes are presented.

4.5.1 DRY DEPOSITION

Dry deposition refers to any physical removal process that does not involve precipitation. Three dry deposition mechanisms are discussed next: gravitational settling, impaction, and absorption.

Gravitational Settling

For particulate atmospheric chemicals, especially those larger than 1 μm in radius, gravitational settling is a significant removal mechanism. The rate of gravitational settling of solid particles through a fluid can be estimated using Stokes' law, which expresses settling velocity as a function of the size and density of the particles and the viscosity of the fluid. (Recall that Stokes' law was presented for particles settling in surface waters in Eq. [2-20].) The two major differences between the settling of particles in water and the settling of particles in air are (1) air viscosity is much lower than water viscosity, and (2) air density is usually a negligible fraction of particle density, whereas water density is often a significant fraction of particle density. Stokes' law is applicable to particles that are spherical, under conditions of *viscous* fluid flow at the scale of the particle (i.e., the particle does not create turbulence, and the rate at which a particle settles depends on the rate at which fluid near the particle is deformed by the passage of the particle). Stokes' law typically is applicable for particles having radii less than approximately 100 μm, but is demonstrably invalid for large and fast-falling objects, such as bricks, artillery shells, or hailstones.

As a minor variant on Eq. [2-20], Stokes' law can also be expressed as

$$w_f = \frac{(2/9)g \cdot r^2 \cdot \Delta\rho}{\mu_f}, \qquad [4\text{-}23]$$

where w_f is the settling velocity [L/T], g is the acceleration due to gravity [L/T^2], r is the particle radius [L], $\Delta\rho$ is the difference between the particle density and the fluid density [M/L^3], and μ_f is the dynamic viscosity of the fluid [M/LT].

Dynamic viscosity also is called the coefficient of viscosity. Note that dynamic viscosity divided by density gives the kinematic viscosity which is used in Eq. [2-20]. For most gases, the dynamic viscosity increases with increasing temperature. At 18°C, the dynamic viscosity of air is 1.83×10^{-4} g/(cm · sec), or *poise*. If particle radius is expressed in micrometers, values for g and μ_f are substituted, and $\Delta\rho$ is set to 1 g/cm^3, Stokes' law reduces to a convenient rule of thumb for air:

$$w_f \approx 10^{-2}\, r^2 \text{ (cm/sec)}. \qquad [4\text{-}24]$$

Note that there is no consistency of dimensions or units in this equation as written; the user must remember the proper units.

For rapidly settling particles between approximately 10 and 100 μm in radius, Stokes' law often is applied to correct the predictions of Gaussian plume models for settling; the plume centerline is assumed to tilt downward from the horizontal at an angle whose value is given by the arc tangent of the settling velocity divided by the wind speed.

For particles having densities of approximately 1 g/cm^3 and radii less than 1 μm, gravitational settling often may be neglected, because it is usually slower than other removal processes; the Stokes' settling velocity for such particles is on the order of 10 m/day or less. Although such small particles may not settle directly, they may be incorporated into larger particles by collision and *coagulation* (sticking together). Small particles also may be removed by impaction or diffusive transport, as discussed later. Alternatively, small particles may be incorporated into raindrops, hailstones, or other forms of precipitation and settle out as wet deposition (Section 4.5.2).

EXAMPLE 4-11

Unintentional venting of an underground nuclear bomb test results in deposition of particles into the atmosphere at a height of 1.5 km over the site. If these soil particles have a density of 2.3 g/cm^3, and the wind speed is 1.5 m/sec, what would be the size of the particles that can be expected to fall predominantly *outside* the border of the test site, which is 200 km downwind?

First, calculate the travel time to the test site border, based on the wind speed:

$$\frac{200{,}000 \text{ m}}{1.5 \text{ m/sec}} \approx 1.3 \times 10^5 \text{ sec.}$$

Second, calculate the maximum settling velocity particles can have if they are to land on or outside the border of the test site:

$$\frac{150{,}000 \text{ cm}}{1.3 \times 10^5 \text{ sec}} = 1.1 \text{ cm/sec.}$$

Third, use Stokes' law in Eq. [4-23] to estimate particle radius, making the assumption that $\Delta\rho \approx 2.3$ g/cm^3, because $\rho_{air} \approx 10^{-3}$ g/cm^3, which is much smaller than $\rho_{particle}$:

$$1.1 \text{ cm/sec} = \frac{(2/9)(981 \text{ cm/sec}^2)(r^2)(2.3 \text{ g/cm}^3)}{1.8 \times 10^{-4} \text{ g/cm} \cdot \text{sec}}$$

$$r = 6.4 \times 10^{-4} \text{ cm} = 6.4\ \mu\text{m.}$$

Therefore, particles with radii smaller than approximately 6 μm may be expected to fall outside the test site.

Note that if the rule of thumb of Eq. [4-24] is used, a radius of 11 μm is estimated. This estimate varies by approximately a factor of 2 from the full solution because $\Delta\rho$ is approximated in the rule of thumb as 1 g/cm^3, whereas in this example, $\Delta\rho$ is 2.3 g/cm.^{3}

FIGURE 4-30 Particle deposition by impaction. In this example, airflow is from front to back past the car. A particle following streamline A, which curves gently, may be deflected sufficiently to avoid contact with the obstruction. Particles with sufficient mass and density that travel along streamline B, which has very sharp turns, may be carried to the surface of the object by their own momentum. Although the process is illustrated for a large object (the car) and a centimeter-sized particle (the moth), the same phenomenon also occurs for much smaller particles and objects.

Impaction

A second physical mechanism for dry deposition is *impaction* (Fig. 4-30). Impaction occurs when air containing particles moves past stationary objects, such as vegetation or buildings, and some of the airborne particles collide with the objects and stick. As moving air is deflected around an obstruction, some particles in the fluid also are deflected; the force required to deflect the particles is transmitted via the fluid. For particles having larger mass, or trajectories that must follow sharp turns of small radii to avoid contact with the object, fluid forces may be inadequate to provide the necessary deflection, and the particles contact the object. Contact is most likely for particles that pass through the *stagnation point,* a point in front of the object at which fluid velocity is zero.

Particle removal by impaction is evident in many common situations: one example is the accumulation of dirt on the leading edges of blades of an electric room fan. The dustier the house and the less frequently the fan is cleaned, the more pronounced this buildup will be, up to some maximum thickness where the mechanical strength of the dirt layer becomes insufficient to resist aerodynamic forces. Other examples of impaction on a larger scale are crushed insects on automobile windshields and headlights (see Fig. 4-30) and the accumulation of snow on the upwind side of a tree trunk during a snowstorm.

Particles also may stick to solid surfaces after diffusing through a stagnant air boundary layer above the solid. Diffusive transport results from the random motion of particles, often called Brownian motion (see Section 2.2.5, Fig. 2-11). This process may occur for particles that are too small to deposit effectively by impaction.

Accurate estimates of particle deposition are difficult to obtain because of the complexity of natural surfaces and the diversity of atmospheric particles. The long-term deposition of particles on vegetative surfaces often is estimated using a deposition velocity model in which the flux density of particles is calculated as the product of a modeled *deposition velocity* and a measured concentration of particles in the air (Hicks *et al.*, 1987). A deposition velocity is used in a manner similar to a piston velocity in air–water gas exchange (Section 2.3). In a deposition velocity model, the flux density of atmospheric particles to the surface is

$$J = V_d \cdot C_a, \qquad [4\text{-}25]$$

where J is the flux density of particles to the surface [M/L^{2}T], V_d is the deposition velocity [L/T], and C_a is the particle concentration in the air [M/L^3]. Deposition velocities are typically in the range of 0.1 to 1 cm/sec for many atmospheric particles.

EXAMPLE 4-12

Measured sulfate aerosol concentration over a forest averages 10 μg/m^3. Estimate dry deposition of sulfate to the forest if the deposition velocity is 1 cm/sec.

First, convert the sulfate concentration to units of grams per cubic centimeter:

$$\frac{10\ \mu\text{g}}{\text{m}^3} \cdot \frac{1\ \text{m}^3}{10^6\ \text{cm}^3} \cdot \frac{1\ \text{g}}{10^6\ \mu\text{g}} = 10^{-11}\ \text{g/cm}^3.$$

Then use Eq. [4-25]:

$$J = 1\ \text{cm/sec} \cdot 10^{-11}\ \text{g/cm}^3 = 10^{-11}\ \text{g/(cm}^2 \cdot \text{sec)},$$

or

$$J \approx 10^{-6}\ \text{g/(cm}^2 \cdot \text{day)} \approx 3\ \text{g/(m}^2 \cdot \text{year)}.$$

This deposition rate could represent a significant source of acidity to some forests.

Absorption of Gases

A third physical mechanism for dry deposition is absorption. Many atmospheric gases are absorbed, either by liquid surfaces (as discussed in Section 2.3) or by solid surfaces such as soil and vegetation. The absorption of atmospheric gases by solid surfaces is usually described by a thin film model analogous to that used for air–water transfer. Figure 4-31 illustrates the surface of an absorbing material, with an adjacent stagnant boundary layer of air. By assuming that any gas that diffuses through the stagnant air boundary layer and contacts the surface is absorbed, the flux of vapor into the surface is given by

$$J = (C_a - C_s) \cdot D/\delta, \quad [4\text{-}26]$$

where J is the flux density of vapor [M/L^{2}T], C_a is the vapor concentration in the bulk air [M/L^3], C_s is the vapor concentration in the air at a solid surface [M/L^3], D is the diffusion coefficient in air [L^2/T], and δ is the thickness of the stagnant air boundary layer [L]. The quantity D/δ is again a piston velocity, conceptually identical to the piston velocity described in Section 2.3. As in the case of impaction, the piston velocity describing the deposition of a chemical from the air into a liquid or solid surface by absorption is usually called a *deposition velocity*.

The deposition velocity V_d for vegetative surfaces is commonly estimated by an empirical modeling approach (often called an inferential modeling

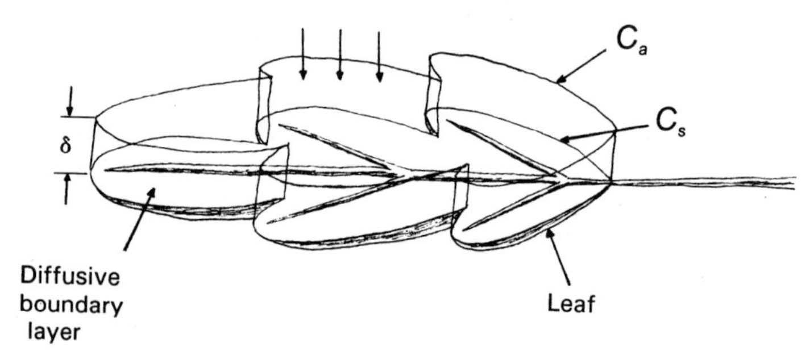

FIGURE 4-31 Dry deposition by absorption. C_a and C_s are chemical concentrations in bulk air and at the leaf surface, respectively. The chemical flux is determined in part by the thickness of the stagnant air boundary layer (shown for simplicity as having uniform thickness). This model is essentially the same as that applied to gas exchange between air and water for the case in which transport resistance is dominated by the air side (Section 2.3).

approach), which uses meteorological data and information on the surface characteristics of the vegetation (Brook *et al.*, 1997; Hicks *et al.*, 1987). In the ADOM model, for example, the dry deposition velocity to vegetation is estimated as the inverse of the sum of three resistances: aerodynamic, laminar-layer or diffusive, and canopy. The aerodynamic resistance is estimated based on the intensity of atmospheric turbulence, which depends on both mechanical mixing and buoyancy, and therefore is a function of atmospheric stability. The laminar-layer resistance depends on the thickness of the laminar layer and on the molecular diffusivity of the species of interest. The canopy resistance is estimated from the ratio of leaf area to ground surface area (known as the leaf area index, LAI) and from the stomatal resistance of the leaves. Because the stomata open and close to regulate water potential, there is a diurnal fluctuation in V_d, as shown in Figure 4-32. Other models take into account components of the leaf resistance due to diffusion across the covering of the leaf (cuticle) and diffusion through the mesophyll tissue in the leaf interior, as well as diffusion through the stomata. Figure 4-32 shows estimates of the diurnal fluctuation of V_d for ozone from four models, as well as the measured V_d.

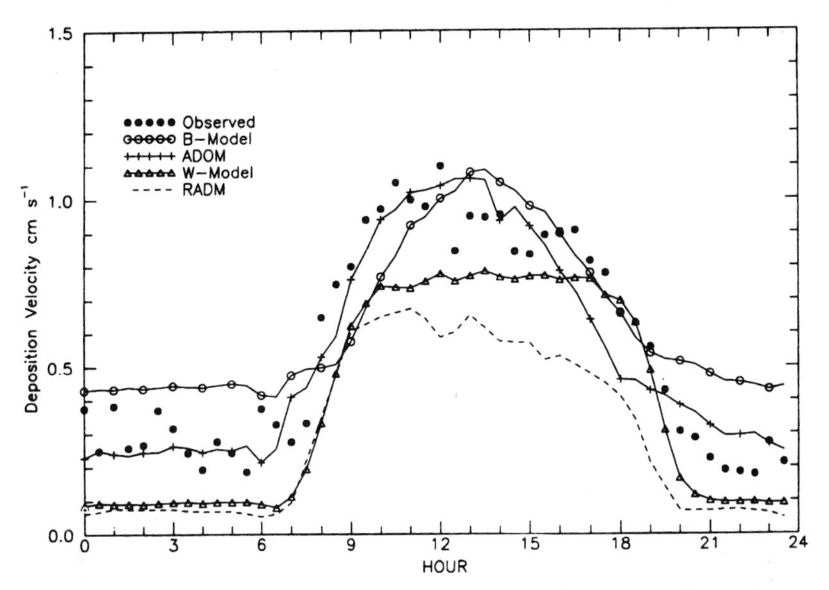

FIGURE 4-32 Diurnal cycle of V_d for ozone for a deciduous forest, averaged over the period July 7 to August 30, 1988. Observed values of V_d and values estimated by four models are shown. (Reprinted from *Atmospheric Environment,* Vol. 30, L. Zhang, J. Padro, and J. L. Walmsley, "A Multi-Layer Model vs. Single-Layer Models and Observed O_3 Dry Deposition Velocities," pp. 339–345, Copyright 1996, with permission from Elsevier Science.)

Despite the different mechanisms involved, deposition velocities for absorption often are found to be similar in magnitude to values for impaction (i.e., approximately 0.1 to 1 cm/sec for gases whose absorption into a surface is limited by the rate of Fickian transport through a stagnant boundary layer (Table 4–9). For additional information on the estimation of dry deposition velocities, the reader is referred to Brook *et al.* (1996, 1997), Hicks *et al.* (1987), Wesely (1989), and Bennett (1988).

4.5.2 Wet Deposition

Wet deposition refers to processes in which atmospheric chemicals are accumulated in rain, snow, or fog droplets and are subsequently deposited onto Earth's surface. Wet deposition removes from the atmosphere many chemicals, including gases, whose rates of gravitational settling, impaction, or absorption are slow or even zero. When incorporation of chemicals into water droplets occurs within a cloud, the process is called *rainout*. When incorporation occurs beneath a cloud, as precipitation falls through the air toward Earth's surface, the process is called *washout*.

Wet Deposition of Gases and Vapors

Gases and vapors in the atmosphere can be removed by dissolving into raindrops. At equilibrium, the chemical concentration in the raindrops is

TABLE 4-9 Dry Deposition Velocities for Several Gases

Chemical	Mean deposition velocity (cm/sec)	Site
Ammonia (NH_3)[a]	2	Forest
Nitric acid (HNO_3)[b]	0.8–1.5	Rural and metropolitan sites
Ozone (O_3)[c]	0.002–2	Vegetative surfaces
Peroxyacetyl nitrate (PAN)[d]	0.5	Cornfield at night
Sulfur dioxide (SO_2)[b]	0.3–0.4	Rural and metropolitan sites

[a] Andersen and Hovmand (1995).
[b] Pratt *et al.* (1996).
[c] Padro (1996).
[d] Schrimpf *et al.* (1996).

given by Henry's law (see Section 1.8.2):

$$C_{water} = C_{air}/H, \quad [4\text{-}27]$$

where C_{water} is the concentration of the chemical dissolved in the raindrops [M/L^3], C_{air} is the concentration of the chemical in the air [M/L^3], and H is the Henry's law constant (dimensionless). Chemical equilibrium may generally be assumed if the rain forms in contact with the air (rainout) or travels through several tens of meters in contact with an air mass (washout). In the context of wet deposition, the ratio C_{water}/C_{air}, which is equal to the reciprocal of the dimensionless Henry's law constant, is sometimes called the *washout ratio*, W_r. The resulting flux density of dissolved gas or vapor, J, associated with the falling rain is thus,

$$J = C_{water} \cdot I = C_{air} \cdot \frac{I}{H} = C_{air} \cdot W_r \cdot I, \quad [4\text{-}28]$$

where I is the rainfall rate [L/T] and W_r is the washout ratio (dimensionless). Note that the quantity I/H has dimensions of [L/T] and may be regarded as a deposition velocity (i.e., the flux density is equal to the gas concentration in the air multiplied by I/H).

Table 4-10 presents values for washout ratios for many common atmospheric gases.

In the case of washout of sulfur dioxide (SO_2), a precursor of acid rain, the high solubility and the chemical reactivity of aqueous SO_2 result in nonattainment of equilibrium. Thus, semiempirical models have been proposed for

TABLE 4-10 Washout Ratios for Several Chemicals[a]

Gas or vapor	Washout ratio
Carbon monoxide (CO)	0.027
Methane (CH_4)	0.039
Nitrous oxide (N_2O)	0.79
Carbon dioxide (CO_2)	1.1
Sulfur dioxide (SO_2)	130
Ammonia (NH_3)	2100
Mercury (Hg)	2.1
Aroclor 1260	3.4
Aroclor 1254	9.1
DDT	620
Aldrin	1900
Dieldrin	130,000

[a] Adapted from Slinn *et al.* (1978).

the SO_2 concentration in air beneath the rain-forming cloud. A first-order decay constant for the SO_2 concentration, λ, varies with the rainfall characteristics (rainfall rate and size of raindrops). Boubel *et al.* (1994) suggest a value of λ equal to

$$\lambda = 10^{-4} \cdot I^{0.53}, \qquad [4\text{-}29]$$

where λ is the *scavenging* or *washout coefficient* (per sec) and I is the rainfall rate in millimeters per hour (mm/hr). Values of λ for SO_2 have been observed in the range of 0.4×10^{-5} to 6×10^{-5}/sec (Boubel *et al.*, 1994). (Note that this is an empirical equation, and consistency of units does not hold; the user must use the units given.) For a further discussion of washout processes and how they are modeled, the reader is referred to NAPAP (1990).

EXAMPLE 4-13

Estimate the rate (per square meter of area) of wet deposition of sulfur by washout of airborne SO_2 to a watershed. Assume that the SO_2 concentration is 20 μg/m^3 (as sulfur) at the beginning of each rainstorm and that the watershed receives 1 m of precipitation per year, in 50 equals storms of 10-hr duration each, from clouds 2000 ft above the ground.

First, estimate a washout coefficient λ for SO_2 by using Eq. [4-29]:

$$\lambda = 10^{-4} \cdot I^{0.53} \text{ (per sec)}.$$

Note that this purely empirical expression for λ necessarily takes into account the net effect of *all* processes affecting SO_2 removal from air (i.e., dissolution into water droplets, hydration, oxidation, and ionization). Given that I must be expressed in millimeters per hour in Eq. [4-29]:

$$I = \frac{2 \text{ cm}}{\text{storm}} \cdot \frac{1 \text{ storm}}{10 \text{ hr}} = \frac{0.2 \text{ cm}}{\text{hr}} = \frac{2 \text{ mm}}{\text{hr}}.$$

The washout coefficient then can be estimated as

$$\lambda = (10^{-4})\left(\frac{2 \text{ mm}}{\text{hr}}\right)^{0.53} = 1.4 \times 10^{-4}/\text{sec}.$$

Given that washout is approximated in this model as a first-order removal process, at the end of each storm the concentration of SO_2 in the air can be estimated using a first-order decay equation:

$$C_{air} = \left(\frac{20\ \mu g}{m^3}\right) e^{(-1.4\times10^{-4}/sec)\cdot(10hr)\cdot(3600\ sec/hr)}$$

$$C_{air} = \frac{0.11\ \mu g}{m^3}.$$

Therefore, the amount of SO_2 washed out and deposited by rain on the watershed is

$$20\,\frac{\mu g}{m^3} - 0.11\,\frac{\mu g}{m^3} = 19.9\ \mu g/m^3.$$

An estimate of the annual wet deposition of sulfur to the watershed thus is

$$\left(\frac{19.9\ \mu g}{m^3\cdot storm}\right)\left(\frac{50\ storms}{year}\right)\left(\frac{2000\ ft}{1}\right)\left(\frac{1\ m}{3.28\ ft}\right)\left(\frac{1\ g}{10^6\ \mu g}\right) = \frac{0.61\ g}{m^2\cdot year}$$

This idealized problem does not reflect the variability of the storms and the air found over an actual watershed. It does show, however, that the expected rate of wet deposition of sulfur is significant.

Wet Deposition of Particles

Particulate chemicals also may be removed from the atmosphere through wet deposition processes. The major mechanism by which particles are incorporated into precipitation is by serving as *nucleation sites* for condensation at the onset of water droplet or ice crystal formation. [If nucleation sites were absent, water droplets would not form unless the air temperature were significantly below the dew point. In the rainmaking practice of cloud seeding, condensation nuclei in the form of silver iodide (AgI) crystals are dispersed into air masses to increase the formation of ice crystals, which in turn accumulate additional water from the cloud and thus promote precipitation.] Particles also can be incorporated into already-formed water droplets within a cloud by collision.

In theory, the removal of fine particles by collision processes results in chemical concentrations in the air mass following an exponential decay law. In actual clouds, both collision and nucleation may contribute to particle removal, and their combined effects often are lumped into a single scavenging coefficient, $\lambda\,[T^{-1}]$:

$$\frac{dC_{air}}{dt} = -\lambda \cdot C_{air}. \qquad [4\text{-}30a]$$

Thus,

$$C_{air,t} = C_{air,0} \cdot e^{-\lambda t}, \quad [4\text{-}30b]$$

where $C_{air,0}$ is the initial chemical concentration in the air [M/L^3] and $C_{air,t}$ is the chemical concentration [M/L^3] after some time t. λ is analogous in its use to the scavenging coefficient discussed previously for SO_2 washout. Although the λs are used in the same fashion, note that λ for particle removal arises from nucleation and collision processes instead of dissolution of SO_2. λ is usually poorly known because it depends on particle size distributions and cloud dynamics that are difficult to measure in any given situation. Typical values of λ range from 0.4×10^{-5} to 3×10^{-3}/sec (Yamartino, 1985).

Removal of particles by rainout is far more effective than dry deposition of particles, as illustrated in Example 4-14.

EXAMPLE 4-14

Assume the dry deposition velocity over a New England forest for particles emanating from a midwestern U.S. power plant averages 0.5 cm/sec on a certain day. The scavenging coefficient in a rain-forming cloud over the forest is 10^{-3}/sec. If the cloud is 500 m in vertical extent, what is the equivalent "wet deposition velocity," V_w, for rainout of the particles?

The wet deposition flux density (flux per square meter of ground area) is equal to the mass flux leaving each square meter of cloud area:

$$J = (500 \text{ m}) \cdot (10^{-3}/\text{sec} \cdot C_{air}) = 0.5 \cdot C_{air} \text{ m/sec}$$

$$V_w = \frac{J}{C_{air}} = 0.5 \text{ m/sec} = 50 \text{ cm/sec}.$$

In this situation, the wet deposition velocity is 100 times faster than the dry deposition velocity.

Particles, unlike gases, do not partition between air and water in a consistent and thermodynamically predictable way. Nevertheless, some workers have empirically constructed particle washout ratios analogous to those used for wet deposition of gases and vapors, as shown in Eq. [4-28]. These particle washout ratios are calculated as the ratio of the measured particle concentration in near-surface precipitation to the average measured concentration of particles in near-surface air. Examples of such washout ratios for fine atmospheric particles are shown in Table 4-11.

TABLE 4-11 Measured Particle Washout Ratios[a]

Particle (diameter)	Washout ratio
Radioactive fallout	$(1.0 \pm 0.3) \times 10^6$
Inorganic dust	3.2×10^6
Pollen ($\sim$15–50 μm)	$(0.65–3.8) \times 10^6$
Lead ($\sim$0.5 μm)	0.063×10^6
Zinc ($\sim$1 μm)	0.15×10^6
Magnesium (5.7 μm)	0.38×10^6
Chloride	1.08×10^6
Sodium	0.98×10^6
Arsenic	0.22×10^6

[a]Slinn *et al.* (1978).

The amount of chemical mass removed from the atmosphere by rain and snow is more easily measured, by collecting and analyzing precipitation, than is the mass removed by dry deposition. Figure 4-33 shows an apparatus widely used to measure the deposition of atmospheric acids in North America. When a sensor detects the presence of precipitation, a motor moves the small roof away from the bucket, allowing rain or snow to collect in the wet bucket. If the movable roof is omitted, the collector is said to be a *bulk collector*.

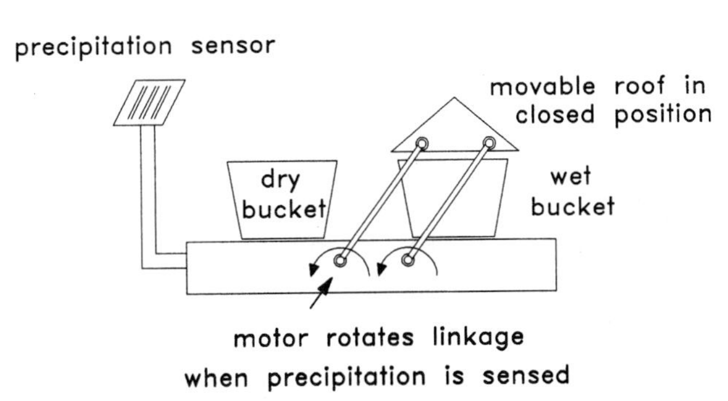

FIGURE 4-33 A wet–dry atmospheric deposition collector commonly used for acid deposition monitoring. The roof over the wet bucket opens only when precipitation is sensed, thereby minimizing the collection of debris, bird droppings, etc. The dry bucket for collecting dry deposition is not a very close approximation to natural surfaces and yields results of uncertain meaning.

4.6 ATMOSPHERIC CHEMICAL REACTIONS

In addition to being removed from the atmosphere by physical processes, atmospheric chemicals can be removed by chemical transformations. Chemical transformations also can be *sources* of atmospheric pollutants; a notorious example is the production of urban smog by reactions involving hydrocarbons, nitrogen oxides, and oxygen.

Most atmospheric chemical reactions are driven either directly or indirectly by sunlight. Many direct photochemical reactions in the atmosphere parallel those that occur in illuminated surface waters (see Section 2.7.1); many of the reactive light-produced chemical species, such as the hydroxyl radical ($OH\cdot$), are the same. The *relative* importance of photochemistry in the atmosphere is greater than in surface waters, however, both because there are consistently higher levels of illumination in the atmosphere and because many of the competing transformation processes, such as those that are biologically mediated, are minor in the atmosphere. Furthermore, at higher altitudes there is a significant flux of radiation at the energetic shorter wavelengths below 290 nm. Most air pollutants are oxidized in atmospheric chemical transformations, as would be expected in a medium containing 20% oxygen; this contrasts with surface waters and groundwater, which, in addition to oxidizing zones, often have anoxic, reducing zones (e.g., in the lower layers of groundwaters and stratified surface waters and in bottom sediments).

In addition to chemical reactions brought about directly by light energy, there are other atmospheric chemical reactions that can occur in the dark, but that typically involve reactive chemical species previously produced photochemically (cf. indirect photolysis in surface waters). The rates of such *thermal*, or *dark*, reactions depend on the temperature and on the concentrations of the reactive chemicals involved. For example, consider the rate of the dark reaction between the photochemically produced compounds ozone (O_3) and nitric oxide (NO). The rate is proportional to the concentration of each reactant

$$O_3 + NO \longrightarrow NO_2 + O_2 \qquad [4\text{-}31a]$$

$$\frac{d[O_3]}{dt} = \frac{d[NO]}{dt} = -[O_3] \cdot [NO] \cdot k, \qquad [4\text{-}31b]$$

where k is a temperature-dependent second-order rate constant equal to 24/(ppm · min) (Boubel *et al.*, 1994). Note that Eq. [4-31b] is of the same general form as the equations used to describe the kinetics of chemical reactions in water (i.e., as in Section 2.7.2, a rate of change in concentration is equal to the product of the reactant concentrations and an intrinsic rate constant).

Water droplets and particulate matter often influence the rates of chemical transformations in the atmosphere. Whereas *homogeneous* reactions involve only gaseous chemical species in the atmosphere, reactions involving a liquid phase or a solid surface in conjunction with the gas phase are called *heterogeneous* reactions. Reactions that occur much more rapidly in water than in air may occur primarily in droplets, even though the droplets constitute only a small fraction of the total atmospheric volume. Solid surfaces also can catalyze reactions that would otherwise occur at negligible rates; specific examples are discussed in the following sections on acid deposition and stratospheric ozone chemistry.

4.6.1 Oxidants in the Troposphere

Although the atmosphere contains approximately 20% oxygen, individual oxygen molecules usually do not react directly with anthropogenic chemicals at ordinary atmospheric temperatures. Instead, photochemical reactions often incorporate oxygen into highly reactive intermediate species, which in turn oxidize other chemicals by donating oxygen to them (Fig. 4-34).

Ozone (O_3) is one of the best known atmospheric oxidants; it exists in both the troposphere and the stratosphere, but plays very different roles in each. At ground level, O_3 is associated with lung irritation and plant damage; it has a major deleterious impact on agriculture. Ozone also causes degradation of many materials, such as rubber windshield wiper blades, which harden and crack rapidly in polluted urban air. In the stratosphere, however, ozone is beneficial; it blocks harmful amounts of ultraviolet radiation from reaching Earth's surface. In the troposphere, ozone is produced mainly by a photochemical cycle involving oxides of nitrogen, as discussed in the next section.

Although ozone is quite reactive, most photochemical oxidations in the atmosphere (as in surface waters) involve still more reactive *free radicals*. Free radicals are species that contain an unpaired electron; therefore, the reaction of any free radical with a chemical other than another free radical still leaves an electron unpaired. The newly formed free radical may readily react with another chemical, forming yet another free radical, and so on. An example of such a *free-radical chain reaction* is a combustion flame. A free radical is destroyed only when reaction with another free radical causes the two unpaired electrons to pair with each other.

Examples of free radicals in the atmosphere include the hydroxyl radical ($OH\cdot$), hydroperoxy radical ($HO_2\cdot$), and alkoxy and alkyl peroxy radicals ($RO\cdot$ and $RO_2\cdot$, respectively, where R is an alkyl group). The most important of these free radicals in the oxidation of atmospheric chemicals is the hydroxyl

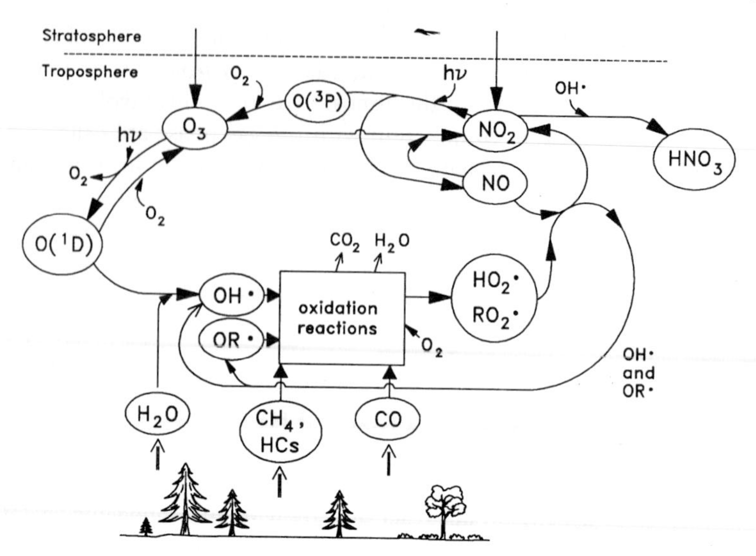

FIGURE 4-34 A simplified diagram of the tropospheric chemical system. One of the net effects of this system is the oxidation of reduced gases (methane and hydrocarbons) emitted by biota and human activities. Some of the oxidized species such as nitrate (in the form of nitric acid, HNO_3) may be deposited to the surface. Note the O_3–NO_x cycle in the top of the diagram (this is later shown in more detail in Fig. 4-35) and the routes for formation of the highly reactive OH· radical. [The atomic oxygen species $O(^1D)$ and $O(^3P)$ are singlet and triplet, respectively (see Section 2.7.1), with D and P referring to the angular momentum of the species.]

radical. Over the rural United States, a typical summer daytime average hydroxyl radical concentration in the convectively mixed layer under clear sky conditions is approximately 3×10^6 molecules per cubic centimeter (NAPAP, 1990). Globally averaged hydroxyl radical concentrations have been estimated to range between 0.6×10^6 and 0.8×10^6 molecules per cubic centimeter (NAPAP, 1990). The hydroxyl radical can form by the photodissociation of ozone at light wavelengths less than 315 nm and the subsequent reaction of the resultant excited oxygen species with water vapor. The overall reaction is

$$O_3 + H_2O \xrightarrow{h\nu} 2OH\cdot + O_2. \quad [4\text{-}32]$$

Many reduced gases and vapors released by both human activities and natural ecosystems are eventually oxidized in the atmosphere by free-radical chain reactions initiated by OH·. Rate constants for the reaction of OH· with several organic vapors are shown in Table 4-12.

TABLE 4-12 Chemical Reaction Rates with OH·[a]

Chemical	Reaction rate (cm^3/molecule/sec)
Ethane	2.74×10^{-13}
Acetylene	7.8×10^{-13}
Ethylene dichloride	2.8×10^{-13}
Ethylene dibromide	2.4×10^{-13}
p-Dichlorobenzene	4.3×10^{-13}
Carbon disulfide	29×10^{-13}

[a] Amts *et al.* (1989).

EXAMPLE 4-15

Carbon disulfide is emitted continuously from the stacks of a manufacturing facility in a generally rural area at a rate of 1 mg/sec. The resulting plume of carbon disulfide in the air is to be modeled using ISCLT to estimate pollutant concentrations in a neighborhood 5 mi from the facility. ISCLT can take into account the removal of a pollutant through photodegradation processes. By assuming that oxidation by OH· is the primary process by which carbon disulfide is photodegraded in the atmosphere, what first-order decay rate should be used in the model for daytime hours? Will the photodegradation of carbon disulfide be a significant mitigating factor in the long-term average exposure of nearby residents? Assume average wind speed is 7 mph.

From Table 4-12, the reaction rate of carbon disulfide with OH· is 29×10^{-13} cm^3 per molecule per second. Assuming a concentration of OH· in ambient air of 3×10^6 molecules per cubic centimeter during daylight hours under clear skies, the first-order decay rate can be estimated as

$$\frac{29 \times 10^{-13}\ \text{cm}^3}{\text{molecules} \cdot \text{sec}} \cdot \frac{3 \times 10^6\ \text{molecules}}{\text{cm}^3} = 8.7 \times 10^{-6}/\text{sec}.$$

To estimate the long-term average half-life of carbon disulfide, based on oxidation by OH·, the fact that photodegradation is occurring only during daylight hours must be taken into account. On average, over a 24-hr period, the photodegradation rate would be half that calculated earlier (i.e., 4.4×10^{-6}/sec). Based on that rate, carbon disulfide has a half-life in air of approximately 1.8 days; see Eq. [1-20]. Given that it takes on average less than 1 hr for carbon disulfide to reach the neighborhood, this half-life is not sufficiently short to appreciably decrease the carbon disulfide concentration in the neighborhood air.

4.6.2 Production of Photochemical Smog: The Ozone–NO_x–Hydrocarbon Connection

One of the first atmospheric chemical cycles to be well documented involves the production of high levels of ozone and photochemical smog. This cycle is ubiquitous in polluted air in any urban area, but was first studied in detail in Los Angeles, California, where the problem is particularly bad due to the high vehicle density and low ventilation rate of the Los Angeles Basin. The set of all possible atmospheric reactions that occur is complex, so only the basic O_3–NO_x cycle and one representative hydrocarbon pathway are presented here.

Figure 4-35 shows the steps in the O_3–NO_x cycle, which occurs in the absence of hydrocarbons but is enhanced in their presence. The species involved are nitrogen dioxide (NO_2), nitric oxide (NO), molecular oxygen (O_2), and ozone (O_3). In urban areas, nitrogen oxides (NO_x) are largely of automotive and industrial origin. Of the four species involved in the O_3–NO_x cycle, the only one that absorbs light is nitrogen dioxide, a gas sometimes

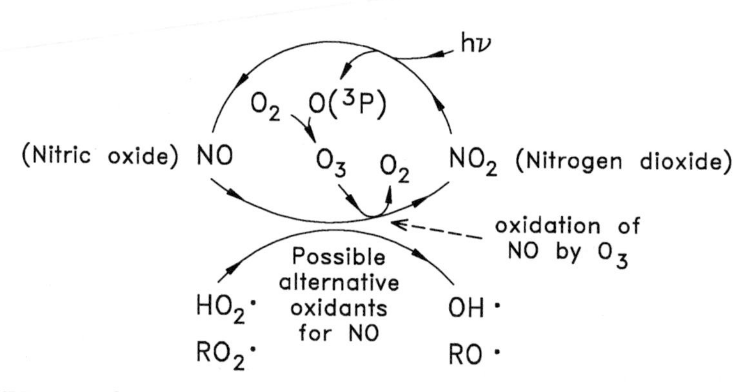

FIGURE 4-35 The O_3–NO_x cycle. The cycle is driven by sunlight; brown-colored nitrogen dioxide gas (NO_2) absorbs a photon and dissociates into nitric oxide (NO) and a highly reactive oxygen atom, which combines with an oxygen molecule to form ozone (O_3). The ozone can be reduced back to O_2 by reaction with nitric oxide. The amount of O_3 formed in this cycle cannot exceed the amount of NO_2 initially present in the air unless alternative means of regenerating NO_2 from NO exist; such means can be provided by free radicals such as $HO_2\cdot$ or $RO_2\cdot$.

visible as a brownish haze overlying urban areas. Nitrogen dioxide can be dissociated into nitric oxide and a free oxygen atom when it absorbs the energy of a photon:

$$NO_2 \xrightarrow{h\nu} NO + O. \quad [4\text{-}33]$$

The rate constant for this reaction depends on light intensity and can be as high as 0.55/min (Boubel *et al.*, 1994). The resultant highly reactive free oxygen atom may combine with molecular oxygen to form a molecule of ozone:

$$O + O_2 \longrightarrow O_3. \quad [4\text{-}34]$$

The reaction shown in Eq. [4-34] also occurs in the stratosphere, where ozone molecules can be formed via the direct action of short-wave ultraviolet light (wavelengths less than 242 nm) on O_2 molecules; see Eq. 4-36b. Ozone is capable of reoxidizing nitric oxide to nitrogen dioxide by the following thermal (dark) reaction. In so doing, nitrogen dioxide and oxygen molecules are recreated, and O_3 is destroyed:

$$O_3 + NO \longrightarrow NO_2 + O_2. \quad [4\text{-}35]$$

When the atmosphere is illuminated, tropospheric ozone levels increase until the rate of ozone destruction equals the photochemical production rate of ozone. During the night, ozone levels decrease because only the dark reaction—Eq. [4-35]—can proceed. Because one molecule of NO_2 is consumed for every molecule of O_3 produced, the maximum amount of O_3 that is produced by the reactions shown in Eqs. [4-33], [4-34], and [4-35] can never be greater than the initial amount of NO_2 present in the air.

Measured concentrations of ozone in polluted urban areas, however, commonly exceed the concentration predicted by the reactions of the O_3–NO_x cycle. These data reflect the fact that other photochemical oxidants, formed from hydrocarbons, reoxidize NO to NO_2 without consuming ozone. As illustrated in Fig. 4-35, a free radical such as a hydroperoxy radical ($HO_2\cdot$), or an alkyl peroxy radical ($RO_2\cdot$), can oxidize nitric oxide to nitrogen dioxide while producing a free radical such as RO· or OH·. The RO· or OH· radicals in turn can react with hydrocarbons in the atmosphere, incorporating molecular oxygen and regenerating species such as $RO_2\cdot$ and $HO_2\cdot$. Figure 4-36 shows one of several possible pathways involving the well-studied atmospheric photochemistry of the hydrocarbon propene.

One effect of introducing hydrocarbons, therefore, is accelerated oxidation of nitric oxide to NO_2, which in turns reacts in the light to produce more ozone. Atmospheric oxygen is the source of oxidation capacity for both this reaction and the transformation of propene to more oxidized compounds. Although the ultimate fate of the hydrocarbon is oxidation to carbon dioxide,

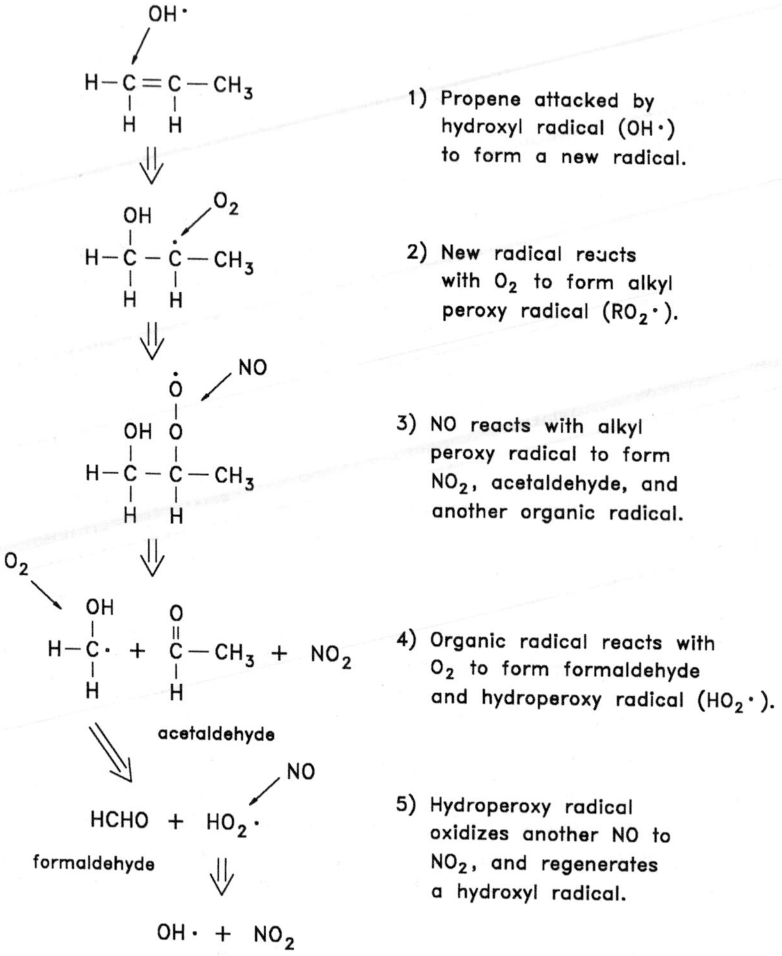

OH·

H—C=C—CH$_3$ (with H, H below)

1) Propene attacked by hydroxyl radical (OH·) to form a new radical.

⇓ O_2

H—C(OH)(H)—Ċ(H)—CH$_3$

2) New radical reacts with O_2 to form alkyl peroxy radical (RO_2·).

⇓ NO

H—C(OH)(H)—C(O—Ȯ)(H)—CH$_3$

3) NO reacts with alkyl peroxy radical to form NO_2, acetaldehyde, and another organic radical.

⇓

O_2 → H—Ċ(OH)(H)· + C(=O)(H)—CH$_3$ + NO_2

acetaldehyde

4) Organic radical reacts with O_2 to form formaldehyde and hydroperoxy radical (HO_2·).

HCHO + HO_2· ← NO

formaldehyde

5) Hydroperoxy radical oxidizes another NO to NO_2, and regenerates a hydroxyl radical.

⇓

OH· + NO_2

FIGURE 4-36 Simplified photochemistry of propene. In this diagram, a few of the pathways by which propene can be oxidized are shown. Numerous reactive free-radical intermediates are produced; these free radicals can react with molecular oxygen and also can oxidize NO to NO_2, thus enhancing ozone production.

many of the intermediates, such as formaldehyde (HCHO) and PAN (peroxyacetyl nitrate), are capable of causing health effects and damage to materials.

Among the models that have attempted to quantify the relationships among NO_x, hydrocarbons, and ozone concentrations in urban air, perhaps the most widely used are those based on the Empirical Kinetic Modeling Approach (EKMA). These models initially were developed from physical models in which different mixtures of NO_x and typical hydrocarbon compounds were irradiated and the ozone concentration was recorded over time. The models then were improved to incorporate additional assumptions, such as the amount of insolation and ventilation, to give more site-specific predictions of urban air quality. Several generations of such models have been produced, each incorporating chemical processes in an increasingly realistic manner. From such models, isopleths of maximum daily ozone concentrations as a function of initial concentrations of NO_x and hydrocarbons (the latter designated as *nonmethane hydrocarbons,* NMHC, or *nonmethane organic carbon,* NMOC) are generated. (Methane is excluded due to its relatively low reactivity.)

Figure 4-37 shows ozone isopleths generated by an EKMA model. In the lower right portion of the figure, ozone concentrations are primarily limited by the amount of NO_x available. The most effective management strategy for limiting ozone in such a situation is to focus on controlling NO_x. For higher NO_x concentrations, the model results suggest that controlling NMHC would be more effective to limit ozone levels. There is even a region in which the model predicts that daily maximum ozone levels would *decrease* with increasing NO_x concentrations. Such a decrease could be attributed to several reasons. One reason could be increased rates of reactions that destroy free radicals, such as the oxidation of NO_2 (itself a free radical, although the dot is customarily omitted) to HNO_3 by the hydroxyl radical. Another reason relates to the timing of peak ozone levels; slower ozone buildups could delay the peak until late in the day when solar radiation is decreasing. Whether any municipality should attempt to control ozone by failing to control NO_x emissions is debatable, however; not only is NO_x a pollutant in its own right, but also regions downwind, outside the model domain, may suffer increased ozone levels as a result.

4.6.3 Acid Deposition

Acid deposition, the deposition of acidic substances from the atmosphere onto Earth's surface, is another phenomenon associated with atmospheric pollution. *Acid rain* refers to the low pH of precipitation that has been observed in

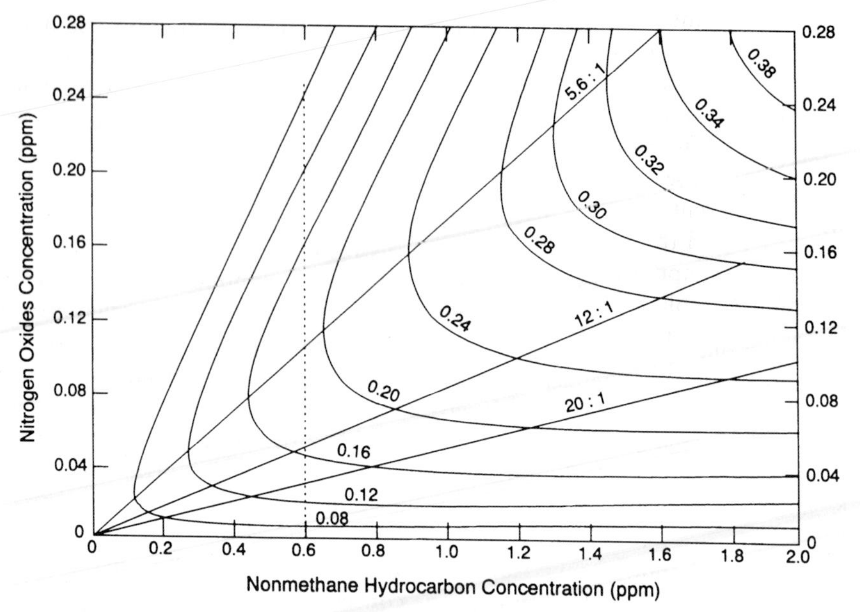

FIGURE 4-37 Ozone isopleths generated with an empirical kinetic modeling approach for a high-oxidant urban area. Straight lines passing through the origin refer to ratios of initial NMHC concentration to initial NO_x concentration (US EPA, 1996b).

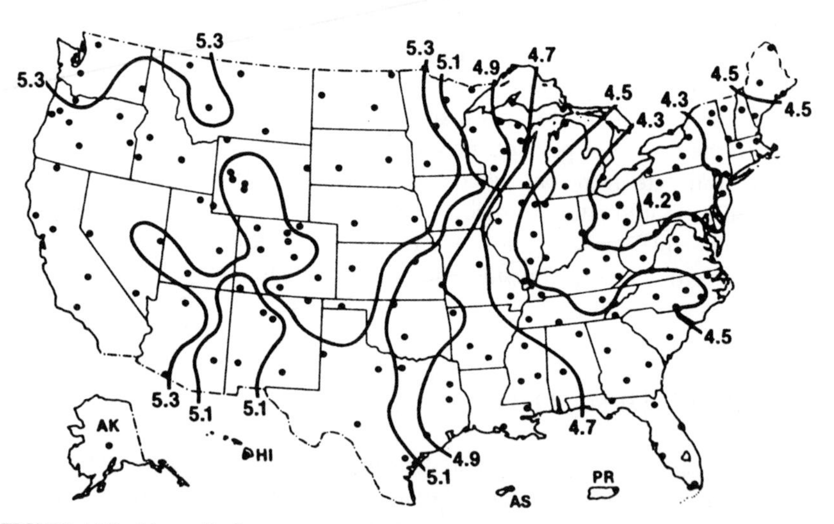

FIGURE 4-38 Mean pH of precipitation of the United States in 1990, based on weekly measurements at 170 sites (adapted from National Atmospheric Deposition Program, 1990).

certain regions (Fig. 4-38). Acid deposition received widespread public attention throughout the 1980s, due to concern about its effects on freshwater ecosystems, forests, structures and materials, human health, and crops. Acid deposition has caused the acidification of some lakes and streams, and consequent loss of fish populations, especially in soft water lakes in acid-sensitive regions such as the northeastern United States and southern Scandinavia. Severe forest damage has been attributed at least in part to acid deposition, especially in central Europe; the term *Waldsterben* (literally, forest death) was coined to describe the phenomenon. Scientific debates raged on many aspects of acid deposition, especially the quantitative relationship between acid deposition rates and changes in surface-water chemistry. This relationship still is not fully understood. Because of the large costs associated with decreasing the emissions of acid-producing pollutants, the debates involved political and economic considerations as well, and were characterized by a high degree of polarization. The primary purpose of this section is not to address the effects of acid deposition, but instead to give an overview of the atmospheric transport and transformation processes that lead to its existence.

Sources of Acid

The two acids that contribute the most to atmospheric acid deposition are sulfuric acid (H_2SO_4) and nitric acid (HNO_3). Other acids, such as hydrochloric acid (HCl) and a variety of organic acids, also have been found in precipitation samples. Most acids are not directly emitted to the atmosphere; instead, emissions contain *acid precursors,* typically oxides of sulfur (SO_x) and oxides of nitrogen (NO_x), which can be oxidized further in the atmosphere to form acids (Fig. 4-39). SO_x originate from sulfur-containing impurities in fuels, notably coal and residual fuel oils. NO_x has two sources: nitrogen-containing impurities in fuels, and reactions between atmospheric nitrogen and oxygen at elevated temperatures in fuel-burning equipment, such as industrial boilers, stationary power plants, and automobile engines. In addition to their roles as acid precursors, both SO_x and NO_x are toxic and irritating air pollutants and also are precursors for fine particles.

Although some sulfuric acid is emitted directly by fuel-burning equipment, most of the sulfur in fuel is oxidized to and emitted as sulfur dioxide (SO_2). Sulfur dioxide contains sulfur in the (+IV) oxidation state and dissolves in water to form sulfurous acid (H_2SO_3), a relatively weak acid. In the presence of hydroxyl radicals in the atmosphere, sulfur dioxide is oxidized to sulfur trioxide (SO_3), which contains sulfur in the (+VI) oxidation state. Sulfur trioxide reacts with water to form H_2SO_4, a strong acid.

Typically, the bulk of sulfur oxidation in the atmosphere occurs by heterogeneous processes in water droplets. The homogeneous oxidation of SO_2 to

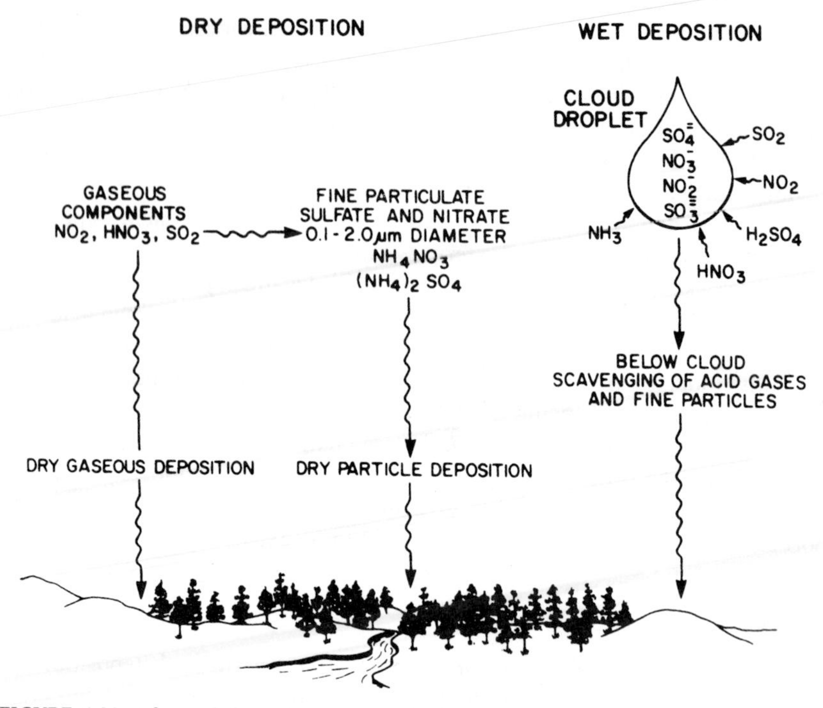

FIGURE 4-39 The acid deposition process. Acid precursors, notably oxides of nitrogen and sulfur, are emitted to the atmosphere, primarily by fuel-burning equipment. Acid precursors are oxidized in the atmosphere to nitric and sulfuric acids by a variety of homogeneous and heterogeneous reactions. The acids are deposited by precipitation-related processes such as washout and rainout, by sorption of nitric acid vapor, and by dry deposition of acidic particulate material such as ammonium sulfate aerosol. (Stern *et al.*, 1984.)

H_2SO_4 has an average daily rate of approximately 0.01/hr, whereas heterogeneous oxidation, in the presence of clouds, fog, or sea salt aerosols, can proceed at a rate greater than 0.3/hr (Luria and Sievering, 1991). Both rainout and washout processes contribute to the incorporation of H_2SO_4 into precipitation.

The concentration of sulfuric acid produced within water droplets can be surprisingly high. *Acid fogs,* in which the pH of the water reaches 3 or lower, have been documented both at low elevations (e.g., the Los Angeles Basin) and in clouds at higher elevations (e.g., Whiteface Mountain, New York). Such acid fogs are believed to be one reason why forest damage often is more pronounced at higher elevations, where the vegetation is within clouds a larger fraction of the time.

TABLE 4-13 Selected Ion Budgets for Bickford Watershed, July 1981 through June 1983[a,b]

	Ca^{2+}	Mg^{2+}	K^+	Na^+	NH_4^+	Cl^-	SO_4^{2-}	NO_3^-
Input	87	42	27	122	111	207	579	295
Output	776	286	94	860	5	757	1186	9
Net output	689	244	67	738	−106	550	607	−286

[a]Input is measured in bulk precipitation, while output is measured in streamflow from the watershed. Positive net output means the watershed has other sources of the ion besides precipitation or is being depleted of its storage of the element. Ca^{2+}, Mg^{2+}, K^+, and Na^+ are being depleted by mineral weathering; Na^+ and Cl^- are believed to have a significant road salt source. Both NH_4^+ and NO_3^- are retained as nutrients. Sulfate is not believed to have major sources on the watershed; its excess of output over input is believed to be a measure of dry deposition. Units are equivalents per hectare per year.

[b]Eshleman and Hemond (1988).

The deposition of nitric acid is similar to the deposition of sulfuric acid: both processes begin with the emission of acid precursors that are subsequently oxidized in the atmosphere to form strong acids. Most anthropogenic NO_x emissions occur as NO, which contains nitrogen in the (+II) oxidation state. NO is subsequently oxidized in the atmosphere to NO_2 [(+IV) oxidation state] by O_3, $HO_2\cdot$, or $RO_2\cdot$. NO_2 then reacts with $OH\cdot$ to form HNO_3, which contains nitrogen in the (+V) oxidation state. Nitric acid also can be formed at night, through reactions involving the nitrate radical (NO_3). The nitrate radical is formed through the reaction of NO_2 and O_3; the nitrate radical subsequently forms HNO_3 through reactions with aldehydes (RCHO) or possibly water. Both rainout and washout processes contribute to the incorporation of HNO_3 into precipitation.

Accurately estimating the total amount of acid deposited on a receptor, such as a lake, or a forest, is problematic because acids are deposited by difficult-to-quantify dry processes as well as wet processes. The wet deposition of both sulfuric and nitric acids is believed to account for only about half of the total deposition of these acids on surface waters, soils, and vegetation. Data from whole watershed mass balance studies (e.g., Table 4-13) support the hypothesis that total deposition of sulfate considerably exceeds what is measured in the form of wet deposition alone. A significant amount of H_2SO_4 is deposited as sulfate aerosols, such as ammonium sulfate [$(NH_4)_2SO_4$]. The direct absorption of SO_3, followed by oxidation of SO_3 to H_2SO_4 at the absorbing surface, is another deposition mechanism. Dry deposition of nitric acid includes sorption of nitric acid vapor onto surfaces, as well as deposition of ammonium nitrate aerosol (NH_4NO_3) that forms from the reaction of NO_3^- with ammonia in the atmosphere. The dry deposition processes of nitric acid are very difficult to measure on a whole ecosystem basis, because of the several other sources of nitrogen that exist within an ecosystem and the multiplicity of pathways that interconvert nitrogen species. For example, the dry deposition of nitric acid can be masked by incorporation of the nitrate into living organisms, or by denitrification. Denitrification results in release of nitrogen into the atmosphere, mainly as N_2, with some small amount of N_2O; see Eq. [2-53] in Section 2.4.3.

Environmental Fate of Acid Deposition

Two important questions concerning the fate of acid deposition are (1) what happens to the acidic chemicals deposited on a watershed, and (2) to what extent does acid deposition change the pH of surface waters? Available data suggest that there are multiple answers to these questions, due to variability from region to region, and even from watershed to watershed. In certain areas, such as the southern Appalachian Mountains of the United States, sulfate appears to be absorbed chemically, primarily by iron sesquioxides (Fe_2O_3) in the soil; in the Adirondacks of New York and in New England, however, sulfate appears to pass through the soils with little sorption. Waters draining from highly sorptive soils do not experience a large drop in pH until the sorption capacity of the soil is filled; only after excess sulfate starts to appear in the surface waters does pH decrease. Sulfate reduction—(see Eq. [2-55])—also can remove sulfate from the water, sequestering sulfur either permanently (e.g., as iron sulfide or organic sulfur in continuously saturated sediments) or seasonally (e.g., in locations where anoxic sediments dry out and become aerated, leading to reoxidation of sulfur species).

The behavior of nitric acid (and nitrate anions) in ecosystems is more complex than that of sulfuric acid, because nitric acid is a frequently limiting plant nutrient as well as an acid. (Sulfur also is a necessary plant nutrient, but is typically present at levels in excess of the needs of most ecosystems, even in unpolluted environments.) In some ecosystems nitrate appears to be taken up completely, and there is seemingly no acidifying effect of HNO_3 deposition on the ecosystem or its associated waters (e.g., Eshleman and Hemond, 1988). Other ecosystems have large amounts of nitrate in their waters, suggesting that the waters have been acidified by HNO_3 deposition. Natural nitrification processes on the watershed, however, also may be responsible for the occurrence of nitrate in stream waters. The nitric acid deposited in ecosystems is not necessarily the same nitric acid that appears in stream waters; instead, the nitric acid input can alter the internal nitrogen cycle and lead indirectly to increased nitrate in runoff. As a nutrient, excess nitric acid deposition also may stimulate growth in both aquatic and terrestrial ecosystems. A study of the Chesapeake Bay suggested that nitrate from acid rain and dry deposition contributed 25% of the anthropogenic nitrogen load to the bay (Fisher and Oppenheimer, 1991), raising concern that nitrate deposition is accelerating eutrophication of the bay—a problem very different from acidification.

Even if acids remain in water as it percolates through soils and ultimately drains to streams and lakes, other processes can partially mitigate the acidifying influence. At lowered pH, the rate of soil weathering is increased, releasing additional base cations such as calcium, which, by contributing to alkalinity, lessen the acidifying effect of atmospheric acids. This process is difficult to describe quantitatively, because the kinetics of weathering of natural soils remain rather poorly understood. Moreover, possible interactions between soil water chemistry and the varied hydrologic pathways of water through the soil remain essentially undescribed; in many cases, it is not known what portions of the soil profile actually come in contact with moving water. There is, however, evidence that base cation production from some soils partially offsets the increase in acidity that would otherwise result from the addition of atmospheric acids to the ecosystem.

Some workers have proposed the use of a quantity called the *F-factor* to describe the degree of acidification that actually has occurred in water bodies, relative to the degree of acidification that would have occurred in the absence of such offsetting influences as increased cation weathering. The F-factor was originally defined as $\Delta C_B/\Delta SO_4$, implicitly assuming that all acidification was due to sulfate deposition and that if a watershed had no internal processes to mitigate acid inputs, the alkalinity (C_B) of its waters would decrease by an amount equal to the change in sulfate concentration, measured in equivalents (NAPAP, 1990). Given the importance of nitric acid deposition, it probably would be useful to define a second F-factor based on alkalinity change in response to a change in nitrate deposition. F-factors are purely empirical, and only can be directly measured if long-term records of both water chemistry and atmospheric deposition are available. However, they provide useful terminology for the purpose of discussion. Some researchers believe that typical F-factors for the northeastern United States average approximately 0.7 or even less.

4.6.4 Stratospheric Ozone Chemistry

In the stratosphere, ozone is produced when ultraviolet light with wavelengths between 200 and 240 nm breaks the bond between two oxygen atoms in a molecule of oxygen:

$$O_2 + h\nu \longrightarrow 2O. \quad [4\text{-}36a]$$

In turn, the highly reactive oxygen atom reacts with other molecules of oxygen—as in Eq. [4-34]—to form O_3:

$$O + O_2 \longrightarrow O_3. \quad [4\text{-}36b]$$

Ozone, in turn, can be destroyed by interaction with another photon that breaks it into an oxygen molecule (O_2) and an oxygen atom (O). Stratospheric ozone also can be destroyed by reaction with other species, such as nitric oxide (NO)—as in Eq. [4-35], and chlorine atoms (from CFCs). The net concentration of ozone is established by the rates of both the production and destruction reactions.

Stratospheric ozone is produced at maximum rates in equatorial regions, where solar radiation is most intense. Ozone does not really occur as a layer, but instead as a broadly distributed gas whose peak concentration occurs in midstratosphere. The total amount of ozone present in the atmosphere is small, typically between 200 and 400 *Dobson units.* A Dobson unit is the amount of ozone that, if gathered together in a thin layer covering Earth's surface at a pressure of 1 atm, would occupy a thickness of 1/100 of a millimeter (10 μm). The entire ozone shield, which protects life on Earth from damage by the UV-B radiation of the Sun (ultraviolet radiation in the 280–320 nm range), is equivalent to a layer of ozone only 2 to 4 mm thick at sea level pressure.

Concern about the diminishment of stratospheric ozone began more than two decades ago. Molina and Rowland (1974) proposed that the release of CFCs through human activity played a major role in O_3 depletion. In 1978, the United States banned the use of CFCs as propellants in aerosol sprays. In 1987, the Montreal Protocol on Substances that Deplete the Ozone Layer was established to halt the production by industrialized countries of most O_3-destroying CFCs by 2000. The Montreal Protocol was subsequently amended to change the date to 1996 (Ham, 1993).

Concrete evidence for a dramatic decrease in stratospheric ozone over the South Pole was first obtained in 1985, when a team from the British Antarctic Survey, using ground-based instrumentation, reported that average ozone concentrations over Antarctica for the month of October were decreasing. In the early 1970s, the October average was approximately 300 Dobson units, whereas in 1984 it was 180 Dobson units. Based on this finding, American investigators reanalyzed data obtained from the Nimbus 7 satellite from 1978 to 1986 (Stolarski *et al.*, 1986) and confirmed the existence of the Antarctic *ozone hole.* The existence of the ozone hole had not been recognized earlier

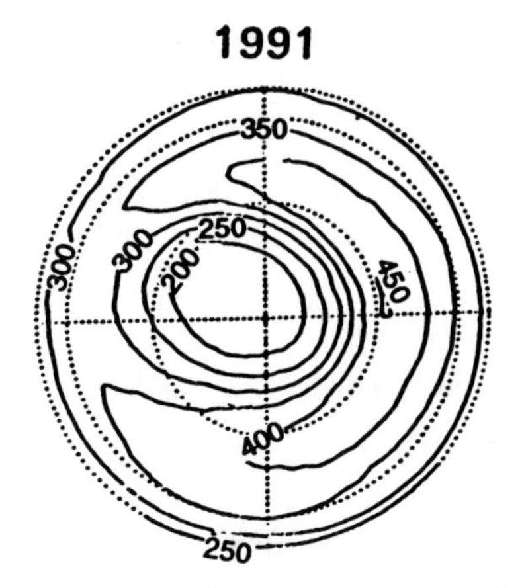

FIGURE 4-40 The Antarctic ozone hole, October 1991, as measured by the total ozone mapping spectrometer (TOMS) from the Nimbus 7 satellite. The hole covers the Antarctic continent and extends as far north as the tip of South America. (Reprinted with permission from R. Stolarski, R. Bojkov, L. Bishop, C. Zerefos, J. Staehelin, and J. Zawodny, 1992, "Measured Trends in Stratospheric Ozone," *Science* 256: 342–349. Copyright ©1992, American Association for the Advancement of Science.)

from the satellite data because the entire analysis was done by computer and programmers had written the code to prefilter the data; all values less than 190 Dobson units were rejected as being impossibly low and therefore in error! Direct observations of Antarctic atmospheric ozone levels (Fig. 4-40) were made during the Airborne Antarctic Ozone Experiment in 1987, using ER-2 research aircraft (a derivative of the famous U-2 spy plane, capable of flying at an altitude of 70,000 ft) and a lower flying DC-8 aircraft.

The current Antarctic ozone hole, an ozone-depleted region between 12 and 25 km in altitude, now is acknowledged to be a result of the catalytic degradation of ozone by chlorine, which is released into the atmosphere through human production and use of CFCs. CFCs are able to affect stratospheric chemistry because their chemical stability and resultant long lifetime in the troposphere allow them to mix into the stratosphere (Section 4.4.5). [The majority of anthropogenic vapors and gases (including numerous chlorine-containing species) are removed by tropospheric mechanisms before mixing appreciably into the stratosphere.] In the stratosphere, under the action of short-wave ultraviolet light, CFCs are degraded, and chlorine is released. Much of the chlorine becomes incorporated in so-called *reservoir compounds,* such as hydrochloric acid (HCl) and chlorine nitrate ($ClONO_2$), which are not directly reactive with ozone. The free chlorine atom, however, reacts directly with ozone molecules to form chlorine oxide and molecular oxygen:

$$Cl + O_3 \longrightarrow ClO + O_2. \quad [4\text{-}37]$$

A chlorine oxide molecule in turn can react with an oxygen atom to form more molecular oxygen plus a free chlorine atom,

$$ClO + O \longrightarrow Cl + O_2, \quad [4\text{-}38]$$

or it may react with another chlorine oxide molecule,

$$ClO + ClO \longrightarrow Cl_2O_2 \longrightarrow Cl_2 + O_2. \quad [4\text{-}39]$$

The chlorine molecule, Cl_2, is readily photodissociated into additional chlorine atoms, ready to attack more ozone. A single Cl atom thus can destroy as many as 100,000 ozone molecules during its residence in the stratosphere (Levi, 1988).

The preceding set of reactions occurs throughout the stratosphere and is responsible for depletion of ozone on a global scale. The extreme ozone depletion at the South Pole, however, did not at first seem to be consistent with the kinetics of the known ozone-destroying reactions and the known free chlorine concentrations. The discrepancy turned out to be due to heterogeneous chemistry (Fig. 4-41). Under wintertime conditions when polar stratospheric temperatures drop below $-85°C$, the condensation of water forms stratospheric clouds, even though the partial pressure of water vapor in the stratosphere is extremely low. On the surface of the ice crystals in these clouds, two heterogeneous reactions involving reservoir compounds can occur:

$$HCl + ClONO_2 \longrightarrow HNO_3 + Cl_2, \quad [4\text{-}40]$$

and

$$H_2O + ClONO_2 \longrightarrow HNO_3 + HOCl. \quad [4\text{-}41]$$

Polar stratospheric clouds thus catalyze the release of chlorine molecules from reservoir compounds—Eq. [4-40]—and the eventual formation, when springtime sunlight arrives, of damaging chlorine atoms. The polar stratospheric clouds also may have a second effect; it has been suggested that nitric acid (HNO_3) condenses onto polar stratospheric ice particles, which subsequently settle out of the lower stratosphere. Nitrogen oxides are thereby removed, preventing them from capturing chlorine atoms in the reservoir compound $ClONO_2$.

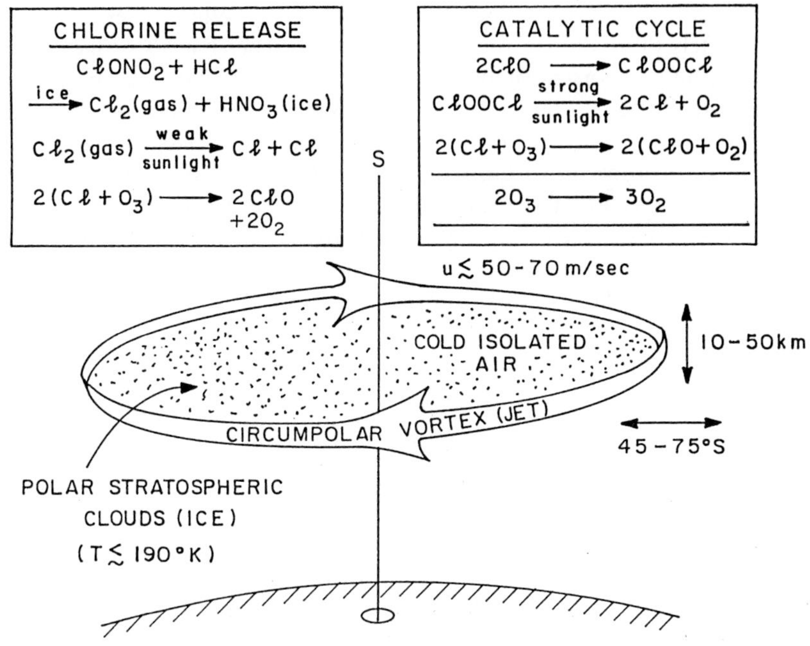

FIGURE 4-41 Key chemical processes of the Antarctic ozone hole. Of particular importance are the polar stratospheric clouds, which catalyze the release of ozone-destroying chlorine from chlorine nitrate and hydrogen chloride (Prinn and Hartley, 1992).

4.7 GLOBAL CLIMATE CHANGE: THE GREENHOUSE EFFECT

4.7.1 Radiation Balance of Earth

The temperature of Earth is determined by the balance between solar radiation input (see Fig. 2-30) and reflection and reradiation of energy from Earth and its atmosphere back into space. Recall from Section 4.3.1 that the solar constant is approximately 1400 W/m^2. The projected area intercepting the solar radiation is πr^2, where r is Earth's radius. Given that Earth's surface area is $4\pi r^2$, the average solar radiation received outside Earth's atmosphere is approximately one-fourth the solar constant, or 350 W/m^2.

Not all this solar radiation reaches Earth's surface, however. Some of the solar energy is absorbed by the atmosphere and clouds, and some is reflected from the atmosphere and clouds back into space. When skies are clear, approximately 80 to 85% of the solar radiation reaches Earth's surface; when skies are cloudy, approximately 50% reaches Earth's surface. Earth's surface itself has an albedo that on average is approximately 0.35. Thus, about a third of the solar radiation that reaches Earth's surface is reflected. The remaining solar energy is absorbed by Earth; consequently, Earth's surface temperature increases, and its rate of radiation in the long-wave, infrared spectrum also increases. For a *blackbody*, an object that absorbs all incident energy at all wavelengths, the rate of reradiation (the flux density J over all wavelengths) is given by the Stefan–Boltzmann law,

$$J = \sigma \cdot T^4, \qquad [4\text{-}42]$$

where σ is the Stefan–Boltzmann constant, 5.7×10^{-12} W/(cm$^2 \cdot$ K^4) and T is the absolute temperature. In the case of radiation from an actual surface that does not behave as an idealized blackbody, the radiated energy is adjusted by an emissivity factor, ϵ, which is about 0.97 for a water surface, but is less for many other surfaces.

The wavelengths of emitted long-wave, infrared radiation occupy a broad band, with the wavelength of maximum emission given by *Wien's displacement law*. Wien's displacement law states that the product of the absolute temperature of a radiating blackbody and the wavelength corresponding to maximum energy (λ_{max}) is a constant (i.e., if the temperature of the radiating blackbody increases, λ_{max} decreases). When λ_{max} is in centimeters, Wien's displacement law can be expressed as

$$\lambda_{max} \cdot T = 0.29, \qquad [4\text{-}43]$$

where T is absolute temperature and 0.29 is Wien's displacement constant for λ_{max} in centimeters. In the infrared portion of the spectrum, absorbance data are usually expressed as a function of the reciprocal of wavelength, $1/\lambda$ or *wave number*. Maximum emission from a water surface at 298 K (25°C) is at a wave number of 1030/cm; the same radiation expressed in wavelength is 9.7 μm.

EXAMPLE 4-16

What average surface temperature would be predicted for Earth if it had no atmosphere? Assume it behaves as a blackbody. What would be the wavelength at which maximum energy is reradiated by Earth at that temperature? What is the corresponding wave number?

Approximately 350 W/m^2 of solar radiation would be received on average at Earth's surface if atmospheric effects are ignored. Earth reflects approximately 35% of received radiation. Use the Stefan–Boltzmann law—Eq. [4-42]—to solve for *T*, equating the average solar radiation that would be absorbed by Earth's surface under these conditions to the rate of reradiation:

$$\frac{350\ \text{W}}{\text{m}^2} \cdot (1 - 0.35) = \frac{5.7 \times 10^{-12}\ \text{W}}{\text{cm}^2 \cdot \text{K}^4} \cdot \text{T}^4 \cdot \frac{(100\ \text{cm})^2}{\text{m}^2}$$

$$\text{T}^4 = 4.0 \times 10^9$$

$$\text{T} = 251\ \text{K or } -22°\text{C}.$$

This estimated average surface temperature for Earth, if it did not have an atmosphere, is much lower than what actually exists.

The wavelength corresponding to the maximum energy reradiated by Earth at 251 K can be estimated using Wien's displacement law, Eq. [4-43]:

$$\lambda_{max} = \frac{0.29}{251\ \text{K}} = 0.0012\ \text{cm or } 12\ \mu\text{m}.$$

The corresponding wave number for this λ_{max} is

$$\frac{1}{0.0012\ \text{cm}} = 870/\text{cm}.$$

As can be inferred from Example 4-16, not all energy reradiated by Earth escapes back into space; if it did, the present life-supporting temperature of Earth, approximately 15°C (annual average global mean temperature), would not exist. Although atmospheric gases and vapors scatter some of the incoming solar radiation, they also trap a portion of the long-wave infrared radiation emitted from Earth's surface and reradiate a fraction of the absorbed energy back to Earth, in the *greenhouse effect*. The more strongly the atmosphere absorbs long-wave radiation, the warmer Earth's average surface temperature is for a given input of solar short-wave radiation.

4.7.2 Greenhouse Gases

There are five major greenhouse gases that occur naturally: water vapor, carbon dioxide, ozone, methane, and nitrous oxide. Human activity has in-

TABLE 4-14 **Properties of Several Greenhouse Gases**[a]

Gas	1994 Mixing ratio	Radiative forcing (W/m^2)	Atmospheric residence time (year)	Rate of increase
Carbon dioxide	358 ppm(v)	1.6[f]	50–200[b]	1.5 ppm(v)/year
Methane	1720 ppb(v)	0.47[f]	12	10 ppb(v)/year
Nitrous oxide	312 ppb(v)[c]	0.14[f]	120	0.8 ppb(v)/year
CFC-11	268 ppt(v)[c,d]	0.06[f]	50	0 ppt(v)/year
CFC-12	503 ppt(v)[d,e,f]	0.14[f]	102[f]	7 ppt(v)/year[f]

[a]IPCC (1996), unless otherwise noted.
[b]Different uptake rates by different sink processes result in a range of atmospheric residence times.
[c]Estimated from 1992 and 1993 data.
[d]ppt(v) = part per trillion by volume.
[e]1992 concentration.
[f]Prather *et al.* (1996).

creased the atmospheric mixing ratios of most of these gases, and also has released other greenhouse gases, such as CFCs, which are entirely anthropogenic in origin. [Note that one chlorinated greenhouse gas, methyl chloride (CH_3Cl) is produced by some marine algae as well as by human activity.] These increases in greenhouse gas concentrations are causing an *enhanced greenhouse effect,* due to more long-wave radiation emitted by Earth's surface being trapped and reradiated back to Earth. This change in energy available to Earth has been called *radiative forcing* (IPCC, 1996). Radiative forcing can affect Earth's climate through changes in available energy or the distribution of energy among the atmosphere, land, and oceans, thus resulting in *global warming* or *global climate change.* Table 4-14 provides information on several greenhouse gases, including their atmospheric mixing ratios and radiative forcing. Figure 4-42 shows the absorbance spectra, including the wave numbers at which peak absorbance occurs, for several gases. The absorbance spectra in Fig. 4-42 were weighted by the concentration of each gas in the atmosphere in the mid-1970s to show the approximate percentage of long-wave infrared energy absorbed by (or transmitted through) each gas or vapor at each wavelength.

It is immediately evident from Fig. 4-42 that water vapor is a primary greenhouse gas. Currently, about 65% of the greenhouse effect is due to water vapor (Mann and Lazier, 1996). Although on a global scale water vapor concentration is not greatly influenced by direct anthropogenic emissions, it may be altered indirectly via global-scale, human-induced climate change.

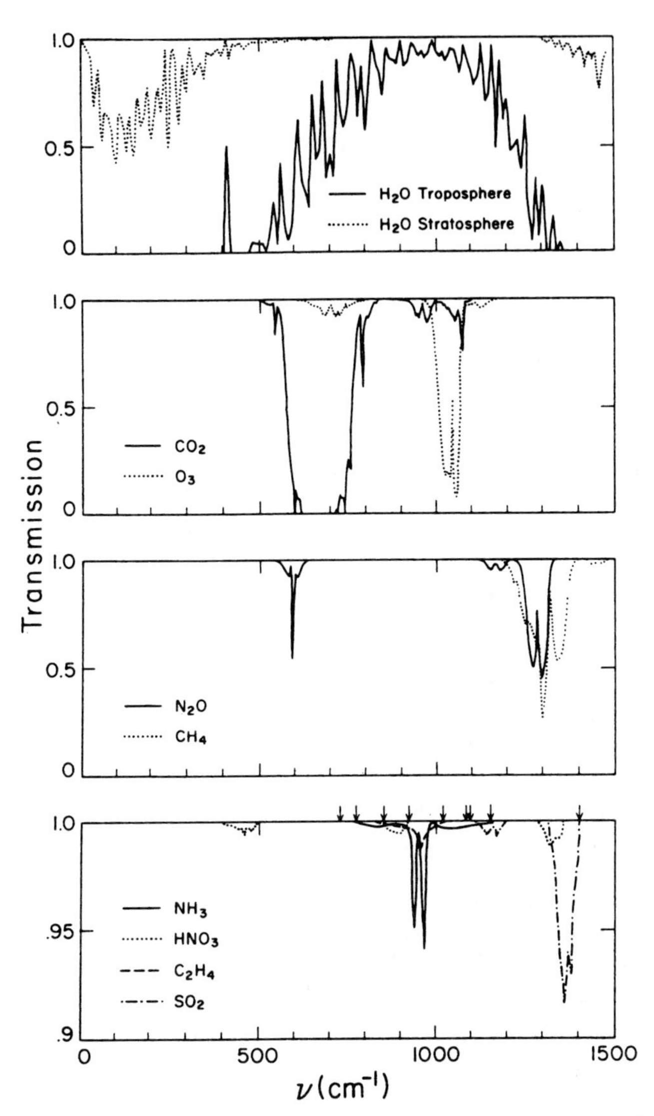

FIGURE 4-42 Atmospheric absorption of radiation by major greenhouse gases, including water, carbon dioxide, ozone, methane, and nitrous oxide. Location of peak absorbances from several CFCs also are shown by arrows in the bottom panel. These spectra are in the thermal infrared region (i.e., in the range of wavelengths emitted by objects at ambient environmental temperatures) and give transmission data for long-wave radiation emitted by Earth's surface. In this figure, ν is the wave number, that is, the reciprocal of wavelength λ. (Reprinted with permission from W. C. Wang, Y. L. Yung, A. A. Lacis, T. Mo, and J. E. Hansen, 1976, "Greenhouse Effects Due to Man-Made Perturbations of Trace Gases," *Science* 194:685–690. Copyright ©1976, American Association for the Advancement of Science.)

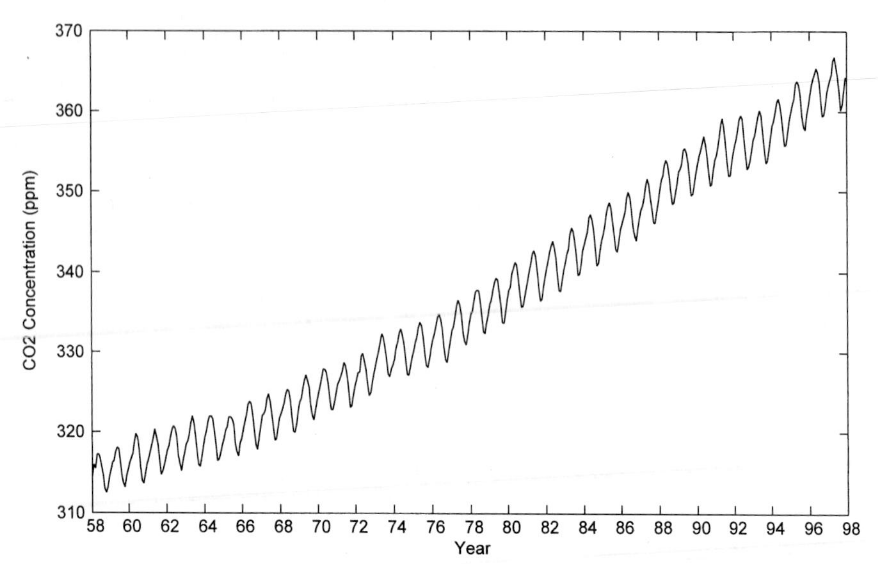

FIGURE 4-43 Increase in the mixing ratio of carbon dioxide in Earth's atmosphere as measured since March 1958 at the Mauna Loa Observatory (data from Keeling and Whorf, 1998).

Carbon Dioxide

As shown in Fig. 4-42, carbon dioxide (CO_2) absorbs the second largest amount of long-wave infrared radiation in the atmosphere (about 32%; Mann and Lazier, 1996). Over Earth's history, the predominant natural source of CO_2 in the atmosphere has been volcanic eruptions, and the vast majority of that CO_2 is now stored in ocean sediments and rocks derived from those sediments (Mann and Lazier, 1996). If Earth did not have oceans, the concentration of CO_2 in Earth's atmosphere would be far higher than it is currently.

Carbon dioxide concentrations in the atmosphere have risen steadily since the industrial period began in about 1750. The current atmospheric mixing ratio of CO_2 is approximately 3.6×10^5 ppb(v). Figure 4-43 shows the atmospheric mixing ratio of CO_2 measured since March 1958 at the Mauna Loa Observatory in Hawaii. Seasonal fluctuations are primarily due to summertime maxima and wintertime minima in the net rate of photosynthesis, which removes CO_2 from the atmosphere. The long-term trend, however, is clearly upward, and reflects anthropogenic releases of carbon dioxide from fossil fuel combustion processes, land alteration (primarily the clearing of forest for agriculture), and, to a lesser extent, cement production. The contribution of land-use conversion is difficult to quantify because carbon dioxide

TABLE 4-15 The Global Carbon Dioxide Budget[a,b]

Storage, source, or sink of carbon dioxide	Magnitude (GtC/year)[c]
Rate of atmospheric increase (storage in the atmosphere)	3.3 ± 0.2
Emissions from fossil fuel combustion and cement production	5.5 ± 0.5
Net emissions from changes in tropical land use	1.6 ± 1.0
Global net respiration	60
Ocean uptake	2.0 ± 0.8
Uptake by Northern Hemisphere forest regrowth	0.5 ± 0.5
Other inferred terrestrial sinks[d]	1.3 ± 1.5
Global net primary production (photosynthesis)	61.3

[a] Schimel *et al.* (1996).
[b] Average annual budget for 1980 to 1989. Error limits correspond to an estimated 90% confidence interval.
[c] GtC/year = gigatonnes of carbon per year. 1 Gt equals 10^{15} g.
[d] Such sinks could include enhanced photosynthesis due to CO_2 and/or nitrogen fertilization, as well as effects resulting from climate change. These are inferred sinks whose magnitude is that required to complete the carbon balance.

release occurs not only from the burning of forest, but also from the subsequent enhanced decomposition of soil organic material. Table 4-15 shows estimates of the major sources and sinks of atmospheric carbon dioxide.

It is important to recognize two features of the global carbon dioxide budget. First, the uncertainties in the budget are considerable, in part because of the difficulty of measuring the temporally and spatially varying fluxes of CO_2 on a global scale. Current budgets have a *missing carbon problem;* the budgets do not balance, thereby requiring the inference of one or more global-scale terrestrial carbon sinks (see Table 4-15). Second, anthropogenic CO_2 fluxes are a significant fraction of the natural ones; human activities therefore can have an immense effect on the atmospheric CO_2 budget. Given the economic, political, and social forces operating in both heavily industrialized and newly industrializing countries, the task of reducing anthropogenic fluxes of CO_2 through energy conservation, utilization of energy sources that do not require combustion, and forest conservation is daunting. Most researchers consider a continued rise in atmospheric CO_2 to be almost inevitable, but the rate of increase may be somewhat modified by actions such as the adoption of the Kyoto Protocol.

Considerable uncertainty surrounds the short- and long-term response of biota to altered temperatures and carbon dioxide concentrations in the atmosphere. Although enhanced photosynthesis is observed in some species under increased atmospheric CO_2 concentrations, unpredictable species changes and replacements are likely under altered climate. If the net effect of biotic change

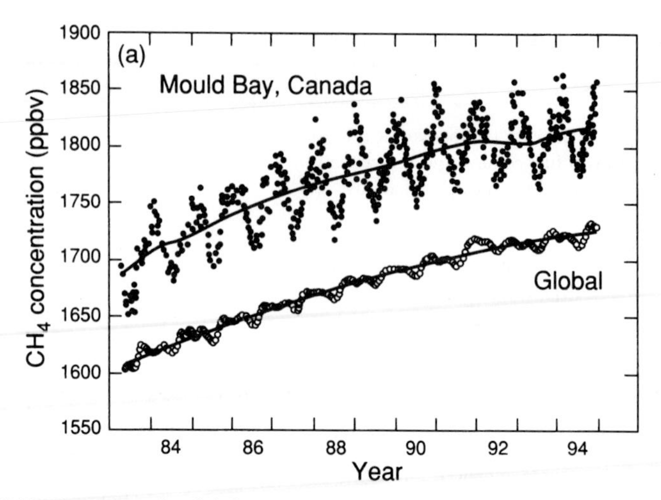

FIGURE 4-44 Increase in the global average mixing ratio of methane in Earth's atmosphere and increase in CH_4 atmospheric mixing ratios observed at Mould Bay, Canada. Individual points represent actual measurements and show the seasonal modulation of the CH_4 concentration, probably due in part to a seasonal pattern of release from ecosystems and possibly to seasonal changes in the atmospheric oxidation rate of CH_4. The solid lines drawn through the two data sets represent the long-term trends in CH_4 atmospheric mixing ratios (Prather *et al.*, 1996).

were to store additional carbon in the form of organic material, then the negative feedback created would lessen the effects of anthropogenic CO_2 releases on climate. Other mechanisms, however, may provide destabilizing *positive feedback*; for example, warmer temperatures favor decomposition of organic material, thereby further increasing CO_2 release to the atmosphere.

Methane

After water vapor and CO_2, methane (CH_4) is the third most important greenhouse gas. Each additional molecule of CH_4 added to the atmosphere absorbs about 20 times as much long-wave infrared radiation as does a molecule of carbon dioxide. This occurs in part because some of the absorption spectrum of methane lies in *windows* in the carbon dioxide absorption spectrum (see Fig. 4-42); therefore, methane absorbs wavelengths that are not already being highly attenuated by carbon dioxide. Currently, the global concentration of methane in the atmosphere is approximately 1.7 ppm and is increasing at an annual rate of approximately 0.01 ppm per year (Table 4-14). The seasonal fluctuations shown in Fig. 4-44 may correspond to seasonal

TABLE 4-16 Global Methane Sources (1978 to 1988)[a,b]

Sources	Budget No. 1	2	3	4	5	6	7	8	9	10	11
Ruminants	160	160	90	60	120	115	85	60	85	75	80
Rice paddy fields	280	210	39	45	95	165	53	160	120	54	110
Biomass burning			60	70	25	68	75	45	78	65	55
Other anthropogenic	33	193	150	20	80	81	55	67	72	103	120
Swamps and marshes	245	250	39	125	150	155	38	80	48	82	115
Other natural	24	27	832	150	83	256	14		25	15	55
Anthropogenic	473	563	339	195	320	429	268	332	354	297	365
Natural	269	277	871	275	233	411	52	80	73	97	170
Anthropogenic/total (%)	64	67	28	41	58	51	84	81	83	75	68
Total	741	840	1210	470	553	840	319	412	427	394	535

[a] Simplified to reflect the major sources. The emissions are given in teragrams per year. Each column represents a different published CH_4 budget.
[b] Adapted from Crutzen *et al.* (1989).

production of CH_4 in ecosystems such as peatlands, or to seasonal changes in the strength of the atmospheric sink, primarily attack by OH·.

As discussed in Section 2.4.3, methane is produced naturally from the anaerobic decomposition of organic material by microbes. Peatlands and bogs are major sources of CH_4 because they contain vast amounts of organic material undergoing anaerobic decomposition. The decomposition pathways are complex and multiple in number; one common pathway involves the breakdown of organic material into small organic molecules, such as acetic acid (CH_3COOH), and the subsequent fermentation of the acetic acid into carbon dioxide and methane by microbes known as *methanogens*.

Methane also is produced as a by-product of anthropogenic activities. The release of natural gas by the petroleum industry during production and transportation is a significant source of CH_4, as is the venting of gases from municipal landfills in which anaerobic decomposition of waste occurs. These sources of CH_4 are thought to be smaller on a global scale than the release of CH_4 from the rumina of cattle and from rice cultivation. Although these latter two processes are, in a sense, natural processes, they are under the direct management of humans and therefore can be thought of as anthropogenic sources of methane. Several estimates of the magnitude of global CH_4 sources are shown in Table 4-16.

The atmospheric methane budget appears in Table 4-17. As is the case with the carbon dioxide budget, many of the sink and source strengths are

TABLE 4-17 The Global Methane Budget[a,b]

Storage, source, or sink of methane	Magnitude (Tg/year)[c]
Rate of atmospheric increase (storage in the atmosphere)	37
Natural sources[d]	160
Fossil fuel related	100
Biospheric sources managed by humans[e]	275
Reaction with tropospheric OH·	490[f]
Transport into the stratosphere	40
Soils[g]	30

[a] Prather *et al.* (1995), unless otherwise noted.
[b] Estimated sources and sinks for 1980 to 1990.
[c] Tg/year = teragrams of methane per year. 1 Tg equals 10^{12} g.
[d] Includes wetlands, termites, oceans, and other natural sources.
[e] Includes enteric fermentation in cattle, sheep, and buffalo; rice paddies; biomass burning; landfills; animal waste; and, domestic sewage.
[f] Prather *et al.* (1996).
[g] Methane from the atmosphere can be consumed (oxidized) by microbial communities in soils.

only approximately known. Current estimates of the global methane budget do not account for the rapidly rising concentrations in the atmosphere; there is a missing source term of approximately 60 Tg/year.

Climate changes that result in increased surface temperatures will influence natural sources and sinks of CH_4 and potentially lead to positive feedback mechanisms. For example, at higher temperatures the anaerobic decomposition rates of organic matter in northern wetlands may accelerate, resulting in the release of more CH_4 to the atmosphere, which would then lead to further temperature increases. (Conversely, water tables could drop under warmer conditions, leading to a decrease in net CH_4 emissions.) There also exist large CH_4 reservoirs in methane *clathrate,* a crystalline icelike structure of water and CH_4 that exists in areas of permafrost and beneath certain marine sediments (Fig. 4-45). Methane clathrate also could release significant quantities of CH_4 to the atmosphere under conditions of warmer surface temperatures.

Nitrous Oxide

Nitrous oxide (N_2O), another greenhouse gas, is stable in the troposphere, but oxidizes to ozone-reactive NO in the stratosphere. Molecule for molecule, N_2O is currently about 200 times more effective in absorbing long-wave infrared radiation than is carbon dioxide, and its atmospheric concentration

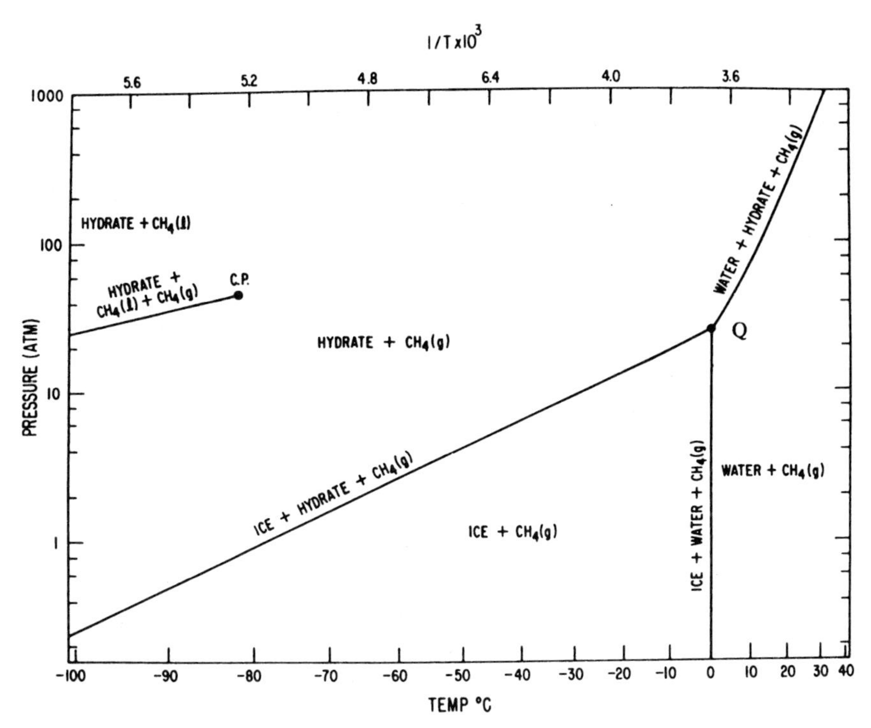

FIGURE 4-45 The stability diagram of methane hydrate (clathrate). Stable region is in the upper left of the diagram; the hydrate becomes unstable with both warming and decreases in pressure. Methane clathrate is of interest because of the possibility of releasing large amounts of methane by global warming (Miller, 1974).

is increasing by 0.8 ppb(v)/year (Fig. 4-46). As is the case with methane and carbon dioxide, nitrous oxide has both natural and anthropogenic sources. The largest sources of N_2O are associated with the complex cycle of nitrogen in natural systems (Haynes, 1986), but N_2O also occurs as a combustion by-product. Two estimates of the major sources of nitrous oxide appear in Table 4-18.

Most nitrous oxide produced in terrestrial and aquatic ecosystems is associated with the processes of nitrification and denitrification. *Nitrification* is the oxidation of ammonium (NH_4^+) to nitrate, and is carried out by *chemoautotrophic* bacteria in oxygenated environments, where ammonium is present (Haynes, 1986):

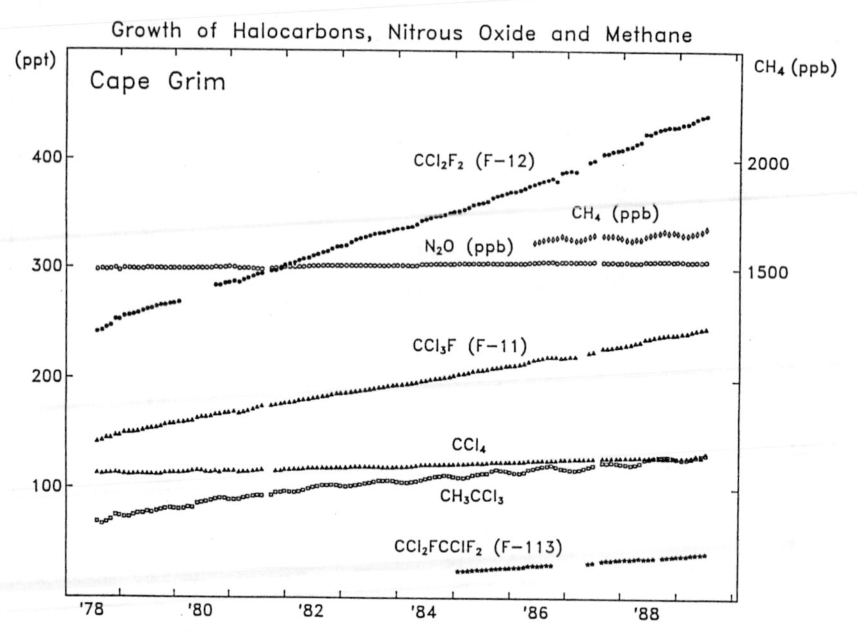

FIGURE 4-46 Increase in the mixing ratios of nitrous oxide and several other greenhouse gases (mostly CFCs) in the atmosphere. For N_2O, use the left scale, but with units of ppb (adapted from Prinn and Hartley, 1992).

TABLE 4-18 Two Estimates of Global Sources of Nitrous Oxide[a,b]

Oceans	9	3
Land and soils	13.4	12
Anthropogenic	6.6	9
Total	29	24
Anthropogenic/total (%)	23	38

[a]Emissions in teragrams per year.
[b]Crutzen *et al.* (1989).

$$NH_4^+ \longrightarrow NH_2OH \longrightarrow [HNO] \longrightarrow NO_2^- \longrightarrow NO_3^-. \quad [4\text{-}44]$$

$$[HNO] \downarrow N_2O, \quad NO_2^- \to N_2O$$

The chemoautotrophic bacteria use the energy released by ammonium oxidation to reduce CO_2 to organic matter, much as green plants use energy captured from light to reduce CO_2. The fraction of ammonium converted to nitrous oxide during nitrification is usually less than 1%, and is strongly influenced by pH and oxygen concentration. Maximum N_2O production usually occurs at low concentrations of oxygen (less than 0.01 atm). Human activities can influence N_2O production; it is stimulated in heavily fertilized agricultural systems, especially where nitrogen fertilizer is added in the form of ammonium. Many scientists believe that fertilized agriculture is at least partly responsible for rising levels of N_2O in the atmosphere. Because nitrous oxide production is widely distributed, time varying, and heterogeneous over space, however, estimates of how much extra N_2O production occurs due to ammonium fertilization are very uncertain.

Denitrification, the process in which nitrate is used as an oxidant by organisms in the degradation of organic material, also leads to the production of N_2O. Nitrous oxide is an intracellular intermediate in the overall process of denitrification:

$$NO_3^- \longrightarrow NO_2^- \longrightarrow NO \longrightarrow N_2O \longrightarrow N_2. \quad [4\text{-}45]$$

The release of N_2O in this process is enhanced under certain conditions, such as low pH, low temperatures, and high nitrate concentrations; these factors inhibit, or at least do not favor, the last step of the preceding sequence of reactions. As in the case of nitrification, agricultural management practices may influence both the total rate of denitrification and the ratio of nitrous oxide to N_2 produced.

Chlorofluorocarbons

In Section 4.6.4, the role of CFCs in stratospheric ozone destruction was discussed. CFCs also are of concern because they are radiatively active in portions of the infrared spectrum not strongly attenuated by water vapor, CO_2, CH_4, or N_2O. Currently, a CFC molecule added to the atmosphere absorbs about 10,000 times as much long-wave infrared radiation as does a CO_2 molecule. CO_2 has a radiative forcing of 1.8×10^{-5} W/(m$^2 \cdot$ ppb(v)), whereas CFCs range from 0.22 to 0.32 W/(m$^2 \cdot$ ppb(v)) (Prather *et al.*, 1996). CFCs also have long atmospheric residence times, ranging from 50 to 1700 years. The locations of some CFC absorbance bands are shown in Fig. 4-42.

Unlike the several radiatively active trace gases that have both natural and

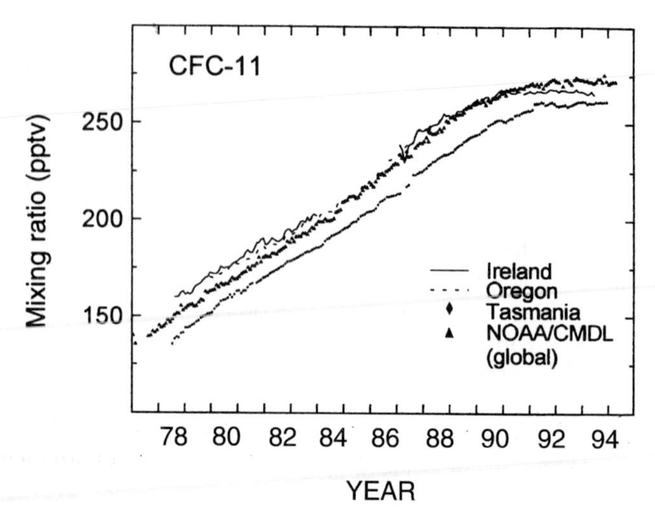

FIGURE 4-47 Global average mixing ratios and selected station mixing ratios of CFC-11. The effect of the CFC phaseout under the Montreal Protocol on atmospheric mixing ratios of CFC-11 can clearly be seen (Prather *et al.*, 1996).

anthropogenic sources, the source of CFCs is exclusively anthropogenic. The two CFCs with the highest atmospheric concentrations are CFC-11 (CCl_3F) and CFC-12 (CCl_2F_2), shown in Fig. 4-46. These CFCs have been used as foam-blowing agents, as refrigerants, and as aerosol propellants in spray cans. It has been estimated that CFC-11 and CFC-12 contributed about a third as much to greenhouse warming as did CO_2 during a 10-year period beginning in the late 1970s (Rogers and Stephens, 1988). Currently, CFCs are being phased out under the Montreal Protocol. The effect of this phaseout on atmospheric mixing ratios of CFC-11 already is beginning to be seen, as shown in Fig. 4-47. Table 4-14 includes the current radiative forcing and atmospheric residence times for CFC-11 and CFC-12.

Inexpensive and safe alternative compounds to CFCs are difficult to find; hydrochlorofluorocarbons (HCFCs) are being used as temporary replacements in some applications, leading to sharply increasing atmospheric mixing ratios of HCFCs, as shown in Fig. 4-48. HCFCs contain at least one hydrogen to render them more susceptible to oxidative degradation in the troposphere (their atmospheric residence times range from 1.4 to 18.4 years), but they are not a long-term solution as CFC substitutes. They still contribute to the enhanced greenhouse effect with a radiative forcing of 0.14 to 0.28 $W/(m^2 \cdot ppb(v))$ (Prather *et al.*, 1996) and they still destroy stratospheric ozone. HCFCs also have potential adverse health effects (Ham, 1993).

Hydrofluorocarbons (HFCs), another class of potential CFC substitutes,

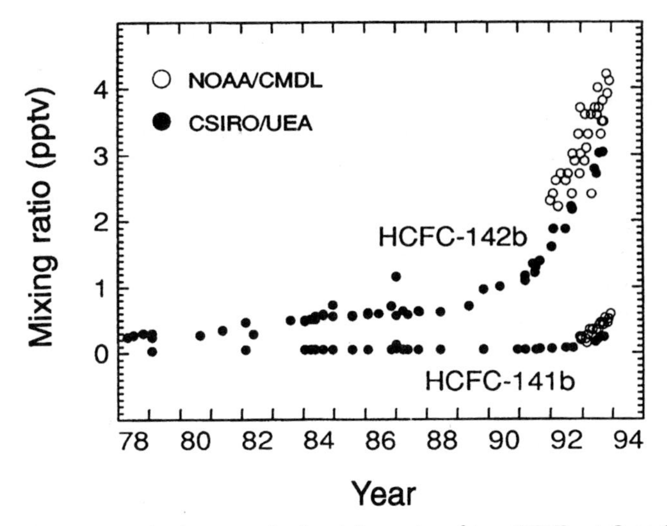

FIGURE 4-48 Increases in the atmospheric mixing ratios of two HCFCs at Cape Grim, Tasmania. (HCFC-141b is CH_3CFCl_2 and HCFC-142b is CH_3CF_2Cl; plot is a composite of two data sources.) As CFCs are phased out under the Montreal Protocol, atmospheric mixing ratios of substitutes are increasing (Prather *et al.*, 1996).

do not contribute to stratospheric ozone destruction because they do not contain chlorine. Their stability, however, ensures their atmospheric presence for many decades to come; their atmospheric residence times may exceed 250 years. Although the infrared absorbance of HFCs is typically less than that of the CFCs they replace, the potential contribution of HFCs to global warming is still significant, with estimates of radiative forcing of 0.02 to 0.35 $W/(m^2 \cdot ppb(v))$ (Prather *et al.*, 1996).

Even when CFC production is completely halted, CFCs still will be released to the atmosphere when the equipment in which they were used, such as refrigerators and air conditioners, are "junked" and the fluid systems are breached. Some municipalities (e.g., the city of Cambridge, Massachusetts) now have ordinances requiring the recovery of CFC refrigerants from appliances and air conditioners before the appliances are discarded.

4.8 CONCLUSION

Chemical transport in the atmosphere has many parallels with transport in water; both advection and Fickian processes are important. However, the velocities and the Fickian transport coefficients tend to be larger in the atmosphere, and the distances over which pollutant sources have influence can be greater. Additional factors that affect both advection and Fickian transport in the atmosphere include the compressibility of air (which creates a pronounced adiabatic lapse rate) and changes in the stability of the air due to water evaporation and condensation. Because of the large spatial scales over which air circulation occurs, the Coriolis force also must be considered; in contrast, this effect is negligible in groundwater and in small water bodies (although it is significant in large lakes and in oceanic circulation).

Atmospheric chemical processes are dominated by photochemistry; biota, which compete with photochemistry as agents of chemical transformation in surface waters, are few in the air. Also, light intensities in the atmosphere are higher than those in surface waters, and the presence of ultraviolet light of shorter wavelengths contributes to an expanded suite of photochemical reactions that may occur (e.g., the photodegradation of CFCs in the stratosphere).

Chemical contamination in the atmosphere is important for different reasons at several distinctly different scales; on the indoor and local scales, the major concern is that of human health. On regional scales, as pollutants threaten forests, lakes, and agriculture, the health of natural ecosystems is the major issue. On the global scale, maintenance of a viable planetary environment is of the utmost importance and must be treated in an international context.

EXERCISES

1. An unvented kerosene space heater releases CO at the rate of 1.0 g/hr into a room having a volume of 50 m^3 and an air change rate of 1/hr.
   a. What is the steady-state CO concentration expected in the room?
   b. How does the CO concentration change with time after the heater's fuel supply is depleted (assume that steady state is reached before it runs out of fuel)?
2. A factory town covers an area of 10 by 10 km, and contains SO_2 sources amounting to a total source strength of 1 metric ton/hr. Given a mixing height of 0.3 km and a wind speed of 0.8 m/sec, what is the approximate steady-state SO_2 concentration you would expect in the air?
3. A smelter releases 10 kg/day of zinc fumes via a stack whose effective height is 120 m. When the wind velocity is 10 mph from the southwest, measurably elevated atmospheric zinc levels occur in residential areas east and north of the smelter.
   a. Estimate maximum steady-state zinc concentrations in the air at ground level as a function of distance from the stack (these maximum concentrations will occur along the plume centerline).
   b. Estimate steady-state concentrations of zinc at ground level 10 km from the smelter, at distances of 100 m, 300 m, and 1 km from the plume centerline. Use stability category B.

4. On a sunny midwinter day, outside London, the wind is from the northwest at 4 m/sec. A plume of smoke is streaming from the stack of a coal-fired power plant. *Effective* height of the stack is 200 m. Over level terrain, how far downwind does the plume travel before it reaches the ground [use two standard deviations (σ_z) as your criterion for "reaching the ground"].
5. What is the mixing height associated with the following temperature profile at 11:00 A.M.?

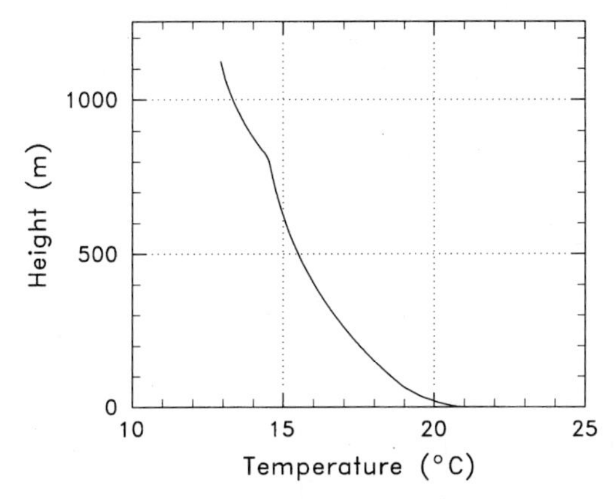

6. A simplified weather map for the United States from October 2, 1992 follows. Note that the symbols indicating weather conditions (rain, sunshine, cloudiness, etc.) have been deleted.

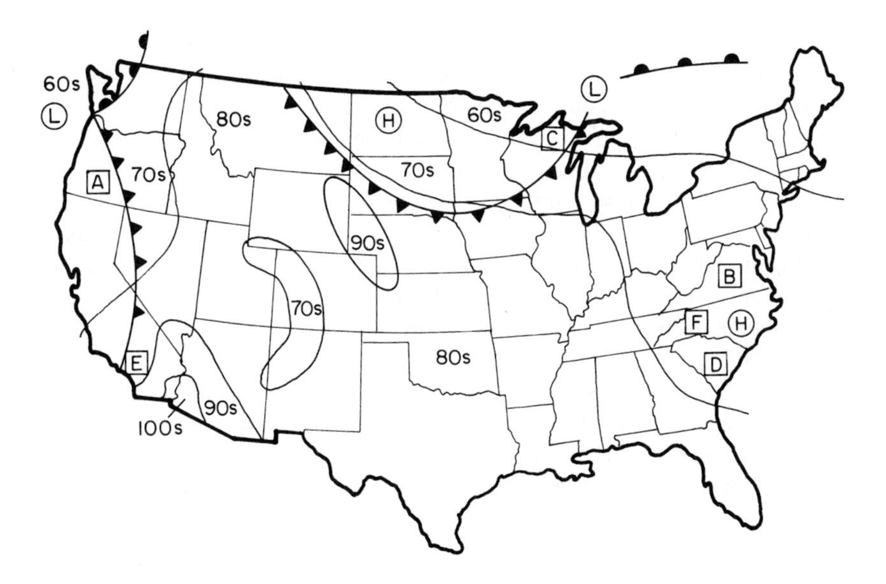

a. Which region is more likely to encounter precipitation—region A (western Oregon) or region B (central Virginia)? Briefly indicate *why*.
b. Predict the wind direction at points C (upper Michigan) and D (southern South Carolina). Justify your reasoning.
c. Which region is more likely to be subject to air pollution under these weather conditions—region E (southern California) or F (North Carolina)? Indicate your reasoning.

7. Assume that the soil under a poorly insulated house emits 2 pCi/($m^2 \cdot$ sec) of radon gas and that all this gas finds its way through the cellar floor and into the house. Assuming outdoor concentrations of radon are negligible and that the ventilation rate of the house is 0.9 air changes per hour, what is the ambient steady-state radon concentration inside the house? Should action be taken to increase the ventilation? Alternatively, if the ventilation rate is lowered to 0.3 air changes per hour to save energy, what is the effect on radon concentration? The house has 100 m^2 floor space, the average ceiling height is 2.6 m, and the radon decay rate is 7.6×10^{-3}/hr.

   Note the EPA recommendations for radon: consider remedial action if $C \geq 4$ pCi/liter; remedial action is strongly recommended if $C \geq 8$ pCi/liter.

8. A plastics factory being proposed for a location 5 km upwind from your neighborhood will emit 5 kg/day of vinyl chloride, a carcinogenic gas. The proposed stack is to be 30 m in height. Winds at the stack may vary from 6 to 12 m/sec blowing toward the neighborhood. Temperature inversions are common in this region; mixing heights of 100 m are not unusual.

   A citizens' group has asked you—a noted environmental expert—to comment as to whether vinyl chloride from this plant will adversely affect human health in your neighborhood. Assume problems may occur if concentrations at the near edge of your neighborhood exceed 1 ppm.

9. An inventor extraordinaire has read that the photocatalytic cycle of nitrogen oxides (the major source of air pollution in his city) begins with the emission of NO, chiefly by automobiles. This has encouraged him to invent a simple, cheap, and efficient device that converts NO in automobile emissions to NO_2. Will this device help or harm local air pollution problems in his city? Would it be preferable to convert the NO to N_2O? to HNO_3? to N_2?

10. a. Make an order-of-magnitude estimate of the dry deposition rate of nitric acid (HNO_3) aerosol on a forest, if atmospheric concentration of HNO_3 is 5 $\mu g/m^3$. How important is this source of acid compared with the annual deposition of HNO_3 in the rain, assuming the nitrate concentration in rain is 60 μM and annual precipitation depth averages 115 cm?

b. Similarly, make an order-of-magnitude estimate of the dry deposition rate of sulfate aerosol on a forest, if the atmospheric concentration is 8 $\mu g/m^3$. Compare this rate with wet input resulting from rain that averages 80 μM SO_4^{2-}.

11. Prevailing wind off the Pacific Ocean carries foggy air at 50°F over a coastal mountain range whose average height is 5000 ft. On the eastern side, the terrain drops to 500 ft above sea level. Estimate the temperature on the leeward side of the mountains, assuming all liquid water is deposited as rain on the windward side.
12. The second-order rate constant for the reaction of ethene with OH· is approximately 2.7×10^{-13} $cm^3/(molecule \cdot sec)$. If [OH·] is maintained at 10^{-12} M in air in a test chamber, and the initial ethene concentration is 100 ppm, what is:
    a. the amount of ethene remaining after 1 hr?
    b. the pseudo-first-order rate constant for ethene under these conditions?
    c. the pseudo-first-order rate constant in the atmosphere, where [OH·] is approximately 0.01 parts per trillion near Earth's surface?
13. The effective height of the stack of a power plant is 125 ft. The plant burns 5 ton of lignite (a low-grade coal) per hour; the lignite has a sulfur content of 1.5%. Wind speed is 6 m/sec, and the Pasquill–Gifford stability category may be taken as D.
    a. Estimate vertical and lateral plume dimensions at a distance 500 m downwind, and at a distance 5 km downwind.
    b. Estimate steady-state SO_2 concentration in a shopping district 5 km downwind.
14. An urban area contains several metalworking industries that rely heavily on chlorinated solvents for degreasing. Despite improved controls, approximately 80 kg of 1,1,1-TCA is still released to the urban atmosphere over a 3-mi radius, primarily by evaporation from solvent vats and from freshly degreased parts, during an 8-hr shift. During a period of atmospheric inversion, the effective mixing height is only 100 m, and wind speed averages 0.7 m/sec.
    a. Estimate average steady-state concentration of 1,1,1-TCA in the urban air during the work day, neglecting any sinks for the chemical.
    b. What would you expect to be the primary sink for this chemical in the air?
15. Assume mean atmospheric concentration of sulfate aerosol in a certain region is 8 $\mu g/m^3$.
    a. Estimate the dry deposition rate of sulfate aerosol onto a forested watershed in this region.
    b. Compare this source of sulfur to measured wet deposition of 30 kg/(ha · year) of sulfate.
    c. Annual surface water runoff from the watershed is 450 mm, and the

mean concentration of sulfate in stream water is 75 μM. Is the watershed acting as a net source or sink of sulfate?

16. Combustion produces oxides of nitrogen, which can be oxidized to nitrate (NO_3^-) in the atmosphere; the nitrate subsequently may be deposited by both wet and dry processes. Cite three specific regional or global-scale problems, or potential problems, that may be exacerbated by increased levels of nitrate deposition in precipitation.
17. A WWI fighter pilot is trying to reach friendly territory 35 km distant. The battle-damaged aircraft is still flying at 30 m/sec, but is losing altitude at 80 m/min. Fortunately (?), the pilot encounters a *thermal,* a rising column of air below a cumulus cloud, which has an upward velocity of 120 m/min, and begins to circle within the thermal to gain altitude. A thermometer and a hygrometer aboard the plane indicate a temperature of 20°C and an RH of 75%, respectively, at a height of 500 m.
    a. Can the pilot make it back to friendly territory on reaching the cloud base?
    b. The pilot decides to attempt to gain additional altitude by circling upward *within* the cloud. How high can the aircraft go before encountering ice? Can the pilot reach friendly territory from this altitude?
18. Compare the mechanisms leading to precipitation associated with each of the following meterological features:
    a. a cold front
    b. a low-pressure area
    c. orographic lifting.
19. Although some automotive catalytic converters are intended only to catalyze the oxidation of CO and hydrocarbons in engine exhaust, "mixed-bed" catalysts can be built to simultaneously decrease nitrogen oxide emissions by *reducing* the nitrogen oxides to nitrogen gas. Such catalysts are more expensive and also seem to require more accurate control of the fuel–air mixture in the engine (e.g., too much excess air might make it impossible to reduce the nitrogen oxides).
    a. How many equivalents of reducing capacity are required to reduce each milligram of nitric oxide (NO)?
    b. Where would this reducing capacity come from?
20. A certain cargo container is designed to be nearly airtight when sealed, to minimize possible salt and humidity effects when sensitive equipment is shipped. To test the container tightness, it is proposed to inject radon-222 ($t_{1/2}$ is 3.8 days) and then observe its concentration in the sealed container over a 2-week period. An initial concentration of 85 pCi/liter inside the container is established.
    a. What is the concentration after 2 days if the container is totally airtight?

b. What is the concentration after 2 days if the container has a ventilation rate of 0.05 air changes per hour (ACH)?

c. What is steady-state concentration if, instead of making a single injection, you place a radon source in the container? The source, a radium salt, emits 5 pCi/sec into the 60 m^3 container. Assume the container is fully airtight.

21. In a predawn traffic accident, a tanker truck overturns, and several tens of gallons of gasoline are spilled on the roadway. The fuel forms a pool in a small depression of the roadway. Pool area is 10 m^2. Wind speed is 2.5 m/sec, and the initial vapor pressure of this gasoline at the temperature of the pavement (15°C) is 45 mm Hg (this changes slowly with time as the more volatile compounds preferentially evaporate). The density of the liquid gasoline is approximately 0.7 g/cm^3 and average molecular weight is 120 g/mol.

a. Estimate the rate at which the depth of gasoline in the pool decreases.

b. What is the gasoline vapor concentration in air at 2-m height (breathing height), at a location 200 m directly downwind? Assume that the Pasquill–Gifford dispersion model is applicable, that it is overcast, and that the source dimensions are small compared with plume width.

22. The pilot of an IFR-equipped Cessna 172, which has adequate instrumentation to fly in the clouds, is planning a flight from eastern New York into Pennsylvania. The plane has no deicing equipment. The flight will require a minimum enroute altitude of 2500 m to maintain required terrain clearance over the Allegheny Mountains. Surface temperature at the departure airfield 200 m above sea level is 12°C, and RH is 60%. If the air along the route is characterized by an adiabatic lapse rate:

a. Will the pilot need to fly by instrumentation (i.e., in clouds) to clear the mountains?

b. Is airframe icing a consideration on the flight (i.e., what is the temperature at 2500 m)? Icing can occur, for example, when subcooled water droplets in a cloud, which have a temperature below 0°C, but have not yet changed to a solid phase, freeze on contact with the wings and fuselage.

23. A theme park in Florida hires you to help prepare an environmental impact statement for proposed nightly fireworks displays. One concern is the metals used to color the displays (Cu for green, Sr for red, K for violet, etc.), which may be deposited on the surrounding subtropical marshland. Although the metal releases occur in pulses, you decide to treat the problem as a continuous average release of 1 kg/day of fine (submicrometer-sized) Cu particulate at an effective height of 100 m.

When the nightly displays are conducted, assume typical conditions are cloudy, and the trade winds provide a wind speed of 3.5 m/sec.

a. Estimate the average concentration of Cu in air at ground level, 10 km downwind of the display site.
b. Estimate the deposition rate of Cu per unit area of marshland at this point.

24. A smelter emits SO_2 at a rate of 1 g/min via a stack whose effective height is 80 m.
    a. Estimate maximum SO_2 concentration that would be expected at ground level, at a point located 2.0 km directly downwind, when wind speed is 2.5 m/sec, and insolation is slight.
    b. By approximately what factor is SO_2 concentration expected to change if the wind shifts by 5 degrees?
25. A swimming pool in the neighborhood of a smelter receives atmospheric SO_2 deposition; the average concentration over the pool is 300 $\mu g\ SO_2/m^3$. If SO_2 deposition velocity is 1.2 cm/sec and all SO_2 deposited on the pool is transformed to H_2SO_4, how much NaOH must be added to the pool (moles per square meter of surface area) weekly to neutralize the H_2SO_4 and prevent the pool from becoming acidified?
26. Although NO_x is not depleted in the basic NO_x–O_3 cycle, mechanisms do exist for the conversion of NO_x to other compounds, such as HNO_3, a component of acid rain that is effectively removed by rainout.

    For the following reaction, the rate constant is 1.1×10^{-11} cm^3/(molecule · sec):

    $$OH\cdot + NO_2 \longrightarrow HNO_3.$$

    a. If this were the only reaction involving NO_2, what would be the half-life of NO_2 in a certain urban atmosphere where average $[OH\cdot]$ is 3×10^6 molecules per cubic centimeter?
    b. What would be the half-life of NO_2 if photolysis of NO_2 to NO and O (rate constant approximately 0.53/min) is included in addition to hydroxyl radical attack?
    c. Why is there measurable NO_2 in most urban atmospheres, given your result in (b)?
27. Air temperature at the base of a certain late summer thunderstorm (altitude of 500 m) is 68°F; yet, hail (ice) up to 0.1 cm in diameter is produced by the storm.
    a. Estimate the altitude at or above which the hail forms.
    b. What would be the minimum updraft strength in the storm for hail of this size to be carried upward in the cloud? (A hailstone has a layered structure reflecting the deposition of ice at several discrete time periods; hail may cycle through a cloud several times as it falls, is caught in updrafts, falls, etc.)
28. The concentration of carbonyl sulfide, COS, in the troposphere is approx-

imately 500 parts per trillion. It is removed by reaction with the OH· radical (k is 9×10^{-15} cm^3/(molecule · sec)) and by mixing into the stratosphere (first-order rate constant for tropospheric depletion of COS by this mixing process is approximately 0.04/year).

a. Write an appropriate mass balance expression for COS in the troposphere, on a per-liter basis. State your assumptions; simplify by assuming the troposphere is at a reasonable average temperature and pressure throughout its volume.

b. What is your estimate of the COS source strength (per liter of troposphere)? Use [OH·] equal to 10×10^6 molecules per cubic centimeter.

29. Many towns in the United States once operated open dumps, which required little maintenance. When refuse piled up too deeply, it was often ignited. A dump fire could burn for days, adding an eerie glow to the nighttime landscape.

    Assume that a certain dump fire emits 10 g/sec of soot particles (less than 0.1 μm in diameter). You decide to model soot transport using the Pasquill–Gifford approach, treating the dump as a zero-height stack. On this day, an outdoor ceremony is planned at city hall, 2 km directly downwind.

    a. If wind speed is 4 m/sec and it is a bright, sunny day, with a high mixing height, what is your prediction for soot concentrations in the ground-level air at city hall?

    b. How does this differ from your prediction for an overcast night with the same wind speed and a mixing height larger than the vertical extent of the plume?

    c. What is your nighttime prediction if there is an inversion at 50 m?

30. Various pesticides and solvents often found their way into open dumps. Based on airborne concentrations of a certain solvent vapor near a dump, you have calculated that concentrations in Jonesville, an hour's travel time downwind, should be 2 ng/m^3. Instead, you measure only 0.2 ng/m^3. What would be the required second-order rate constant if this difference were due to attack by some atmospheric species whose concentration were 3×10^5 molecules per cubic centimeter?

31. Icing can occur when air cools in a carburetor to below the dew point *and* the freezing temperature. Some of the cooling is due to the heat of vaporization of the fuel; some is adiabatic cooling as the air pressure drops in the venturi (the narrow portion of the carburetor where pressure is lowest and fuel is sucked into the airstream). Note that icing can occur even when the ambient air temperature is above freezing.

    Assume air must cool adiabatically to 5°C or lower in a certain engine carburetor before ice can form. A plane is flying at 1500 m, and air pressure in the venturi is equivalent to atmospheric pressure at 4500 m. Can icing potentially occur under these conditions:

a. Ambient air temperature is 10°C and RH is 100%?
b. Ambient air temperature is 15°C and RH is 100%?
c. Explain how air at less than 100% RH might cause icing when air at 100% RH would not.

(For an excellent story involving carburetor ice, read E. Gann's *Fate is the Hunter*.)

32. The second-order rate constant for oxidation of methane via OH· attack is approximately 3×10^{-15} cm³/(molecule · sec).
    a. Estimate a half-life of methane in the troposphere (state your assumptions; you may consider the troposphere as a whole to be relatively well mixed).
    b. Assume that the average half-life of a nonreactive molecule in the troposphere is approximately 8 years before diffusing to the stratosphere. What is your estimate of the relative importance of methane oxidation by hydroxyl radical in the troposphere compared with diffusion to the stratosphere?
33. Estimate an atmospheric half-life of acrolein. The second-order rate constant for reaction with OH· is approximately 2×10^{-11} cm³/(molecule · sec) and the second-order rate constant for reaction with O_3 is 4×10^{-13} cm³/(molecule · sec). Assume [OH·] is approximately 10^6 molecules per cubic centimeter and the O_3 partial pressure is 10^{-7} atm.
34. The boilers for a paper mill in an urban area emit 50 kg of SO_2 per hour.
    a. What concentration of SO_2 would you expect at ground level in the town center 2 km directly downwind, if the paper mill smoke stack is 40 m high (consider this to be effective height), and wind speed is 2.9 m/sec? Assume midday with moderate insolation, and use the formulae recommended by Briggs.
    b. The wind shifts by 30 degrees. What concentration do you expect now in the town center?
35. The temperature profile below was measured near an urban area shortly after sunrise. It is now 11:00 A.M. and surface air temperature has risen to 18°C in the city. Cumulative emission of carbon monoxide (CO) during the morning rush hour is approximately 5 kg/km².
    a. What concentration of carbon monoxide would you expect to find in the city? (Assume very light winds can be neglected.)
    b. In an effort to refine your air pollution model, you want to determine whether a certain pollutant behaves conservatively or is significantly degraded on the time scale of hours. Assume hydroxyl radical concentration is 10^6 molecules per cubic centimeter. What would be the necessary second-order rate constant for reaction with hydroxyl radical if the half-life were 3 hr?

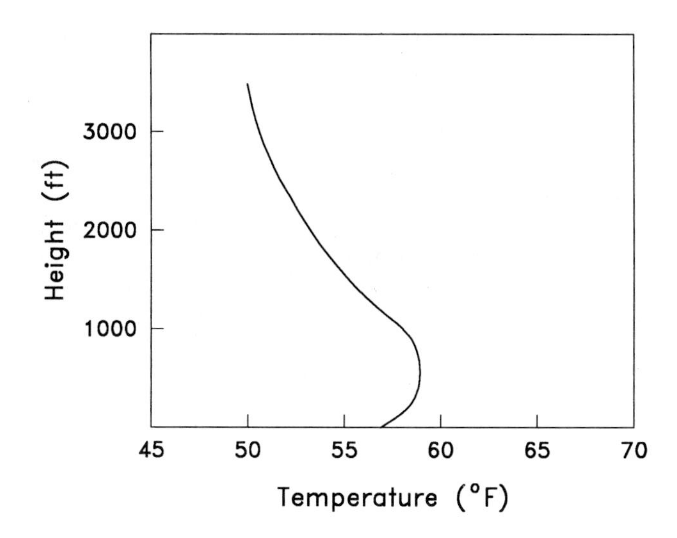

36. A root cellar receives an input of 0.1 μCi of radon-222 ($t_{1/2}$ is 3.8 days) per hour through the dirt floor.
    a. If the volume of the room is 15 m^3, what ventilation rate must be maintained [air changes per hour (ACH)] to keep radon levels below the EPA recommended limit of 4 pCi/liter?
    b. The cellar is closed up in an experiment to see if ethylene, a gas that stimulates ripening and is released by certain fruits such as apples, will accumulate enough to speed the ripening of some fruits. The ventilation rate is decreased to 0.05 ACH. The apples emit 7 μg/hr of ethylene. What is the steady-state ethylene concentration?
    c. What is the steady-state radon concentration in the closed-up cellar (0.05 ACH)?
37. A pilot is flying at an altitude of 6000 m above sea level and is "on instruments" (i.e., flying in a cloud). The outside air temperature is -13°F. The pilot notices the beginnings of ice accumulation on the wings, and prepares to call for a clearance to change either course or altitude (or both). The options are
    a. Immediately reverse course.
    b. Seek a safe lower altitude where ice will not be forming.
    c. Attempt to climb above the clouds.
    d. Continue on course and hope for the best.

    Explain why or why not plan (b) is reasonable. The terrain is mountainous and entirely covered by clouds to an elevation greater than 6000 m, with many mountain peaks as high as 3000 m above sea level. (Collectively, the maximum altimeter errors should be no greater than 75 m in this instance.)

38. Refer to the synoptic chart that follows.
    a. What is the probable origin of the low (point A) near International Falls, Minnesota? Contrast with the low-pressure area near North Carolina (point B).
    b. Suggest a Pasquill–Gifford stability category to use at midday in Kansas City (point C) and indicate how you arrived at your choice.
    c. What are the most probable locations for the development of subsidence inversions?
    d. Why is it clear near Springfield, Illinois (point D) but fully overcast near Indianapolis, Indiana (point E)?

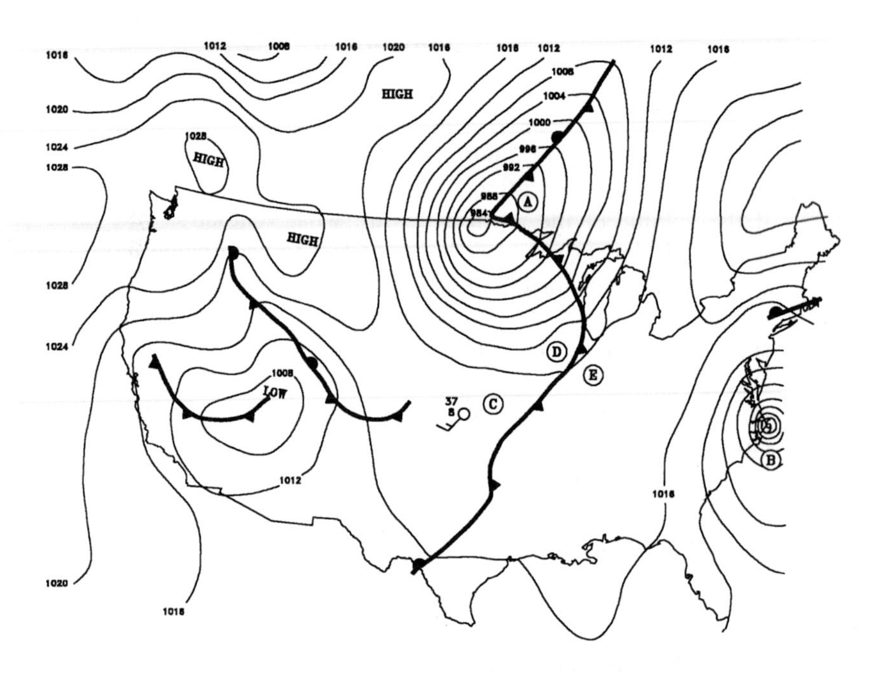

39. The major source of nitric oxide in a certain urban area is auto exhaust. At 6:00 A.M. the NO_x concentration (mixing ratio) is 0.3 ppb(v). During the morning rush hour, automobiles collectively emit 1000 g of NO_x (measured as N) per square kilometer of land area. The 10:00 A.M. temperature profile is shown next. By neglecting possible sinks for NO_x, what mixing ratio do you expect in ambient air at 10:00 A.M.?

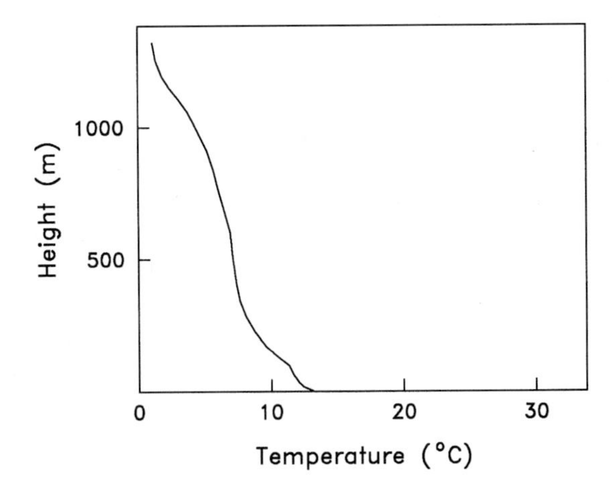

40. An incinerator ship burns waste solvents at sea. The furnace combusts 4,000 liters/hr of hydrocarbons and chlorinated hydrocarbons, having an average chlorine content of 50 g/liter. Most of the chlorine is emitted as HCl. The ship steams slowly through the night into a 15-knot wind, emitting combustion products at an effective height of 15 m.
    a. What is the HCl concentration in the air at sea level, 200 m astern and 1 km astern?
    b. HCl is lost from the atmosphere by absorption into seawater and by rainout.
        i. What is the pH of rain containing 3 ppm of HCl (neglect other solutes except for CO_2)?
        ii. Estimate the pH of seawater to which 3 ppm of HCl is added. (Assume that initially in the seawater, pH is 8 and C_T is 5×10^{-4} *M*. Assume further that HCl dissolves into seawater far more rapidly than CO_2 re-equilibrates with the atmosphere.)
41. Pilots of early-model SBD dive bombers would sometimes make high-speed dives from, perhaps, 10,000 ft to near sea level (while facing enemy gunfire), only to have their bomb sight optics fog over with condensed moisture as they approached the target. At the suggestion of surviving pilots, electric bomb sight heaters were later added to prevent this problem. How much would the heaters have to heat the optics above the outside surrounding air temperature to prevent this problem? Assume a rapid (i.e., optics do not have time to change temperature) dive from 10,000 ft into tropical air (85°F, 95% RH) at 500 ft, a dry adiabatic lapse rate, and an atmosphere of neutral stability (i.e., actual lapse rate equals adiabatic lapse rate).
42. Trichloroethylene (TCE) dissolved in a wastewater stream enters a lake.

The mixed depth of the lake is 3 m, the gas exchange coefficient (water side controlled) is 0.5 m/day, and the primary atmospheric sink is attack by OH· (k is 2.4×10^{-12} cm^3/(molecule · sec)).

a. Estimate the half-life for TCE in the lake.

b. Estimate the half-life for TCE once it escapes from the lake to the atmosphere (assume [OH·] is 10^6 molecules per cubic centimeter).

43. New carpeting is installed in a home. The carpet outgasses formaldehyde at a total rate of 2.3 g/hr and the house has a ventilation rate of 0.3 ACH. If the house has a volume of 750 m^3, what is the expected formaldehyde concentration? If the carpet were taken outdoors, how long would it take for the formaldehyde concentration in the home to decrease to 0.1 ppm?

REFERENCES

Allen, O. E. (1983). *A Planet Earth: Atmosphere.* Time-Life Books, Chicago, IL.

Amts, R. R., Seila, R. L., and Bufalini, J. J. (1989). "Determination of Room Temperature OH Rate Constants for Acetylene, Ethylene Dichloride, Ethylene Dibromide, *p*-Dichlorobenzene, and Carbon Disulfide." *JAPCA* **39**, 453–460.

Andersen, H. V., and Hovmand, M. F. (1995). "Ammonia and Nitric Acid Dry Deposition and Throughfall." *Water, Air, Soil Pollut.* **85**, 2211–2216.

Arya, S. P. (1999). *Air Pollution Meteorology and Dispersion.* Oxford University Press, Oxford.

Bacastow, R. B., Keeling, C. D., and Whorf, T. P. (1985). "Seasonal Amplitude Increase in Atmospheric CO_2 Concentration at Mauna Loa, Hawaii, 1959–1982." *J. Geophys. Res.* **90**(D6), 10,529–10,540.

Bennett, M. (1988). "A Simple Physical Model of Dry Deposition to a Rough Surface." *Atmos. Environ.* **22**(12), 2701–2705.

Boubel, R. W., Fox, D. L., Turner, D. B., and Stern, A. C. (1994). *Fundamentals of Air Pollution,* 3rd ed. Academic Press, San Diego, CA.

Briggs, G. A. (1972). "Discussion: Chimney Plumes in Neutral and Stable Surroundings." *Atmos. Environ.* **6**, 507–510.

Brook, J. R., Sirois, A., and Clarke, J. F. (1996). "Comparison of Dry Deposition Velocities for SO_2, HNO_3 and SO_4^{2-} Estimated with Two Inferential Models." *Water, Air, Soil Pollut.* **87**, 205–218.

Brook, J. R., Di-Giovanni, F., Cakmak, S., and Meyers, T. P. (1997). "Estimation of Dry Deposition Velocity Using Inferential Models and Site-Specific Meterology-Uncertainty Due to Siting of Meterological Towers." *Atmos. Environ.* **31**(23), 3911–3919.

Cole, F. W. (1970). *Introduction to Meterology.* Wiley, New York.

Crutzen, P. J., Gerard J.-C., and Zander, R. (Eds.). (1989). Our Changing Atmosphere. *In Proceedings of the 28th Liège International Astrophysical Colloquium.* University of Liege, Cointe-Ougree, Belgium.

Dobbins, R. A. (1979). *Atmospheric Motion and Air Pollution.* Wiley, New York.

Doiron, S. D., Bluth, G. J. S., Schnetzler, C. C., Krueger, A. J., and Walter, L. S. (1991). "Transport of Cerro Hudson SO_2 Clouds." *EOS* **72**(45), 489–490.

Draxler, R. R. (1979). "Estimating Vertical Diffusion from Routine Meteorological Tower Measurements." *Atmos. Environ.* **13**, 1559–1564.

Eder, B. K., Coventry, D. H., Clark, T. L., and Bollinger, C. E. (1986). "RELMAP: A Regional

Lagrangian Model of Air Pollution User's Guide." EPA 600/8-86/013. US EPA, Research Triangle Park, NC.

Eshleman, K. N. and Hemond, H. F. (1988). "Alkalinity and Major Ion Budgets for a Massachusetts Reservoir and Watershed." *Limnol. Oceanogr.* **33**(2), 174–185.

Federal Aviation Administration (FAA). (1965). Aviation Weather.

Fisher, D. C. and Oppenheimer, M. (1991). "Atmospheric Nitrogen Deposition and the Chesapeake Bay Estuary." *Ambio* **20**(3–4), 102–108.

Flagan, R. C. and Seinfeld, J. H. (1988). *Fundamentals of Air Pollution Engineering.* Prentice-Hall, Englewood Cliffs, NJ.

Friedlander, S. K. (1977). *Smoke, Dust, and Haze: Fundamentals of Aerosol Behavior.* Wiley, New York.

Gifford, F. A. (1976). "Turbulent Diffusion-Typing Schemes: A Review." *Nucl. Saf.* **17**(1), 68–86.

Godish, T. (1989). *Indoor Air Pollution Control.* Lewis, Chelsea, MI.

Ham, D. (1993). "CFC Replacements: Many Problems for Chemists to Solve." *Nucleus* **71**(6), 6–7.

Hanna, S. R. (1985). "Air Quality Modeling over Short Distances." *In Handbook of Applied Meteorology* (D. D. Houghton, Ed.), pp. 712–743. Wiley, New York.

Haynes, R. J. (1986). *Mineral Nitrogen in the Plant-Soil System.* Academic Press, Orlando, FL.

Hess, S. L. (1959). *Introduction to Theoretical Meteorology.* Holt, Rinehart & Winston, New York.

Hicks, B. B., Baldocchi, D. D., Meyers, T. P., Hosker, R. P., Jr., and Matt, D. R. (1987). "A Preliminary Multiple Resistance Routine for Deriving Dry Deposition Velocities from Measured Quantities." *Water, Air, Soil Pollu.* **36**, 311–330.

Holton, J. R., Haynes, P. H., McIntyre, M. E., Douglass, A. R., Rood, R. B., and Pfister, L. (1995). "Stratosphere-Troposphere Exchange." *Rev. Geophys.* **33**(4), 403–439.

Houghton, D. D. (Ed.). (1985). *Handbook of Applied Meteorology.* Wiley, New York.

Intergovernmental Panel on Climate Change (IPCC). (1996). *Climate Change 1995. The Science of Climate Change.* J. T. Houghton, L. G. Meira Filho, B. A. Callander, N. Harris, A. Kattenberg, and K. Maskell (Eds.). University Press, Cambridge, Great Britain.

Kaiser, C. D. (1979). "Examples of the Successful Application of a Simple Model for the Atmospheric Dispersion of Dense, Cold Vapors to the Accidental Release of Anhydrous Ammonia from Pressurized Containers." United Kingdom Atomic Energy Authority Safety and Reliability Directorate, Pub. SRD R150.

Keeling, C. D. and Whorf, T. P. (1998). "Atmospheric CO_2 Concentrations—Mauna Loa Observatory, Hawaii, 1958–1997." Numeric Data Package NDP-001. Carbon Dioxide Information Analysis Center, Oak Ridge, TN.

Khalil, M. A. K. and Rasmussen, R. A. (1983). "Sources, Sinks, and Seasonal Cycles of Atmospheric Methane." *J. Geophy. Res.* **88**(C9), 5131–5144.

Levi, B. G. (1988). "Ozone Depletion at the Poles: The Hole Story Emerges." *Phys. Today* **July**, 17–21.

Luria, M. and Sievering, H. (1991). "Heterogeneous and Homogeneous Oxidation of SO_2 in the Remote Marine Atmosphere." *Atmos. Environ.* **25A**(8), 1489–1496.

Lutgens, F. K. and Tarbuck, E. J. (1992). *The Atmosphere: An Introduction to Meteorology,* 5th ed. Prentice-Hall, Englewood Cliffs, NJ.

Lyons, T. J. and Scott, W. D. (Eds.). (1990). *Principles of Air Pollution Meteorology.* CRC Press, Boca Raton, FL.

Mann, K. H. and Lazier, J. R. N. (1996). *Dynamics of Marine Ecosystems. Biological–Physical Interactions in the Oceans,* 2nd ed. Blackwell Science, Cambridge, MA.

Marks, L. S. (Ed.). (1951). *Mechanical Engineers' Handbook.* McGraw-Hill, New York.

Masters, G. M. (1991). *Introduction to Environmental Engineering and Science.* Prentice-Hall, Englewood Cliffs, NJ.

Mészáros, E. (1981). *Atmospheric Chemistry: Fundamental Aspects.* Elsevier, New York.

Miller, S. L. (1974). "The Nature and Occurrence of Clathrate Hydrates." *In Natural Gases in Marine Sediments* (I. R. Kaplan, Ed.). Plenum Press, New York.

Miller, A. and Thompson, J. C. (1975). *Elements of Meteorology,* 2nd ed. Charles E. Merrill, Columbus, OH.

Molina, M. J. and Rowland, F. S. (1974). "Stratospheric Sink for Chlorofluoromethanes: Chlorine Atom-Catalysed Destruction of Ozone." *Nature (London)* **249**, 810–812.

National Acid Precipitation Assessment Program (NAPAP). (1990). "Acidic Deposition: State of Science and Technology." Vol. I. *Emissions, Atmospheric Processes, and Deposition.* Government Printing Office, Washington, DC.

National Atmospheric Deposition Program (NADP) (1990). *NADP/NTN Annual Data Summary: Precipitation Chemistry in the United States 1990.* Government Printing Office, Washington, DC.

Oke, T. R. (1978). *Boundary Layer Climates.* Methuen, London.

Padro, J. (1996). "Summary of Ozone Dry Deposition Velocity Measurements and Model Estimates over Vineyard, Cotton, Grass and Deciduous Forest in Summer." *Atmos. Environ.* **30**(13), 2363–2369.

Pasquill, F. (1961). "The Estimation of the Dispersion of Windborne Material." *Meterol. Mag.* **90**(1063), 33–49.

Pasquill, F. (1974). *Atmospheric Diffusion: The Dispersion of Windborne Material from Industrial and Other Sources,* 2nd ed. Ellis Horwood, Chichester.

Prather, M., Derwent, R., Ehhalt, D., Fraser, P., Sanhueza, E., and Zhou, X. (1995). "Other Trace Gases and Atmospheric Chemistry," Section 2. *In Climate Change 1994. Radiative Forcing of Climate Change and An Evaluation of the IPCC IS92 Emission Scenarios* (J. T. Houghton, L. G. Meira Filho, J. Bruce, Hoesung Lee, B. A. Callander, E. Haites, N. Harris, and K. Maskell, Eds.). Cambridge University Press, Cambridge, Great Britain.

Prather, M., Derwent, R., Ehhalt, D., Fraser, P., Sanhueza, E., and Zhou, X. (1996). "Radiative Forcing of Climate Change," Section 2.2. *In Climate Change 1995. The Science of Climate Change* (J. T. Houghton, L. G. Meira Filho, B. A. Callander, N. Harris, A. Kattenberg, and K. Maskell, Eds.). Cambridge University Press, Cambridge, Great Britain.

Pratt, G. C., Orr, E. J., Bock, D. C., Strassman, R. L., Fundine, D. W., Twaroski, C. J., Thornton, J. D., and Meyers, T. P. (1996). "Estimation of Dry Deposition of Inorganics Using Filter Pack Data and Inferred Deposition Velocity." *Environ. Sci. Technol. 30,* 2168–2177.

Prinn, R. G. and Hartley, D. (1992). "Atmosphere, Ocean, and Land: Critical Gaps in Earth System Models." *In Modeling the Earth System* (D. Ojima, Ed.). University Corporation for Atmospheric Research/Office for Interdisciplinary Earth Studies. Boulder, CO.

Rogers, J. D. and Stephens, R. D. (1988). "Absolute Infrared Intensities for F-113 and F-114 and an Assessment of Their Greenhouse Warming Potential Relative to Other Chlorofluorocarbons." *J. Geophys. Res.* **93**(D3), 2423–2428.

Rosenberg, N. J., Blad, B. L., and Verma, S. B. (1983). *Microclimate: The Biological Environment,* 2nd ed. Wiley, New York.

Sams, Howard W. and Co. Inc. (1977). *Reference Data for Radio Engineers,* 6th ed. Prentice-Hall, Englewood Cliffs, NJ.

Samson, P. J. and Small, M. J. (1984). "Atmospheric Trajectory Models for Diagnosing the Sources of Acid Precipitation." *In Modeling of Total Acid Precipitation Impacts* (J. L. Schnoor, Ed.). Butterworth, Boston.

Schimel, D., Alves, D., Enting, I., Heimann, M., Joos, F., Raynaud, D., and Wigley, T. (1996). "Radiative Forcing of Climate Change," Section 2.1. *In Climate Change 1995: The Science of*

Climate Change (J. T. Houghton, L. G. Meira Filho, B. A. Callander, N. Harris, A. Kattenberg, and K. Maskell, Eds.). Cambridge University Press, Cambridge, Great Britain.

Schrimpf, W., Lienaerts, K., Müller, K. P., Rudolph, J., Neubert, R., Schüßler, W., and Levin, I. (1996). "Dry Deposition of Peroxyacetyl Nitrate (PAN): Determination of Its Deposition Velocity at Night from Measurements of the Atmospheric PAN and ^{222}Radon Concentration Gradient." *Geophys. Res. Lett.* **23**(24), 3599–3602.

Seinfeld, J. H. (1975). *Air Pollution: Physical and Chemical Fundamentals.* McGraw-Hill, New York.

Seinfeld, J. H. (1986). *Atmospheric Chemistry and Physics of Air Pollution.* Wiley, New York.

Stern, A. C., Boubel, R. W., Turner, D. B., and Fox, D. L. (1984). *Fundamentals of Air Pollution,* 2nd ed. Academic Press, Orlando, FL.

Sobel, D. (1995). *Longitude. The True Story of a Lone Genius Who Solved the Greatest Scientific Problem of His Time.* Walker, New York.

Stolarski, R. S., Krueger, A. J., Schoeberl, M. R., McPeters, R. D., Newman, P. A., and Alpert, J. C. (1986). "Nimbus 7 Satellite Measurements of the Springtime Antarctic Ozone Decrease." *Nature (London)* **322**, 808–811.

Stolarski, R., Bojkov, R., Bishop, L., Zerefos, C., Staehelin, J., and Zawodny, J. (1992). "Measured Trends in Stratospheric Ozone." *Science* **256**, 342–349.

US EPA (1986). "Radon Reduction Techniques for Detached Houses." EPA/625/5-86/109.

US EPA (1995). "User's Guide for the Industrial Source Complex (ISC3) Dispersion Models." Volume I-User Instructions. EPA-454/B-95-003a; Volume II-Description of Model Algorithms. EPA-454/B-95-003b. September.

US EPA. (1996a). "Air Quality Criteria for Particulate Matter." Volume I of III. EPA/600/P-95/001aF. April. National Center for Environmental Assessment, Office of Research and Development, U.S. Environmental Protection Agency. Research Triangle Park, NC.

US EPA. (1996b). "Air Quality Criteria for Ozone and Related Photochemical Oxidants." Volume I of III. EPA/600/P-93/004aF. July. National Center for Environmental Assessment, Office of Research and Development, U.S. Environmental Protection Agency. Research Triangle Park, NC.

US EPA. (1997). "Regulatory Impact Analyses for the Particulate Matter and Ozone National Ambient Air Quality Standards and Proposed Regional Haze Rule." Office of Air Quality Planning and Standards, Research Triangle Park, NC.

US EPA. (1998). "Toxic Chemical Release Inventory System. Frozen ADABAS File for Reporting Year 1996."

Venkatram, A., Karamchandani, P. K., and Misra, P. K. (1988). "Testing a Comprehensive Acid Deposition Model." *Atmos. Environ.* **22**, 737–747.

Viegle, W. J. and Head, J. H. (1978). "Derivation of the Gaussian Plume Model." *JAPCA* **28**(11), 1139–1141.

Wang, W. C., Yung, Y. L., Lacis, A. A., Mo, T., and Hansen, J. E. (1976). "Greenhouse Effects Due to Man-Made Perturbations of Trace Gases." *Science* **194**, 685–690.

Weast, R. C. (Ed.). (1990). *CRC Handbook of Chemistry and Physics,* 70th ed. 2nd printing. CRC Press, Boca Raton, FL.

Wesely, M. L. (1989). "Parameterization of Surface Resistances to Gaseous Dry Deposition in Regional-Scale Numerical Models." *Atmos. Environ.* **23**(6), 1293–1304.

Wilson, R. and Spengler, J. D. (eds.). (1996). *Particles in Our Air: Concentrations and Health Effects.* Harvard University Press, Cambridge, MA.

Yamartino, R. J. (1985). "Atmospheric Pollutant Deposition Modeling." *In Handbook of Applied Meteorology* (D. D. Houghton, Ed.), pp. 754–766. Wiley, New York.

Zhang, L., Padro, J., and Walmsley, J. L. (1996). "A Multi-Layer Model vs. Single-Layer Models and Observed O_3 Dry Deposition Velocities." *Atmos. Environ.* **30**(2), 339–345.

APPENDIX

Dimensions and Units for Environmental Quantities

Three fundamental dimensions of measurement, mass [M], length [L], and time [T], form the basis for most environmental quantities. A good understanding of these dimensions and of the way in which they are combined to form various units of measurement clarifies many problems of chemical fate and transport.

A.1 FUNDAMENTAL DIMENSIONS AND COMMON UNITS OF MEASUREMENT

Mass

Mass refers to an amount of substance such as a kilogram of soil, a ton of road salt, or a microgram of soot particles. The gram (g) or the kilogram (kg) are commonly used units of mass in the Système International (SI) measurement system. Mass is frequently determined by measuring the force exerted on it by a gravitational field (i.e., weighing it), but it also can be determined on the basis of its inertia. In the English measurement system, the unit of the

TABLE A-1 Common Mass Units and Interconversion Factors[a]

	Gram	Kilogram	Pound[b]	Slug	Ton (short)[b]	Ton (metric)
Gram	1	10^{-3}	2.205×10^{-3}	6.852×10^{-5}	1.102×10^{-6}	10^{-6}
Kilogram	1000	1	2.205	6.852×10^{-2}	1.102×10^{-3}	0.001
Pound[b]	453.6	0.4536	1	3.108×10^{-2}	5×10^{-4}	4.536×10^{-4}
Slug	1.459×10^{4}	14.59	32.17	1	1.609×10^{-2}	14.59×10^{-3}
Ton (short)[b]	9.072×10^{5}	907.2	2000	62.16	1	0.9072
Ton (metric)	10^{6}	1000	2204	68.50	1.102	1

[a]Weast, R. C. (Ed.) (1990). *CRC Handbook of Chemistry and Physics.* CRC Press, Boca Raton, FL.
[b]Really a unit of force—the amount of mass that weighs 1 lb or 1 short ton.

pound (lb), often loosely used to refer to mass, is actually a unit of force, as described in the next section; a pound of soil is that amount of soil on which Earth's gravity exerts 1 lb of force. See Table A-1 for common units of mass and interconversion factors for both the SI and English systems.

Length

In the SI system, the meter (m) and the centimeter (cm) are common units of length. An English unit, the foot (ft), still is used to some extent in fields such as hydrology and meteorology. Table A-2 gives common units of length and interconversion factors. Table A-3 presents multiplier prefixes for use with SI units.

Time

The SI base unit for time is the second (sec). Often it is convenient to use units considerably longer than the second, such as hours (hr), days, and

TABLE A-2 Common Length Units and Interconversion Factors[a]

	Centimeter	Foot	Inch	Kilometer	Meter	Mile
Centimeter	1	0.03281	0.3937	10^{-5}	10^{-2}	6.214×10^{-6}
Foot	30.48	1	12	3.048×10^{-4}	0.3048	1.894×10^{-4}
Inch	2.54	0.08333	1	2.54×10^{-5}	2.54×10^{-2}	1.578×10^{-5}
Kilometer	10^{5}	3281	39,370	1	1000	0.6214
Meter	10^{2}	3.281	39.37	10^{-3}	1	6.214×10^{-4}
Mile	1.609×10^{5}	5280	63,360	1.609	1609	1

[a]Weast (1990).

TABLE A-3 Multiplier Prefixes for Use with SI Units[a]

10^{-18}	atto-	a
10^{-15}	femto-	f
10^{-12}	pico-	p
10^{-9}	nano-	n
10^{-6}	micro-	μ
10^{-3}	milli-	m
10^{-2}	centi-	c
10^{-1}	deci-	d
10^{1}	deca-	da
10^{2}	hecto-	h
10^{3}	kilo-	k
10^{6}	mega-	M
10^{9}	giga-	G
10^{12}	tera-	T
10^{15}	peta-	P
10^{18}	exa-	E

[a] Weast (1990).

years. Because these units of time are not decimally related, one must be prepared to make unit interconversions with a hand calculator.

A.2 DERIVED DIMENSIONS AND COMMON UNITS

Area and Volume

Among the simplest derived dimensions are area [L^2] and volume [L^3]. While the square meter (m^2) is a common SI unit of area, numerous other units [e.g., square foot (ft^2), acre, hectare (ha), and square mile (mi^2)] are in use, as shown in Table A-4. A common SI unit of volume is the cubic meter (m^3), although for environmental chemical problems the liter is often used. Note that liter is spelled out throughout this book, whereas [L] refers to the dimension length. Table A-5 presents common volume units and interconversion factors.

Velocity and Acceleration

Velocity, V or v, conveys both the speed at which something is traveling and the direction in which it is moving; it has dimensions of [L/T]. Common units include m/sec and cm/sec. Acceleration, a, is the rate at which velocity changes with time. Like velocity, it has both a magnitude and a direction. The

TABLE A-4 Common Area Units and Interconversion Factors[a]

	Acre	Hectare	Square foot	Square kilometer	Square meter	Square mile
Acre	1	0.4047	43,560	4.047×10^{-3}	4047	1.563×10^{-3}
Hectare	2.471	1	1.076×10^{5}	0.01	10^{4}	3.861×10^{-3}
Square foot	2.296×10^{-5}	9.290×10^{-6}	1	9.290×10^{-8}	0.0929	3.587×10^{-8}
Square kilometer	247.1	100	1.076×10^{7}	1	10^{6}	0.386
Square meter	2.471×10^{-4}	10^{-4}	10.764	10^{-6}	1	3.861×10^{-7}
Square mile	640	259	2.788×10^{7}	2.590	2.590×10^{6}	1

[a]Weast (1990).

fundamental dimensions of acceleration are [L/T^2]; acceleration may be expressed, for example, as (cm/sec)/sec or cm/sec^2. Note that these units for velocity and acceleration convey only the speed or magnitude aspect. In most places in this text, the directional aspect of a problem is conveyed with a diagram or figure. In situations where the geometry of a problem is complex, the directional aspect of acceleration or velocity may be more conveniently expressed by the use of vector notation.

Force

Force, F, is defined with the help of Newton's laws of motion. Force has fundamental dimensions of [ML/T^2]. These dimensions result from taking the product of mass (m) and acceleration (a):

$$F = m \cdot a.$$

A dyne (dyn) is the amount of force that gives a gram of mass an acceleration

TABLE A-5 Common Volume Units and Interconversion Factors[a]

	Acre-foot	Cubic foot	Cubic inch	Cubic meter	Liter	U.S. gallon
Acre-foot	1	43,560	7.53×10^{7}	1233	1.233×10^{6}	3.26×10^{5}
Cubic foot	2.30×10^{-5}	1	1728	0.0283	28.3	7.48
Cubic inch	1.33×10^{-8}	5.79×10^{-4}	1	1.64×10^{-5}	1.64×10^{-2}	4.33×10^{-3}
Cubic meter	8.11×10^{-4}	35.3	61,000	1	1000	264
Liter	8.11×10^{-7}	0.0353	61	10^{-3}	1	0.264
U.S. gallon	3.07×10^{-6}	0.134	231	3.79×10^{-3}	3.79	1

[a]Linsley, R. K., Jr., Kohler, M. A., and Paulhus, J. L. H. (1982). *Hydrology for Engineers.* McGraw-Hill, New York.

TABLE A-6 Common Force Units and Interconversion Factors[a]

	Dyne	Kilogram-force	Newton	Pound
Dyne	1	1.020×10^{-6}	1×10^{-5}	2.248×10^{-6}
Kilogram-force	9.807×10^{5}	1	9.807	2.205
Newton	10^{5}	0.102	1	0.225
Pound	4.448×10^{5}	0.454	4.448	1

[a]Weast (1990).

of 1 cm/sec^2; thus, it has units of g · cm/sec^2. The unit of force required to accelerate a kilogram at a rate of 1 m/sec^2 is called a newton (N). A kilogram-force is 9.8 N, the weight of a mass of 1 kg at Earth's surface. See Table A-6 for common units of force and interconversion factors for both the SI and English measurement systems.

Pressure

Pressure is force per unit area and has dimensions of [M/LT^2]. Pressure is an important measurement in many fields of science, and each field has different traditional units. The SI unit, the pascal [N/m^2 or kg/(m · sec^2)], is commonly used, along with the dyn/cm^2. Other pressure units frequently encountered include the millimeter of mercury (mm Hg), the atmosphere (atm), the bar (10^6 dyn/cm^2), and the pound per square inch (psi). The origin of some of these units is implicit in their names; the millimeter of mercury (also called a torr) is the amount of pressure that causes the mercury in a manometer to rise by 1 mm—an easy unit of measure for the laboratory experimentalist to use. Many of the common units of pressure and their interconversion factors are shown in Table A-7.

TABLE A-7 Common Pressure Units and Interconversion Factors[a]

	Atmosphere	Bar	Kilopascal	Millimeter of Hg	Psi
Atmosphere	1	1.013	101.3	760	14.7
Bar	0.987	1	100	750	14.5
Kilopascal	9.87×10^{-3}	1×10^{-2}	1	7.50	0.145
Millimeter of Hg	1.32×10^{-3}	1.33×10^{-3}	0.133	1	1.93×10^{-2}
Psi	0.0680	0.0689	6.89	51.7	1

[a]Weast (1990).

TABLE A-8 Common Energy Units and Interconversion Factors[a]

	Btu[b]	Calorie	Erg	Foot-pound	Joule	Kilowatt-hour
Btu	1	251.996	1.055×10^{10}	778.169	1055.056	2.931×10^{-4}
Calorie	3.968×10^{-3}	1	4.187×10^{7}	3.088	4.187	1.163×10^{-6}
Erg	9.478×10^{-11}	2.388×10^{-8}	1	7.376×10^{-8}	10^{-7}	2.778×10^{-14}
Foot-pound	1.285×10^{-3}	0.324	1.356×10^{7}	1	1.356	3.766×10^{-7}
Joule	9.478×10^{-4}	0.239	10^{7}	0.738	1	2.778×10^{-7}
Kilowatt-hour	3412.14	8.598×10^{5}	3.6×10^{13}	2.655×10^{6}	3.6×10^{6}	1

[a]Weast (1990).
[b]British thermal unit.

Energy

Energy is the capability to perform work. Energy may occur in mechanical, thermal, chemical, nuclear, or electrical form, and has dimensions of $[ML^2/T^2]$. The basic SI energy unit, the joule (J), is equal to the energy conveyed by a force of 1 N exerted over a distance of 1 m; therefore, it has the units of $kg \cdot m^2/sec^2$. Other units of energy are the British thermal unit (Btu), calorie, erg, foot-pound, and kilowatt-hour, as shown in Table A-8.

Power

Power is the rate at which energy is transferred, and therefore it has dimensions of $[ML^2/T^3]$. One watt (W) is defined as 1 J/sec. Other common units are horsepower and Btu/hr. (Curiously, the metric cheval-vapeur is equivalent to only 735 W, slightly less than the U.S./British horsepower of 746 W.) See Table A-9 for interconversion factors for these units.

TABLE A-9 Common Power Units and Interconversion Factors[a]

	Btu per hour	Horsepower	Joule per second	Watt
Btu per hour	1	3.930×10^{-4}	0.293	0.293
Horsepower	2544.43	1	745.700	745.700
Joule per second	3.412	1.341×10^{-3}	1	1
Watt	3.412	1.341×10^{-3}	1	1

[a]Weast (1990).

TABLE A-10 Atomic Weights of Some Common Elements

Chemical	Symbol	Atomic weight (g/mol)	Chemical	Symbol	Atomic weight (g/mol)
Aluminum	Al	26.98	Mercury	Hg	200.59
Argon	Ar	39.95	Molybdenum	Mo	95.94
Arsenic	As	74.92	Neon	Ne	20.18
Barium	Ba	137.34	Nickel	Ni	58.70
Beryllium	Be	9.01	Nitrogen	N	14.01
Bromine	Br	79.90	Oxygen	O	16.00
Cadmium	Cd	112.40	Phosphorus	P	30.97
Calcium	Ca	40.08	Platinum	Pt	195.09
Carbon	C	12.01	Potassium	K	39.10
Chlorine	Cl	35.45	Radium	Ra	226.03
Chromium	Cr	52.00	Radon	Rn	222.00
Cobalt	Co	58.93	Selenium	Se	78.96
Copper	Cu	63.55	Silicon	Si	28.09
Fluorine	F	19.00	Silver	Ag	107.87
Helium	He	4.00	Sodium	Na	23.00
Hydrogen	H	1.01	Sulfur	S	32.06
Iodine	I	126.90	Tin	Sn	118.69
Iron	Fe	55.85	Titanium	Ti	47.90
Lead	Pb	207.20	Uranium	U	238.03
Magnesium	Mg	24.30	Zinc	Zn	65.38
Manganese	Mn	54.94			

Atomic and Molecular Weight

The atomic weight of an element is numerically equal to the mass, in grams, of a mole (6.023×10^{23} atoms) of the element. Molecular weight (MW) of a molecule is the sum of the atomic weights of its constituent atoms. Table A-10 gives atomic weights for many common elements.

Gas Constant

The gas constant R appears in numerous expressions, including the ideal gas law—Eq. [1-29]—and the equation relating an equilibrium constant of a reaction to its free energy—Eq. [1-12]. Equivalent values of R are

$8.3145\ \text{J/(mol} \cdot K)$
$0.082058\ (\text{L} \cdot \text{atm})/(\text{mol} \cdot K)$
$1.9859\ \text{cal/(mol} \cdot K)$
$8.3145 \times 10^{7}\ \text{erg/(mol} \cdot K)$

INDEX

B

C

D

E

F

G

H

I

Q

R

S

ACKNOWLEDGMENTS

The authors thank the reviewers of individual sections of this second edition, whose insightful comments have strengthened the book: Lisa Akeson, Peter Eglinton, Benjamin Levy, and Don McCubbin. The authors also acknowledge Lynn Roberts' critique of the entire text of the first edition. Thanks are also due to John MacFarlane, who assisted with the graphics. The authors greatly appreciate the support for this endeavor that was provided to E. Fechner-Levy by Abt Associates Inc., and to H. Hemond by the William E. Leonhard Chair in the Department of Civil and Environmental Engineering at MIT.